Lecture Notes in Mathematics

2009

Editors:
J.-M. Morel, Cachan
F. Takens, Groningen
B. Teissier, Paris

W0080347

FONDAZIONE
CIME
ROBERTO CONTI
CENTRO INTERNAZIONALE MATEMATICO ESTIVO
INTERNATIONAL MATHEMATICAL SUMMER CENTER

C.I.M.E. means Centro Internazionale Matematico Estivo, that is, International Mathematical Summer Center. Conceived in the early fifties, it was born in 1954 and made welcome by the world mathematical community where it remains in good health and spirit. Many mathematicians from all over the world have been involved in a way or another in C.I.M.E.'s activities during the past years.

So they already know what the C.I.M.E. is all about. For the benefit of future potential users and co-operators the main purposes and the functioning of the Centre may be summarized as follows: every year, during the summer, Sessions (three or four as a rule) on different themes from pure and applied mathematics are offered by application to mathematicians from all countries. Each session is generally based on three or four main courses (24–30 hours over a period of 6–8 working days) held from specialists of international renown, plus a certain number of seminars.

A C.I.M.E. Session, therefore, is neither a Symposium, nor just a School, but maybe a blend of both. The aim is that of bringing to the attention of younger researchers the origins, later developments, and perspectives of some branch of live mathematics.

The topics of the courses are generally of international resonance and the participation of the courses cover the expertise of different countries and continents. Such combination, gave an excellent opportunity to young participants to be acquainted with the most advanced research in the topics of the courses and the possibility of an interchange with the world famous specialists. The full immersion atmosphere of the courses and the daily exchange among participants are a first building brick in the edifice of international collaboration in mathematical research.

C.I.M.E. Director
Pietro ZECCA
Dipartimento di Energetica "S. Stecco"
Università di Firenze
Via S. Marta, 3
50139 Florence
Italy
e-mail: zecca@unifi.it

C.I.M.E. Secretary
Elvira MASCOLO
Dipartimento di Matematica "U. Dini"
Università di Firenze
viale G.B. Morgagni 67/A
50134 Florence
Italy
e-mail: mascolo@math.unifi.it

For more information see CIME's homepage: http://www.cime.unifi.it

CIME activity is carried out with the collaboration and financial support of:
– INdAM (Istituto Nazionale di Alta Matematica)

Jean-Louis Colliot-Thélène
Peter Swinnerton-Dyer
Paul Vojta

Arithmetic Geometry

Lectures given at the C.I.M.E. Summer School
held in Cetraro, Italy, September 10–15, 2007

Editors:
Pietro Corvaja
Carlo Gasbarri

 Springer

FONDAZIONE
CIME
ROBERTO CONTI

Authors

Prof. Jean-Louis Colliot-Thélène
Université Paris-Sud XI
CNRS
Labo. Mathématiques
Orsay 91405 CX
Bâtiment 425
France
jlct@math.u-psud.fr

Prof. Peter Swinnerton-Dyer
University of Cambridge
Dept. of Pure Math. & Math. Statistics
Wilberforce Road
CB30WB Cambridge
United Kingdom
H.P.F.Swinnerton-Dyer@dpmms.cam.ac.uk

Prof. Paul Vojta
University of California, Berkeley
Department of Mathematics
970, Evans Hall
Berkeley, CA 94720-3840
USA
vojta@math.berkeley.edu

Editors

Prof. Pietro Corvaja
Università di Udine
Dipto. di Matematica e Informatica
Via delle Scienze 206
33100 Udine
Italy
pietro.corvaja@uniud.it

Prof. Carlo Gasbarri
Université de Strasbourg
Institut de Recherche
Mathématique Avancée
7, rue René Descartes
67084 Strasbourg
France
gasbarri@math.unistra.fr

ISBN: 978-3-642-15944-2   e-ISBN: 978-3-642-15945-9
DOI: 10.1007/978-3-642-15945-9
Springer Heidelberg Dordrecht London New York

Lecture Notes in Mathematics ISSN print edition: 0075-8434
                 ISSN electronic edition: 1617-9692

Library of Congress Control Number: 2010938613

Mathematics Subject Classification (2010): 11G35, 11G25, 11D45, 14G05, 14G10, 14G40, 14M22

Cover design: SPi Publisher Services

Printed on acid-free paper

springer.com

Preface

Arithmetic Geometry can be defined as the part of Algebraic Geometry connected with the study of algebraic varieties over arbitrary rings, in particular over non-algebraically closed fields. It lies at the intersection between classical algebraic geometry and number theory.

In recent years, significant progress has been achieved in this field, in several directions. More importantly, new links between arithmetic geometry and other branches of mathematics have been developed, and new powerful tools from geometry, complex analysis, differential equations and representation theory have been imported into number theory, thus putting arithmetic geometry at the crossroads of most of contemporary mathematics.

Some links between arithmetic geometry and classical algebraic geometry come from the classification of algebraic varieties, an old subject initiated by the Italian school in the case of surfaces and developed at a rapid pace in recent time.

As discovered by Osgood and Vojta about 20 years ago, there is a formal analogy between complex analysis and both diophantine approximation and arithmetic geometry. Such analogy has revealed itself as a fertile source of ideas and problems in both complex analysis and arithmetic geometry, and it has recently led to new achievements.

The algebraic theory of differential equations is also connected to arithmetic geometry, especially with algebraic geometry in positive characteristic; many authors, starting with the founders of transcendental number theory, stressed the role of differential equations in transcendence. Recently, the theory of algebraic foliations showed new relations between these topics and diophantine approximation.

The C.I.M.E. Summer School *Arithmetic Geometry*, held in Cetraro (Cosenza, Italy), September 10–15, aimed at presenting some of the most interesting new developments of arithmetic geometry. It consisted of four courses, given by some of the most eminent contributors to the field.

Here is an overview of the three courses which have been written up.

Section 1 Variétés presque rationnelles, leurs points rationnels et leurs dégénérescences, by Jean-Louis Colliot-Thélène.

This survey addresses the general question: Over a given type of field, is there a natural class of varieties which automatically have a rational point? Fields under consideration here include: finite fields, p-adic fields, function fields in one or two variables over an algebraically closed field, C_i-fields. Classical answers are given by the Chevalley-Warning theorem and by Tsen's theorem. More general answers were provided by a theorem of Graber, Harris and Starr and by a theorem of Esnault. The latter results apply to *rationally connected varieties*.

Colliot-Thélène discusses these varieties from various angles: weak approximation (see also Swinnerton-Dyer's contribution), R-equivalence on the set of rational points, Chow group of zero-cycles.

Loosely speaking, R-equivalence on the set of rational points of a variety defined over a given field is generated by the elementary relation: to be connected by a rational curve defined over the given field. Rationally connected varieties are varieties for which R-equivalence becomes trivial when one extends the ground field to an arbitrary algebraically closed field. Rationally connected varieties play an important rôle in the classification of algebraic varieties.

Ongoing work on "rationally simply connected" varieties over function fields in two variables is also mentioned. A common thread in this report is the study of the special fibre of a scheme over a discrete valuation ring: if the generic fibre has a simple geometry, what does it imply for the special fibre?

Many examples are presented in the course showing that, despite important recent advancements, still many questions remain open, keeping the subject strongly alive.

Section 2 Topics in diophantine equations, by Sir Peter Swinnerton-Dyer.

The notes by Swinnerton-Dyer address the main problem in the theory of diophantine equations: to decide whether a given algebraic equation has solutions in integer or rational numbers.

An obvious necessary condition for the existence of rational solutions to a diophantine equation is its solubility over the reals, and more generally over p-adic completions of $\mathbb{Q}$. Since an effective procedure to decide about solubility over local fields is known, such condition is very useful in many cases. Hence it is natural to ask for which class of diophantine equations the converse also holds:

1. If the equation is soluble over every local completion of the rational number field, is it soluble over the rationals?

This is called the Hasse principle. It is known that its does not hold for an arbitrary equation. An obstruction for its validity was discovered by Manin in the seventies and is nowadays called the Brauer-Manin obstruction. The notes briefly describe this obstruction, and then address the second natural question:

2. Is the Brauer-Manin obstruction the only obstruction to the Hasse principle?

In the case when a given equation is known to be soluble, one may be interested in the distribution of its solutions, i.e., of rational points on the algebraic variety V defined by that equation. When such points are Zariski-dense, one would like to "measure" their density. There are at least two very distinct notions of density. First: for every positive integer H, we let $N(H)$ be the number of rational points of height less than H. We ask:

3. Can one estimate the growth of $N(H)$, for $H \to \infty$, in terms of the geometry of V?

Secondly: embed $V(\mathbb{Q})$ in the product $\prod_p V(\mathbb{Q}_p)$ and consider the corresponding product topology.

4. (Weak approximation) Is the image of $V(\mathbb{Q})$ dense in every finite product as above?

These problems and questions are related with many other aspects of arithmetic and geometry, and the author illustrates these links in the first chapters of his text, which can be viewed as an introduction to most of twentieth century Arithmetic Geometry.

In the second part of the notes, answers are given to the above mentioned questions in many concrete nontrivial cases, especially for surfaces. The methods employed have been pioneered by Swinnerton-Dyer himself and his collaborators in the last ten years; here a panoramic view of these methodologies is given. Also, several new examples are presented for the first time, in particular for the most important case of elliptic and rational surfaces.

Section 3 Diophantine approximation and Nevanlinna theory, by Paul Vojta.

In the eighties, P. Vojta discovered striking analogies between Nevanlinna theory in complex analysis, diophantine approximation, some results on entire curves and the distribution of integral and rational points on algebraic varieties.

Suppose that X is a projective variety defined over a field K of characteristic zero. If K is a number field we are interested in the structure of the set $X(K)$ of its rational points. If $K = \mathbb{C}$ we are interested in the image of analytic maps $f : \mathbb{C} \to X$.

We may ask the following questions in the two cases:

(1_{ar}) Is the set $X(K)$ Zariski dense?
(1_{an}) May we find maps $f : \mathbb{C} \to X$ with Zariski dense image?
(2_{ar}) Is there a finite extension L/K such that $X(L)$ is Zariski dense?
(2_{an}) Is there a finite covering $h : Y \to \mathbb{C}$ with a map $f : Y \to X$ with Zariski dense image?
(3_{ar}) May we control the size of the rational points in $X(K)$ outside of a proper Zariski closed set?
(3_{an}) Is it possible to control the order of growth of a map $f : \mathbb{C} \to X$ with Zariski dense image, in terms of the geometry of X?

Analogous question can be asked for open subsets $Y \subset X$ of algebraic varieties, namely:

(4_{ar}) Let $\mathscr{O}_K$ be the ring of integers of K. Is the set $Y(\mathscr{O}_K)$ Zariski dense?
(4_{an}) Does there exist a map $f : \mathbb{C} \to Y$ with Zariski dense image?

Many other similar questions may be asked.

The notes by P. Vojta begin by formalizing the language needed to attack these questions: In the arithmetic context, the theory of height and Weil functions is described, while in the analytic context, the appropriate Nevanlinna theory is used.

Vojta shows how, using an appropriate dictionary, the two theories have striking similarities. Also he shows how his "dictionary" can be used as a source of problems in both theories. In particular, the analogies between Roth's theorem in diophantine approximation and Nevanlinna's Second Main Theorem, between Schmidt's subspace theorem in diophantine approximation and Cartan's Theorem in Nevanlinna theory are presented, and this leads to the natural analogy between Griffiths' conjecture in complex analysis and his own conjecture on rational points.

After showing the classical results on the distribution of rational and integral points in their historical perspective, he presents some of the recent developments obtained from Schmidt's subspace theorem (and from Cartan theorem in the Nevanlinna setting), to give nontrivial answers to questions (4_{ar}) and (4_{an}) in certain cases. In the last part of the course, he explains the relations of these theories with different versions of the famous *abc* conjecture of Masser and Oesterlé, and gives some ideas on recent developments obtained by McQuillan and Yamanoi on the so-called $1 + \varepsilon$ conjecture, in the function field case. Finally, he formulates some new conjectures in arithmetic, which are strongly inspired by the work of McQuillan on the *abc* conjecture over function fields.

Pietro Corvaja
Carlo Gasbarri

Contents

Variétés presque rationnelles, leurs points rationnels et leurs dégénérescences

Jean-Louis Colliot-Thélène

1 Introduction

Voici une série de résultats classiques.

Toute forme quadratique en au moins trois variables sur le corps fini $\mathbb{F}_p$ (p premier) possède un zéro non trivial (Euler). Toute forme de degré d en $n > d$ variables sur $\mathbb{F}_p$ possède un zéro non trivial (Chevalley-Warning).

Toute forme quadratique en au moins trois variables sur le corps $\mathbb{C}(t)$ des fonctions rationnelles en une variable possède un zéro non trivial (Max Noether). Toute forme de degré d en $n+1 > d$ variables sur une extension finie de $\mathbb{C}(t)$ possède un zéro non trivial (Tsen). Ceci vaut encore sur le corps $\mathbb{C}((t))$ des séries formelles en une variable (Lang).

Sur un corps fini, sur un corps de fonctions d'une variable sur $\mathbb{C}$, sur le corps $\mathbb{C}((t))$, tout espace homogène d'un groupe algébrique linéaire connexe a un point rationnel.

Toute forme de degré d en $n > d$ variables sur le corps p-adique $\mathbb{Q}_p$ possède un zéro non trivial sur une extension non ramifiée de $\mathbb{Q}_p$ (Lang).

Toute forme de degré d en $n > d^2$ variables sur un corps de fonctions de deux variables sur $\mathbb{C}$ possède un zéro non trivial (Lang).

Toute forme quadratique en $n > 2^2$ variables sur un corps p-adique possède un zéro non trivial (Hensel, Hasse).

Toute forme cubique en $n > 3^2$ variables sur un corps p-adique possède un zéro non trivial (Demjanov, Lewis).

Pour d donné, pour presque tout premier p, toute forme possède un zéro non trivial (Ax-Kochen).

Sur un corps p-adique, tout espace homogène principal d'un groupe semi-simple simplement connexe possède un point rationnel (Kneser, Bruhat-Tits).

Sur un type donné de corps, y a-t-il une classe naturelle de variétés algébriques qui sur un tel corps ont automatiquement un point rationnel ?

J.-L. Colliot-Thélène (✉)
CNRS, Mathématiques, Bâtiment 425 Université Paris-Sud, 91405 Orsay, France
e-mail: jlct@math.u-psud.fr

P. Corvaja and C. Gasbarri (eds.), *Arithmetic Geometry*, Lecture Notes
in Mathematics 2009, DOI 10.1007/978-3-642-15945-9_1,
© Springer-Verlag Berlin Heidelberg 2011

Sur les corps de fonctions d'une variable sur $\mathbb{C}$ d'une part, sur les corps finis d'autre part, des progrès décisifs ont été accomplis dans les cinq dernières années, et on peut dans une certaine mesure dire que la situation est stabilisée. La similitude apparente des résultats est trompeuse. Les résultats cités sur les corps finis s'étendent à une classe beaucoup plus large de variétés que les résultats sur un corps de fonctions d'une variable. Les techniques utilisées sur un corps fini relèvent de la cohomologie étale (ou, de la cohomologie p-adique). Les techniques utilisées sur un corps de fonctions sur les complexes relèvent de la cohomologie cohérente : théorie de la déformation, théorèmes d'annulation de Kodaira et généralisations, programme du modèle minimal.

Sur les corps de fonctions de deux variables, la recherche est extrêmement active.

Dans ce rapport, qui ne contient pratiquement pas de démonstrations, j'ai essayé de présenter un instantané de la situation.

Une partie importante du texte suit un fil unifiant les travaux sur les corps de fonctions d'une variable, ceux sur les corps de fonctions de deux variables, et l'étude des variétés sur les corps p-adiques. C'est l'étude des modèles projectifs réguliers au-dessus d'un anneau de valuation discrète et de leur fibre spéciale.

Certains aspects de ce texte ont fait l'objet d'exposés depuis quelques années. Je remercie Esnault, Gabber, Hassett, de Jong, Kollár, Madore, Moret-Bailly, Starr et Wittenberg pour diverses discussions.

J'engage les lecteurs à consulter le rapport récent d'O. Wittenberg [69].

2 Notations, rappels et préliminaires

Soit k un corps. On note k_s une clôture séparable de k et $\overline{k}$ une clôture algébrique de k. Une k-variété est *par définition* un k-schéma séparé de type fini sur k (non nécessairement irréductible, non nécessairement réduit). On note $X(k) = \mathrm{Hom}_{\mathrm{Spec}\,k}(\mathrm{Spec}\,k, X)$ l'ensemble des points k-rationnels d'un k-schéma X. Une k-variété est dite intègre si elle est irréductible et réduite. On note alors $k(X)$ son corps des fonctions. Une k-variété est dite géométriquement intègre si la $\overline{k}$-variété $X \times_k \overline{k}$ est intègre. Une k-variété géométriquement intègre possède un ouvert de Zariski non vide qui est lisse sur k. Si k est un corps de caractéristique zéro, une k-variété intègre X est géométriquement intègre si et seulement si le corps k est algébriquement fermé dans le corps $k(X)$.

Pour la cohomologie galoisienne, et en particulier le groupe de Brauer d'un corps, le lecteur consultera Serre [65]. En plusieurs endroits on fera libre usage de la notion de dimension cohomologique d'un corps.

En quelques endroits on fera aussi usage de certaines propriétés du groupe de Brauer d'un schéma. Le lecteur se reportera aux exposés de Grothendieck [36].

Lemme 2.1 *(Nishimura, Lang) Soient k un corps, Z une k-variété régulière connexe et Y une k-variété propre. Si l'on a $Z(k) \neq \emptyset$ et s'il existe une k-application rationnelle de Z vers Y, alors $Y(k) \neq \emptyset$.*

Lemme 2.2 *Soient k un corps, Z/k une k-variété géométriquement intègre et Y/k une k-variété lisse connexe. S'il existe un k-morphisme $Z \to Y$, alors la k-variété Y est géométriquement intègre.*

Démonstration. La k-variété lisse Y est géométriquement intègre si et seulement si Y_{k_s} est irréductible. Supposons qu'elle ne le soit pas. On dispose alors du k_s-morphisme $Z_{k_s} \to Y_{k_s}$. Le groupe de Galois de k_s sur k permute les composantes de Y_{k_s}. L'image de Z_{k_s} doit se trouver dans chaque composante de Y_{k_s}. Comme Y_{k_s} est lisse, ces composantes ne se rencontrent pas. Donc Y_{k_s} n'a qu'une seule composante. ∎

Remarque 2.3. Comme l'observe Moret-Bailly, cet énoncé est une conséquence de deux résultats généraux. Soit $Z \to Y$ un k-morphisme de k-variétés. Si Z est géométriquement connexe et Y connexe, alors Y est géométriquement connexe. Par ailleurs, si Y est normal et géométriquement connexe, alors Y est géométriquement irréductible.

Obstruction élémentaire

Soient k un corps, k_s une clôture séparable de k, $\mathscr{G} = \mathrm{Gal}(k_s/k)$ le groupe de Galois absolu. Soit X une k-variété lisse géométriquement intègre. L'inclusion naturelle de groupes multiplicatifs $k_s^\times \to k_s(X)^\times$ définit une suite exacte

$$1 \to k_s^\times \to k_s(X)^\times \to k_s(X)^\times / k_s^\times \to 1.$$

La classe $e(X)$ de l'extension de modules galoisiens discrets obtenue est appelée l'obstruction élémentaire à l'existence d'un k-point : si X possède un k-point, alors $e(X) = 0$ (CT-Sansuc, voir [4]). Si $e(X) = 0$, alors pour toute extension finie séparable K/k, l'application naturelle de groupes de Brauer $\mathrm{Br}\,K \to \mathrm{Br}\,K(X)$ est injective.

Construction de grands corps

Soit k un corps de caractéristique zéro. Pour chaque corps K contenant k, donnons-nous une classe $\mathscr{C}_K$ de K-variétés algébriques géométriquement intègres admettant un ensemble E_K de K-variétés représentant toutes les classes de K-isomorphie de la classe. Pour $k \subset K \subset L$ on suppose que le changement de corps de base $K \to L$ envoie $\mathscr{C}_K$ dans $\mathscr{C}_L$.

Pour tout corps K avec $k \subset K$ supposons satisfaite la condition suivante :

(Stab) Si $f : X \to Y$ est un K-morphisme dominant de K-variétés géométriquement intègres, si Y appartient à $\mathscr{C}_K$ et si la fibre générique de f appartient à $\mathscr{C}_{K(Y)}$, alors X appartient à $\mathscr{C}_K$.

Une construction bien connue, utilisée par Merkur'ev et Suslin (cf. [23]) permet alors de construire un plongement de corps $k \subset L$ possédant les propriétés suivantes :

(i) Le corps k est algébriquement fermé dans L.
(ii) Le corps L est union de corps de fonctions de k-variétés dans $\mathscr{C}_k$.
(iii) Toute variété dans $\mathscr{C}_L$ possède un point L-rationnel.

Le principe est le suivant : s'il existe un k-variété X dans $\mathscr{C}_k$ qui ne possède pas de point rationnel, on remplace k par le corps des fonctions de cette variété. Et on itère. Je renvoie à l'article de Ducros [23] pour la construction précise, qui est reprise dans [14] et [10].

Prenons pour $\mathscr{C}_K$ la classe des K-variétés géométriquement intègres. Rappelons qu'un corps L est dit *pseudo-algébriquement clos* (PAC) si toute L-variété géométriquement intègre sur L possède un L-point. La construction ci-dessus montre que tout corps k de caractéristique zéro est algébriquement fermé dans un corps pseudo-algébriquement clos.

En prenant pour $\mathscr{C}_K$ la classe des K-variétés birationnelles à des fibrations successives de restrictions à la Weil de variétés de Severi-Brauer, Ducros [23] montre que tout corps k de caractéristique zéro est algébriquement fermé dans un corps L de dimension cohomologique $cd(L) \leq 1$.

3 Schémas au-dessus d'un anneau de valuation discrète

3.1 A-schémas de type (R), croisements normaux, croisements normaux stricts

Soit A un anneau de valuation discrète, K son corps des fractions, F son corps résiduel. Soit π une uniformisante de A.

Dans la suite de ce texte, on dira qu'un A-schéma $\mathscr{X}$ *est de type* (R) s'il satisfait les conditions suivantes :

(i) Le A-schéma $\mathscr{X}$ est propre et plat sur A.
(ii) Le schéma $\mathscr{X}$ est connexe et régulier.
(iii) La fibre générique $X = \mathscr{X} \times_A K = \mathscr{X}_K$ est une K-variété géométriquement intègre lisse sur F.

On note $K(X)$ le corps des fonctions de X, qui est aussi celui du schéma $\mathscr{X}$. On note $Y = \mathscr{X} \times_A F = \mathscr{X}_F$ la fibre spéciale de $\mathscr{X}/A$. La fibre spéciale Y est le F-schéma associé au diviseur de Cartier de $\mathscr{X}$ défini par l'annulation de π.

Comme $\mathscr{X}$ est régulier donc normal, on a une décomposition de diviseurs de Weil

$$Y = \sum_i n_i Y_i$$

où les Y_i sont les adhérences des points x_i de codimension 1 de $\mathscr{X}$ situés sur la fibre spéciale. L'anneau local de tout tel point x_i est un anneau de valuation discrète de corps des fractions $K(X)$. Si l'on note v_i la valuation sur le corps $K(X)$ associée à un tel x_i, alors $n_i = v_i(\pi)$.

Comme $\mathscr{X}$ est régulier, les Y_i sont des diviseurs de Cartier sur $\mathscr{X}$. Ce sont les composantes réduites de la fibre spéciale. Ce sont des F-variétés intègres mais non nécessairement géométriquement irréductibles ni (si le corps F n'est pas parfait) nécessairement géométriquement réduites.

On dit que $Y \subset \mathcal{X}$ *est à croisements normaux* si partout localement pour la topologie étale sur $\mathcal{X}$ l'inclusion $Y \subset \mathcal{X}$ est donnée par une équation $\prod_{i \in I} x_i^{n_i}$, où les x_i font partie d'un système régulier de paramètres et les n_i sont des entiers naturels.

On dit que $Y \subset \mathcal{X}$ *est à croisements normaux stricts* si la fibre $Y \subset \mathcal{X}$ est à croisements normaux et si de plus chaque composante réduite Y_i de Y est une F-variété (intègre) lisse. Une telle composante n'est pas nécessairement géométriquement irréductible.

On note A^h le hensélisé de A, et l'on note A^{sh} un hensélisé strict de A. On note K^h le corps des fractions de A^h et K^{sh} le corps des fractions de A^{sh}. Les inclusions $A \subset A^h \subset A^{sh}$ induisent $F = F \subset F_s$ sur les corps résiduels, où F_s est une clôture séparable de F.

On note $\hat{A}$ le complété de A. Si les corps K et F ont même caractéristique, alors il existe un corps de représentants de F dans $\hat{A}$: il existe un isomorphisme $\hat{A} \simeq F[[t]]$.

3.2 Quand la fibre spéciale a une composante de multiplicité 1

Proposition 3.1 *Soit $\mathcal{X}$ un A-schéma de type (R). Les propriétés suivantes sont équivalentes :*

(1) Il existe une composante réduite Y_i dont l'ouvert de lissité est non vide et qui satisfait $n_i = 1$.

(2) Il existe un ouvert $U \subset \mathcal{X}$ lisse et surjectif sur $\mathrm{Spec}\,A$.

(3) $\mathcal{X} \to \mathrm{Spec}\,A$ est localement scindé pour la topologie étale.

(4) $\mathcal{X}(A^{sh}) \neq \emptyset$.

(5) $X(K^{sh}) \neq \emptyset$.

(6) $\mathcal{X}(\hat{A^{sh}}) \neq \emptyset$.

(7) $X(\hat{K^{sh}}) \neq \emptyset$.

Démonstration. Laissée au lecteur.

Dans la situation ci-dessus, on dira que Y *a une composante de multiplicité 1*.[1]

Proposition 3.2 *(a) Soient $\mathcal{X}$ un A-schéma lisse connexe fidèlement plat sur A et $\mathcal{X}'/A$ un A-schéma de type (R). S'il existe une K-application rationnelle de $X = \mathcal{X}_K$ dans $X' = \mathcal{X}'_K$, alors la fibre spéciale Y' de $\mathcal{X}'/A$ a une composante de multiplicité 1.*

(b) Soient $\mathcal{X}$ et $\mathcal{X}'$ deux A-schémas de type (R). Si les fibres génériques $X = \mathcal{X}_K$ et $X' = \mathcal{X}'_K$ sont K-birationnellement équivalentes, alors la fibre spéciale Y de $\mathcal{X}$

---

[1] La terminologie adoptée dans ce texte diffère de celle de [5].

a une composante de multiplicité 1 si et seulement si la fibre spéciale Y' de $\mathcal{X}'$ a une composante de multiplicité 1.

Démonstration. Il suffit d'établir (a). L'hypothèse sur $\mathcal{X}/A$ et le lemme de Hensel assurent $\mathcal{X}(A^{sh}) \neq \emptyset$, donc $X(K^{sh}) \neq \emptyset$. Comme la K-variété X est régulière et la K-variété X' propre, d'après le lemme 2.1 l'existence d'un K^{sh}-point sur X implique l'existence d'un K^{sh}-point sur X'.

Remarque 3.3. (Wittenberg) Soient $K = \mathbb{C}(u,v)$ le corps des fractions rationnelles à deux variables et $X \subset \mathbb{P}_K^2$ la conique lisse définie par l'équation homogène $ux^2 + vy^2 = z^2$. Pour tout anneau $A \subset K$ de valuation discrète de rang 1, de corps des fractions K, on a $X(K^{sh}) \neq \emptyset$, où K^{sh} est le corps des fractions d'un hensélisé strict A^{sh} de A. Mais si $p : \mathcal{X} \to S$ est un morphisme propre et plat de variétés projectives lisses connexes de fibre générique X/K, le morphisme p n'est pas localement scindé pour la topologie étale.

3.3 Quand la fibre spéciale contient une sous-variété géométriquement intègre

Proposition 3.4 *(a) Soit $\mathcal{X}$ un A-schéma régulier connexe fidèlement plat sur A et soit $\mathcal{X}'$ un A-schéma propre. Si la fibre spéciale Y/F de $\mathcal{X}/A$ contient une sous-F-variété géométriquement intègre, et s'il existe une K-application rationnelle de $X = \mathcal{X} \times_A K$ dans $X' = \mathcal{X}' \times_A K$, alors la fibre spéciale Y' de $\mathcal{X}'/A$ contient une sous-F-variété géométriquement intègre.*

(b) Soient $\mathcal{X}$ et $\mathcal{X}'$ deux A-schémas de type (R). Si les fibres génériques $X = \mathcal{X}_K$ et $X' = \mathcal{X}'_K$ sont K-birationnellement équivalentes, alors la fibre spéciale Y contient une sous-F-variété géométriquement intègre si et seulement si la fibre spéciale Y' contient une sous-F-variété géométriquement intègre.

Démonstration. Il suffit de démontrer le point (a). On peut supposer la sous-variété intègre $Z \subset Y$ fermée. Soit $Z_{lisse} \subset Z$ l'ouvert de lissité de Z/F. Soit $p : \mathcal{X}_1 \to \mathcal{X}$ l'éclaté de $\mathcal{X}$ le long de Z. L'image réciproque de Z_{lisse} dans $\mathcal{X}_1$ est un fibré projectif sur Z_{lisse}, qui est une F-variété géométriquement intègre. Soit x son point générique. C'est un point régulier de codimension 1 sur $\mathcal{X}_1$. L'application rationnelle $\mathcal{X}_1 \to \mathcal{X}'$ est donc définie au point x. Soit $x' \in \mathcal{X}'$ son image. L'adhérence de x' dans $\mathcal{X}'$ est une sous-F-variété fermée de $\mathcal{X}'$, munie d'une application F-rationnelle dominante d'une F-variété géométriquement intègre. C'est donc une F-variété géométriquement intègre.

Proposition 3.5 *Soit A un anneau de valuation discrète de corps résiduel F et soit $\mathcal{X}$ un A-schéma de type (R). Supposons les composantes réduites Y_i lisses sur F. Les propriétés suivantes sont équivalentes :*

(1) La fibre spéciale Y contient une sous-F-variété géométriquement intègre.

(2) Il existe une F-variété géométriquement intègre Z et un F-morphisme Z → Y.

(3) Il existe une composante réduite Y_i de Y qui est géométriquement intègre.

Si de plus $car(F) = 0$, ces propriétés sont équivalentes aux propriétés suivantes :

(4) Il existe une extension locale d'anneaux de valuation discrète $A \subset B$ telle que le corps résiduel $F = F_A$ de A soit algébriquement fermé dans le corps résiduel F_B de B, et que l'on ait $X(K(B)) = \mathscr{X}(B) \neq \emptyset$.

(5) Si $F \hookrightarrow E$ est un plongement de F dans un corps pseudo-algébriquement clos E dans lequel F est algébriquement fermé, et si $\hat{A} = F[[t]]$, il existe une extension finie totalement ramifiée $L/E((t))$ avec $X(L) \neq \emptyset$.

Démonstration. Soit $f : Z \to Y$ comme en (2). Une telle application se factorise par au moins un morphisme $Z \to Y_i$ pour i convenable, et le lemme 2.2 montre que Y_i est alors géométriquement intègre, on a donc (3). Les autres implications entre (1), (2) et (3) sont évidentes. L'énoncé (5) implique trivialement (4), et (4) implique (5) comme l'on voit en passant aux complétés et en remplaçant $\hat{B} \simeq F_B[[u]]$ dans $E[[u]]$, où $F_B \hookrightarrow E$ est un plongement du corps résiduel F_B dans un corps pseudo-algébriquement clos E dans lequel F_B et donc aussi F_A est algébriquement fermé. De (4) on déduit l'existence d'un F-morphisme $\operatorname{Spec} F_B \to Y$, ce qui implique l'énoncé (2). Soit $Y_i \subset Y$ une composante comme en (3). Soit B l'anneau local du point générique de Y_i sur $\mathscr{X}$. L'inclusion $A \subset B$ satisfait (4).

Dans la situation ci-dessus, on dira que la fibre spéciale Y/F *a une composante (réduite) géométriquement intègre.*

Des deux propositions précédentes il résulte :

Proposition 3.6 *Soient $\mathscr{X}$ et $\mathscr{X}'$ deux A-schémas de type(R), de fibres spéciales respectives Y et Y'. Supposons les composantes réduites de Y et Y' lisses sur F. S'il existe une application K-rationnelle de $X = \mathscr{X}_K$ vers $X' = \mathscr{X}'_K$, et si Y a une composante géométriquement intègre, alors Y' a une composante géométriquement intègre.*

Remarque 3.7. On trouvera dans l'article [24] de Ducros de nombreux compléments et extensions des énoncés ci-dessus.

3.4 *Quand la fibre spéciale a une composante géométriquement intègre de multiplicité 1*

Proposition 3.8 *Soit A un anneau de valuation discrète de corps résiduel F et soit $\mathscr{X}$ un A-schéma de type (R). Les propriétés suivantes sont équivalentes :*

(1) Il existe une composante réduite Y_i qui est géométriquement intègre et pour laquelle $n_i = 1$.

(2) *Il existe un ouvert $U \subset \mathscr{X}$ non vide lisse, surjectif sur Spec A et à fibres géométriquement intègres.*

Pour F de caractéristique zéro, ces propriétés sont équivalentes aux propriétés suivantes :

(3) *Il existe une extension non ramifiée d'anneaux de valuation discrète $A \subset B$ telle que F soit algébriquement fermé dans le corps résiduel de B et que $X(K(B)) = \mathscr{X}(B) \neq \emptyset$.*

(4) *Si $F \hookrightarrow E$ est un plongement de F dans un corps pseudo-algébriquement clos E dans lequel F est algébriquement fermé, et si $\hat{A} = F[[t]]$, on a $X(E((t))) \neq \emptyset$.*

L'existence d'une composante comme en (1) est une condition nécessaire pour l'existence d'un K-point sur X.

Démonstration. L'équivalence de (1) et (2) est claire. L'équivalence de (3) et (4) est aussi claire. Pour l'équivalence entre (1) et (2) d'une part et (3) et (4) d'autre part, et pour la démonstration de la dernière assertion, voir [13], fin de l'argument p. 745.

Dans la situation ci-dessus, on dira que Y *a une composante géométriquement intègre de multiplicité 1.*

Proposition 3.9 *Supposons F de caractéristique zéro.*

(a) *Soit $\mathscr{X}/A$ un A-schéma connexe lisse et surjectif sur Spec A, à fibres géométriquement intègres. Soit $\mathscr{X}'/A$ un A-schéma de type (R). S'il existe une application K-rationnelle de $X = \mathscr{X}_K$ vers $X' = \mathscr{X}_K'$, alors il existe un ouvert $U \subset \mathscr{X}'$ tel que le morphisme induit $U \to $ Spec A soit lisse et surjectif.*

(b) *Soient $\mathscr{X}$ et $\mathscr{X}'$ deux A-schémas de type (R). Si les fibres génériques $X = \mathscr{X}_K$ et $X' = \mathscr{X}_K'$ sont K-birationnellement équivalentes, alors la fibre spéciale Y a une composante géométriquement intègre de multiplicité 1 si et seulement si la fibre spéciale Y' a une composante géométriquement intègre de multiplicité 1.*

Démonstration. Cela résulte immédiatement du lemme 2.1 et de la caractérisation (3) dans la proposition 3.8.

Question 3.10 *Soient k un corps de caractéristique zéro, K un corps de type fini sur k, X une K-variété projective, lisse, géométriquement intègre. Supposons que pour tout anneau de valuation discrète de rang un A contenant k et de corps des fractions K il existe un A-modèle de type (R) de X/K dont la fibre spéciale contient une composante géométriquement intègre de multiplicité 1. Existe-t-il un k-morphisme $\mathscr{X} \to B$ de k-variétés projectives, lisses, géométriquement intègres satisfaisant les propriétés suivantes :*

(a) *le corps des fonctions $k(B)$ de B est K ;*
(b) *la fibre générique de $\mathscr{X} \to B$ est K-isomorphe à X ;*
(c) *il existe un ouvert $U \subset \mathscr{X}$ tel que le morphisme induit $U \to B$ soit lisse surjectif (fidèlement plat) et à fibres géométriquement intègres.*

On comparera cette question avec la remarque 3.3.

Remarque 3.11. On exhibe facilement un morphisme $\mathscr{X} \to Y = \mathbb{P}^2_{\mathbb{Q}}$ de $\mathbb{Q}$-variétés projectives, lisses, géométriquement intègres, de fibre générique une quadrique de dimension 2, tel que la fibre en tout point de codimension 1 de Y soit géométrique–ment intègre sans que pour autant l'hypothèse dans la question ci-dessus soit satisfaite (désingulariser l'exemple de la remarque 13.4).

3.5 Un exemple : quadriques

Discutons le cas des modèles de quadriques de dimension au moins 1. Supposons $2 \in A^\times$. Soit v la valuation de A. Une quadrique lisse dans $\mathbb{P}^n_K$ ($n \geq 2$) peut être définie par une forme quadratique diagonale sur K.

Considérons le cas des coniques. En chassant les dénominateurs et en poussant les carrés dans les variables, on voit que l'équation définissant la quadrique dans $\mathbb{P}^2_K$ peut s'écrire

$$a_0 T_0^2 + a_1 T_1^2 + a_2 T_2^2 = 0$$

avec $a_0, a_1 \in A^\times$ et $v(a_2) = 0$ ou $v(a_2) = 1$. Cette équation définit un modèle régulier $\mathscr{X} \subset \mathbb{P}^2_A$, et la fibre spéciale $Y \subset \mathscr{X}$ est à croisements normaux.

Si $v(a_2) = 0$, alors la fibre spéciale Y/F est une conique lisse, en particulier géométriquement intègre, et $Y \subset \mathscr{X}$ est à croisements normaux stricts.

Si $v(a_2) = 1$, la fibre spéciale Y possède un F-point rationnel évident, $P \in Y(F)$, donné par $T_0 = T_1 = 0$.

Si $v(a_2) = 1$ et si la classe de $-a_0.a_1$ dans F est un carré, la fibre spéciale se décompose sous la forme

$$Y = Y_1 + Y_2$$

avec chaque $Y_i \simeq \mathbb{P}^1_F$. Dans ce cas $Y \subset \mathscr{X}$ est à croisements normaux stricts.

Si $v(a_2) = 1$ et si la classe de $-a_0.a_1$ dans F n'est pas un carré, alors la fibre spéciale Y/F est intègre, mais se décompose sur une extension quadratique de F en deux droites conjuguées se rencontrant en P, donc Y n'est pas lisse, $Y \subset \mathscr{X}$ n'est pas à croisements normaux stricts. Si l'on éclate le point rationnel singulier P sur $\mathscr{X}$, on obtient un modèle $\mathscr{X}'/A$ dont la fibre spéciale Y' se décompose sous la forme

$$Y' = Y'_0 + 2E,$$

où $E \subset \mathscr{X}$ est le diviseur exceptionnel introduit par l'éclatement. La F-courbe Y'_0 est intègre, elle se décompose sur une extension quadratique de F en la somme de deux droites conjuguées ne se rencontrant pas, et rencontrant E transversalement. Donc Y'_0 est lisse, et $Y' \subset \mathscr{X}'$ est à croisements normaux stricts.

En résumé, pour toute conique lisse sur K, on a les propriétés suivantes.

(a) Il existe un modèle régulier $\mathscr{X}$ avec $Y \subset \mathscr{X}$ à croisements normaux dont au moins une composante a multiplicité 1 et admet un ouvert non vide lisse sur F, mais n'est pas nécessairement géométriquement intègre.

(b) Il existe un modèle régulier $\mathscr{X}$ avec $Y \subset \mathscr{X}$ à croisements normaux dont la fibre spéciale contient une sous-F-variété géométriquement intègre.

(c) Il existe un modèle régulier $\mathscr{X}$ avec $Y \subset \mathscr{X}$ à croisements normaux stricts dont une composante est géométriquement intègre (mais pas nécessairement de multiplicité 1). D'après les paragraphes 3.2 et 3.3 les propriétés (a) et (b) valent pour tout A-modèle de type (R) et la propriété (c) vaut pour tout A-modèle dont la fibre spéciale est à croisements normaux stricts.

Considérons le cas des quadriques de dimension au moins 3. On peut définir une telle quadrique dans $\mathbb{P}_K^n$ ($n \geq 4$) par une équation

$$\sum_{i=0}^{n} a_i T_i^2 = 0,$$

avec $a_i \in A^\times$. Dans $\mathbb{P}_A^n$, cette équation définit un modèle intègre, normal et propre sur A. La fibre spéciale Y/F est géométriquement intègre (et en particulier de multiplicité 1). D'après le paragraphe 3.4 cette propriété vaut alors pour tout A-modèle de type (R).

4 Groupe de Brauer des schémas au-dessus d'un anneau de valuation discrète

Pour les démonstrations des résultats énoncés dans ce paragraphe, le lecteur se reportera aux exposés de Grothendieck [36].

Dans cette section, la cohomologie employée est la cohomologie étale, qui sur un corps est la cohomologie galoisienne du corps (c'estt-à-dire de son groupe de Galois absolu). Soit A un anneau de valuation discrète de corps des fractions K et de corps résiduel F parfait. On dispose alors d'une application résidu

$$\partial_A : \mathrm{Br}\, K \to H^1(F, \mathbb{Q}/\mathbb{Z})$$

envoyant le groupe de Brauer de K dans le groupe des caractères du groupe de Galois absolu de F. Plus précisément, on a une suite exacte

$$0 \to \mathrm{Br}\, A \to \mathrm{Br}\, K \to H^1(F, \mathbb{Q}/\mathbb{Z}).$$

La flèche de droite est surjective sur la torsion première à la caractéristique de F.

Lorsque A est hensélien, la flèche naturelle $\mathrm{Br}\, A \to \mathrm{Br}\, F$ est un isomorphisme. Si de plus F est de dimension cohomologique ≤ 1, alors $\mathrm{Br}\, A = 0$ et $\mathrm{Br}\, K \simeq H^1(F, \mathbb{Q}/\mathbb{Z})$.

Soit $A \hookrightarrow B$ un homomorphisme local d'anneaux de valuation discrète à corps résiduels parfaits. Soit $K \subset L$ l'inclusion de corps de fractions correspondante. Soit e l'indice de ramification de B sur A, c'est-à-dire la valuation dans B de l'image d'une uniformisante de A. Soit $F_A \hookrightarrow F_B$ l'inclusion des corps résiduels. On a alors le diagramme commutatif suivant :

$$
\begin{array}{ccc}
\operatorname{Br} K & \xrightarrow{\partial_A} & H^1(F_A, \mathbb{Q}/\mathbb{Z}) \\
\downarrow \operatorname{Res}_{K,L} & & \downarrow e_{B/A}.\operatorname{Res}_{F_A, F_B} \\
\operatorname{Br} L & \xrightarrow{\partial_B} & H^1(F_B, \mathbb{Q}/\mathbb{Z}).
\end{array}
$$

Soit $F_B' \subset F_B$ la fermeture algébrique de F_A dans F_B. Le noyau de

$$
e_{B/A}.\operatorname{Res}_{F_A, F_B} : H^1(F_A, \mathbb{Q}/\mathbb{Z}) \to H^1(F_B, \mathbb{Q}/\mathbb{Z})
$$

s'identifie au noyau de

$$
e_{B/A}.\operatorname{Res}_{F_A, F_B} : H^1(F_A, \mathbb{Q}/\mathbb{Z}) \to H^1(F_B', \mathbb{Q}/\mathbb{Z})
$$

Soit A un anneau de valuation discrète de corps des fractions K de corps résiduel un corps F de caractéristique zéro. Soit $\mathscr{X}$ un A-schéma de type (R). Soit X/K la fibre générique. Soit $Y = \sum_i e_i Y_i$ la décomposition de la fibre spéciale Y en diviseurs intègres. Soit F_i la fermeture algébrique de F dans $F(Y_i)$.

Comme les schémas intègres $\mathscr{X}$ et X sont réguliers, les applications de restriction $\operatorname{Br} \mathscr{X} \to \operatorname{Br} X \to \operatorname{Br} K(X)$ sont injectives. On dispose alors du diagramme commutatif de suites exactes

$$
\begin{array}{ccccc}
0 \to & \operatorname{Br} A & \to \operatorname{Br} K & \xrightarrow{\partial_A} & H^1(F, \mathbb{Q}/\mathbb{Z}) \\
 & \downarrow & \downarrow & & \downarrow e_i.\operatorname{Res}_{F, F(Y_i)} \\
0 \to & \operatorname{Br} \mathscr{X} & \to \operatorname{Br} X & \xrightarrow{\oplus_i \partial_i} & \oplus_i H^1(F(Y_i), \mathbb{Q}/\mathbb{Z})
\end{array}
$$

et de la suite exacte qui s'en déduit

$$
0 \to \operatorname{Ker}[\operatorname{Br} A \to \operatorname{Br} K(X)] \to \operatorname{Ker}[\operatorname{Br} K \to \operatorname{Br} K(X)] \to
$$
$$
\operatorname{Ker}[H^1(F, \mathbb{Q}/\mathbb{Z}) \xrightarrow{\oplus_i e_i.\operatorname{Res}_{F, F_i}} \oplus_i H^1(F_i, \mathbb{Q}/\mathbb{Z})]
$$

Proposition 4.1 *Soit A un anneau de valuation discrète hensélien à corps résiduel F de caractéristique zéro et de dimension cohomologique au plus 1. Soit $\mathscr{X}$ un A-schéma de type (R), de fibre générique X. Avec les notations ci-dessus, les deux propriétés suivantes sont équivalentes :*

(i) L'application $\operatorname{Br} K \to \operatorname{Br} X / \operatorname{Br} \mathscr{X}$ est injective.
(ii) L'application $H^1(F, \mathbb{Q}/\mathbb{Z}) \xrightarrow{\oplus_i e_i.\operatorname{Res}_{F, F_i}} \oplus_i H^1(F_i, \mathbb{Q}/\mathbb{Z})$ est injective.

En particulier, si la fibre spéciale Y possède une composante géométriquement intègre de multiplicité 1, ou plus généralement si le pgcd des entiers $e_i.[F_i : F]$ est égal à 1, alors $\operatorname{Br} K \to \operatorname{Br} X / \operatorname{Br} \mathscr{X}$ est injective, et il en est donc de même de $\operatorname{Br} K \to \operatorname{Br} X$ et de $\operatorname{Br} K \to \operatorname{Br} K(X)$.

Je renvoie à [15] pour une discussion du cas où le corps résiduel F est fini.

5 Corps C_i

Théorème 5.1 *(Tsen, 1933) Soit F un corps algébriquement clos et $K = F(C)$ un corps de fonctions d'une variable sur F. Soit $X \subset \mathbb{P}^n_K$ une hypersurface de degré d. Si l'on a $n \geq d$, alors $X(K) \neq \emptyset$.*

On notera que l'on ne fait aucune hypothèse sur X, qui peut être réductible.

Le cas des coniques ($d = 2, n = 2$) avait été établi par Max Noether par une méthode géométrique.

Soit $i \geq 0$ un entier. On dit qu'un corps K possède la propriété C_i si toute forme à coefficients dans K, de degré d en $n + 1 > d^i$ variables a un zéro non trivial sur K. On dit qu'un corps K possède la propriété C'_i si pour toute famille finie de formes $\{\Phi_j(X_0, \ldots, X_n)\}_{j=1,\ldots,r}$ de degrés respectifs $d_1, \ldots, d_r$ avec $n + 1 > \sum_{j=1}^r d_j^i$ il existe un zéro commun non trivial sur K.

La propriété C_i implique la propriété ci-dessus pour un système de formes $\{\Phi_j\}$ lorsque tous les degrés d_j sont égaux (Artin, Lang, Nagata). On ne sait pas si en général C_i implique C'_i (voir [63]).

Un corps est algébriquement clos si et seulement si il est C_0. Le théorème de Tsen dit qu'un corps de fonctions d'une variable sur un corps algébriquement clos est un corps C_1. Dans sa thèse, suivant des suggestions d'E. Artin, S. Lang généralisa le théorème de Tsen. Son résultat, pour lequel on trouve quelques antécédents dans les textes des géomètres algébristes italiens, fut complété par Nagata. Le résultat général est le suivant.

Théorème 5.2 *(Lang, Nagata) [54] Soit K un corps C_i. Toute extension algébrique de K est un corps C_i. Le corps des fractions rationnelles en une variable $K(t)$ est C_{i+1}. De façon générale, toute extension de degré de transcendance n de K est un corps C_{i+n}.*

Lang établit aussi le théorème suivant.

Théorème 5.3 *(Lang) [54] Soit A un anneau de valuation discrète hensélien de corps des fractions K et de corps résiduel F. Soit $\hat{K}$ le complété de K. Supposons $\hat{K}$ séparable sur K. Si F est algébriquement clos, alors K est un corps C_1.*

En particulier l'extension maximale non ramifiée $\mathbb{Q}_p^{nr}$ du corps p-adique $\mathbb{Q}_p$ est un corps C_1.

Dans la situation considérée au paragraphe 3 (voir la Proposition 3.1), si l'on suppose le corps résiduel F de A parfait, ce théorème assure que la fibre spéciale d'une hypersurface de degré d dans $\mathbb{P}^n_K$ avec $n \geq d$ possède une composante de multiplicité 1.

Théorème 5.4 *(Greenberg)[35] Si F est un corps C_i alors $K = F((t))$ est un corps C_{i+1}.*

Ce théorème ne s'étend pas dans une situation d'inégale caractéristique : les corps p-adiques ne sont pas C_2 (Terjanian). Ils ne sont en fait C_i pour aucun i (Arkhipov et Karatsuba).

Le théorème de Tsen est souvent mis en parallèle avec l'énoncé suivant, qui implique que les corps finis sont des corps C_1.

Théorème 5.5 *(Chevalley, Warning, 1935) Soit $\mathbb{F}$ un corps fini de caractéristique p. Soit $X \subset \mathbb{P}_{\mathbb{F}}^n$ une hypersurface de degré d. Si l'on a $n \geq d$, alors $X(\mathbb{F}) \neq \emptyset$. Plus précisément, le cardinal de $X(\mathbb{F})$ est congru à 1 modulo p.*

On montra plus tard (Ax (1964), Katz (1971)) que si le cardinal de $\mathbb{F}$ est q alors le cardinal de $X(\mathbb{F})$ est congru à 1 modulo q.

Pour le corps fini à p éléments, le cas des coniques avait été établi par Euler [29], dans un article où il établit aussi la formule de multiplication pour les sommes de quatre carrés. La combinaison de ces deux résultats lui permet de montrer que tout rationnel positif est une somme de quatre carrés de rationnels.

Comme pour le théorème de Tsen, le théorème de Chevalley-Warning ne fait aucune hypothèse sur X. On peut voir là l'origine de la conjecture suivante.

Conjecture 5.6 *(Ax) [2] Soient K un corps et $X \subset \mathbb{P}_K^n$ une hypersurface de degré d. Si l'on a $n \geq d$, alors il existe une sous-K-variété $Y \subset X$ qui est géométriquement irréductible.*

Si K est parfait, dans la conclusion on peut remplacer « géométriquement irréductible » par « géométriquement intègre ». Mais comme l'exemple de la forme irréductible $T_0^2 + xT_1^2 + yT_2^2$ sur le corps $K = \mathbb{F}_2(x, y)$ le montre, ceci ne vaut pas sur K corps non parfait.

Le cas $d = 2$ a été discuté plus haut. Le cas $d = 3$ est facile. Le cas $d = 4$ fut établi par Denef, Jarden et Lewis dans [22]. Dans le même article, les auteurs établissent la conjecture lorsque K contient un corps algébriquement clos. La démonstration de ce résultat utilise la théorie des corps hilbertiens.

En caractéristique nulle, la conjecture d'Ax est maintenant un théorème de Kollár (Théorème 7.10 ci-après).

Les corps finis, les corps de fonctions d'une variable, le corps $\mathbb{C}((t))$ sont des corps de dimension ≤ 1 au sens de Serre ([65], II.3.1) : le groupe de Brauer de toute extension finie de k est trivial. De fait, tout corps C_1 est de dimension ≤ 1. On sait (Ax) que la réciproque est fausse (pour des références et d'autres résultats dans cette direction, voir [14]).

Tout espace homogène d'un groupe algébrique linéaire connexe sur un corps parfait de dimension ≤ 1 a un point rationnel (Steinberg, Springer, voir [65]).

Les corps finis et le corps $\mathbb{C}((t))$ ont des groupes de Galois isomorphes au groupe $\hat{\mathbb{Z}}$. Tout espace homogène d'une variété abélienne sur un corps fini possède un point rationnel (Lang) mais ceci n'est pas vrai sur $\mathbb{C}((t))$, comme le montre l'exemple de la courbe de genre 1 donnée dans $\mathbb{P}_{\mathbb{C}((t))}^2$ par l'équation $X^3 + tY^3 + t^2Z^3 = 0$.

6 *R*-équivalence et équivalence rationnelle sur les zéro-cycles

Soient k un corps et X une k-variété. Deux k-points de X sont élémentairement R-liés s'il existe un k-morphisme $U \to X$ d'un ouvert U de $\mathbb{P}^1_k$ tel que les deux points soient dans l'image de $U(k)$. La relation d'équivalence sur $X(k)$ engendrée par cette relation est appelée la R-équivalence [59]. Pour k de caractéristique zéro, l'ensemble $X(k)/R$ est un invariant k-birationnel des k-variétés projectives, lisses, géométriquement intègres [16].

La R-équivalence a été beaucoup étudiée lorsque X est un k-groupe linéaire [16, 31].

Soit X/k une k-variété algébrique. On note $Z_0(X)$ le groupe abélien libre sur les points fermés de X. C'est le groupe des zéro-cycles de X. Le groupe de Chow (de degré zéro), noté $CH_0(X)$, est le quotient du groupe $Z_0(X)$ par le sous-groupe engendré par les zéro-cycles de la forme $\pi_*(\mathrm{div}_C(g))$, où $\pi : C \to X$ est un k-morphisme propre d'une k-courbe C normale intègre, g est une fonction rationnelle sur C et $\mathrm{div}_C(g) \in Z_0(C)$ est son diviseur.

Si X/k est propre, l'application linéaire $\deg_k : Z_0(X) \to \mathbb{Z}$ envoyant un point fermé P sur son degré $[k(P) : k]$ passe au quotient par l'équivalence rationnelle ci-dessus définie. On note $A_0(X)$ le noyau de l'application $\deg : CH_0(X) \to \mathbb{Z}$, et on l'appelle le groupe de Chow réduit de X.

Le groupe de Chow réduit est un invariant k-birationnel des k-variétés projectives, lisses, géométriquement intègres.

Pour X/k propre, l'application évidente $X(k) \to Z_0(X)$ induit une application

$$X(k)/R \to CH_0(X)$$

dont l'image tombe dans l'ensemble des classes de cycles de degré 1.

7 Autour du théorème de Tsen : variétés rationnellement connexes

Dans le programme de classification de Mori est apparue au début des années 1990 la notion de variété rationnellement connexe. Les travaux fondateurs résultent d'une collaboration entre Kollár, Miyaoka et Mori ; certains des résultats sont dus à Campana. Un rôle-clé y est joué par la théorie des déformations, plus précisément par l'étude infinitésimale des schémas Hom, cas particulier des schémas de Hilbert. On consultera les livres [46] et [21], ainsi que les articles [1] et le récent rapport [67].

Tout fibré vectoriel sur la droite projective est isomorphe à une somme directe de fibrés de rang 1, donc de la forme $\mathcal{O}(n), n \in \mathbb{Z}$. Soit k un corps algébriquement clos. Soit X une k-variété algébrique projective, lisse, connexe, de dimension d. Soit T_X son fibré tangent. On dit que X est séparablement rationnellement connexe (SRC) s'il existe un morphisme $f : \mathbb{P}^1 \to X$ très libre, c'est-à-dire tel que dans une (donc dans toute) décomposition du fibré vectoriel

$$f^* T_X \simeq \oplus_{i=1}^{d} \mathscr{O}(a_i),$$

on ait $a_i \geq 1$ pour tout i.

On dit qu'une variété projective lisse et connexe X est rationnellement connexe (RC) si, pour tout corps algébriquement clos Ω contenant k, par un couple général de points de $X(\Omega)$ il passe une courbe de genre zéro, i.e. il existe un Ω-morphisme $\mathbb{P}^1_\Omega \to X_\Omega$ dont l'image contient les deux points.

On dit qu'une variété projective lisse et connexe X est rationnellement connexe par chaînes (RCC) si, pour tout corps algébriquement clos Ω contenant k, tout couple général de points de $X(\Omega)$ est lié par une chaîne de courbes de genre zéro. Cette dernière propriété est équivalente à la condition que tout couple de points de $X(\Omega)$ est lié par une chaîne de courbes de genre zéro. En d'autres termes, l'ensemble $X(\Omega)/R$ est réduit à un élément.

Au lieu de faire les hypothèses ci-dessus pour tout corps algébriquement clos Ω contenant k, il suffit de les faire pour un tel corps non dénombrable.

Toute variété RC est clairement RCC.

Théorème 7.1 (*Kollár-Miyaoka-Mori*) *Toute variété SRC est RC donc RCC. En caractéristique zéro, ces trois propriétés sont équivalentes, et elles impliquent :*

Pour tout corps algébriquement clos Ω contenant k et tout ensemble fini de points $x_1, \ldots, x_n \in X(\Omega)$ il existe un morphisme $f : \mathbb{P}^1_\Omega \to X_\Omega$ très libre tels que tous les x_i soient dans l'image de f.

En dimension 1, une variété est RC si et seulement si elle est une courbe lisse de genre zéro. En dimension 2, une variété est SRC si et seulement si elle est rationnelle, i.e. birationnelle à un espace projectif. Une variété projective et lisse unirationnelle est RC. Une variété projective et lisse séparablement unirationnelle est SRC. Sur k algébriquement clos, le groupe de Chow réduit $A_0(X)$ d'une variété RCC est clairement trivial.

Si k est un corps quelconque, une k-variété est dite rationnellement connexe resp. rationnellement connexe par chaînes, respectivement séparablement rationnellement connexe si elle est géométriquement intègre et si elle est RC, resp. RCC, resp. SRC, après passage à un corps algébriquement clos contenant k.

Les compactifications lisses d'espaces homogènes de groupes algébriques linéaires connexes sont des variétés RC.

En dimension 2, on dispose d'une classification k-birationnelle des k-surfaces SRC, c'est-à-dire des k-surfaces projectives, lisses (géométriquement) rationnelles (Enriques, Manin, Iskovskikh, Mori) : toute telle surface est k-birationnelle à une k-surface de del Pezzo ou à une k-surface fibrée en coniques au-dessus d'une conique lisse. Une surface de del Pezzo X est une surface projective et lisse dont le fibré anticanonique ω_X^{-1} est ample. Le degré d'une telle surface est l'entier $d = (\omega.\omega)$. Il satisfait $1 \leq d \leq 9$.

Le groupe de Chow réduit $A_0(X)$ d'une variété RCC sur un corps k quelconque est clairement un groupe de torsion. On peut montrer (Prop. 11.1) qu'il est annulé par un entier $N = N(X) > 0$.

Une variété de Fano est une variété projective lisse dont le fibré anticanonique est ample. Si $X \subset \mathbb{P}^n$ est une intersection complète lisse connexe définie par des formes $\Phi_j, j = 1, \ldots, r$ de degrés respectifs $d_j, j = 1, \ldots, r$, alors X est de Fano si et seulement si $n \geq \sum_j d_j$. On reconnaît là la condition C_1.

Un théorème difficile est le suivant ([6, 51] ; voir aussi [46, 21]) :

Théorème 7.2 *(Campana, Kollár-Miyaoka-Mori) Une variété de Fano est rationnellement connexe par chaînes.*

Le théorème suivant peut donc être vu comme une généralisation du théorème de Tsen.

Théorème 7.3 *(Graber, Harris, Starr [34] ; de Jong, Starr [42]) Soit F un corps algébriquement clos et $K = F(C)$ un corps de fonctions d'une variable sur F. Soit X une K-variété séparablement rationnellement connexe. Alors $X(K) \neq \emptyset$.*

Ce théorème implique le résultat suivant.

Théorème 7.4 *Soit F un corps algébriquement clos de caractéristique zéro. Soit f : $X \to Y$ un morphisme dominant de F-variétés projectives et lisses, à fibre générique géométriquement intègre. Si Y est rationnellement connexe et si la fibre générique est une variété rationnellement connexe, alors X est une variété rationnellement connexe.*

Le théorème 7.3 a aussi le corollaire suivant, connu des experts.

Théorème 7.5 *Soit R un anneau de valuation discrète hensélien équicaractéristique de corps résiduel F algébriquement clos, de corps des fractions K. Soit $X \subset \mathbb{P}^n_K$ une K-variété séparablement rationnellement connexe. Alors $X(K) \neq \emptyset$.*

Démonstration. Soit S le complété de R et L le corps des fractions de S. Comme X/K est lisse, $X(K)$ est dense dans $X(L)$ pour la topologie définie par la valuation de K ([5], Chap. 3.6, Cor. 10 p. 82). Il suffit donc d'établir le théorème en remplaçant R par S et K par L. L'anneau S admet alors un corps de représentants isomorphe à F, on peut donc identifier $S = F[[t]]$ et $L = F((t))$.

Les lettres R et K étant désormais libres, notons maintenant R le hensélisé de $F[t]$ en $t = 0$ et K le corps des fractions de R. Le corps $L = F((t))$ est le complété de K.

Soit $\mathscr{X}' \subset \mathbb{P}^n_S$ l'adhérence schématique de $X \subset \mathbb{P}^n_L$. C'est un schéma intègre projectif et plat sur S. La F-algèbre S est la limite inductive filtrante de ses F-sous-algèbres de type fini. Il existe donc une F-algèbre de type fini intègre $A \subset S$ et un A-schéma $\mathscr{X}$ projectif et plat tel que $\mathscr{X} \times_A S \simeq \mathscr{X}'$. En particulier la fibre générique de $\mathscr{X}/A$ est une variété séparablement rationnellement connexe. Il existe un ouvert non vide $U \subset Y = \operatorname{Spec} A$, qu'on peut prendre lisse sur F, tel que toutes les fibres du morphisme $\mathscr{X} \to Y$ au-dessus de U sont des variétés séparablement rationnellement connexes (Kollár, Miyaoka, Mori, voir [46] IV.3.11).

Notons $\xi \in Y(S) \subset Y(L)$ le point correspondant à $A \subset S$. Munissons $U(L)$ de la topologie définie par la valuation de L. L'ensemble $U(K)$ est dense dans $U(L)$ ([5], Chap. 3.6, Cor. 10 p. 82). Pour tout entier $n \geq 1$ l'application naturelle $Y(S) \to Y(S/t^n)$ a ses fibres ouvertes dans $Y(S) \subset Y(L)$. On peut donc trouver dans $U(K) \subset Y(K)$ un point ξ_n qui soit dans $Y(S)$ et donc dans $Y(R)$ et qui ait même image que ξ dans $Y(R/t^n) = Y(S/t^n)$. On peut donc pour tout $n \geq 1$ trouver un R-schéma projectif $\mathscr{X}_n$ à fibre générique X_n/K séparablement rationnellement connexe, tel que $\mathscr{X}_n \times_R R/t^n \simeq \mathscr{X} \times_A \times_S S/t^n$ (noter que l'on a $R/t^n = S/t^n$.) Le corps des fractions du hensélisé R de $F[t]$ en $t = 0$ est l'union de corps de fonctions de F-courbes. Le théorème 7.3 assure donc $X_n(K) \neq \emptyset$. Comme $\mathscr{X}_n/R$ est projectif, on a donc $\mathscr{X}_n(R) = X_n(K) \neq \emptyset$, donc $\mathscr{X}(S/t^n) = \mathscr{X}_n(R/t^n) \neq \emptyset$. On a donc $\mathscr{X}(S/t^n) \neq \emptyset$ pour tout entier n. Par un théorème de Greenberg ([35], Thm. 1) ceci implique $\mathscr{X}(S) \neq \emptyset$. On a donc $X(L) \neq \emptyset$.

On ne connaît pas de démonstration de ce théorème qui ne passe pas par le cas global $K = F(C)$.

Le théorème 7.3 admet un théorème « réciproque » :

Théorème 7.6 (*Graber-Harris-Mazur-Starr*)[33] *Soit $k = \mathbb{C}$. Soit S une variété lisse sur $\mathbb{C}$ de dimension au moins 2. Soit $X \to S$ un morphisme projectif et lisse à fibre générique géométriquement intègre. Si la restriction de $X \to S$ à toute courbe $C \subset S$ admet une section, alors il existe une $\mathbb{C}(S)$-variété Z géométriquement intègre et rationnellement connexe et un $\mathbb{C}(S)$-morphisme de Z dans la fibre générique de $X \to S$.*

Les variétés rationnellement connexes sont donc en quelque sorte caractérisées par le fait d'avoir automatiquement un point sur le corps des fonctions d'une courbe sur les complexes.

La classe des variétés rationnellement connexes semble être la classe la plus large de variétés projectives lisses à laquelle on peut étendre le théorème de Tsen. Un exemple explicite de surface d'Enriques sur $K = \mathbb{C}((t))$ sans K-point a été construit par Lafon [53]. Un modèle affine, avec variables x, y, u, z, est défini par le système

$$x^2 - tu^2 + t = (t^2 u^2 - t) y^2$$

$$x^2 - 2tu^2 + (1/t) = t(t^2 u^2 - t) z^2.$$

[Dans la classification des surfaces, les surfaces d'Enriques sont en quelque sorte les plus proches des surfaces rationnelles, une telle surface X satisfait en particulier $H^1(X, O_X) = 0$ et $H^2(X, O_X) = 0$.]

En caractéristique zéro, le théorème 7.5 est équivalent à l'assertion suivante :

Théorème 7.7 *Soit A un anneau de valuation discrète de corps résiduel de caractéristique zéro et soit $\mathscr{X}$ un A-schéma intègre propre régulier. Si la fibre générique est une variété SRC, alors la fibre spéciale possède une composante de multiplicité 1.*

Comme indiqué ci-dessus, le théorème ci-dessus ne vaut déjà plus si la fibre générique est une surface d'Enriques.

Remarque 7.8. On ne sait pas si l'analogue de ce théorème vaut dans le cas d'inégale caractéristique. Par exemple, si X est une variété rationnellement connexe sur le corps p-adique $\mathbb{Q}_p$, a-t-elle un point dans une extension non ramifiée de $\mathbb{Q}_p$? C'est vrai en dimension 1 ou 2. En effet l'extension maximale non ramifiée de $\mathbb{Q}_p$ est un corps C_1 (théorème de Lang) et par inspection de la classification k-birationnelle des k-surfaces rationnelles, on montre (Manin, l'auteur) que toute k-surface rationnelle (projective et lisse) sur un corps k qui est C_1 possède un point k-rationnel.

Motivé par la conjecture 5.6 (Ax), par les énoncés des théorèmes de Tsen et de Chevalley-Warning, et par plusieurs résultats qui seront discutés plus bas, on peut, suivant Kollár [49], envisager les énoncés suivants :

Suggestions 7.9 *Soit A un anneau de valuation discrète de corps résiduel F. Soit $\mathscr{X}$ un A-schéma régulier, propre et plat sur A, à fibre générique X lisse géométriquement connexe, à fibre spéciale un diviseur Y/F à croisements normaux stricts. Si X/K est une variété séparablement rationnellement connexe, alors*

(a) *il existe une composante réduite Y_i de Y qui est géométriquement intègre sur F ;*
(b) *mieux, il existe une F-variété rationnellement connexe Z et un F-morphisme de Z dans Y ;*
(c) *encore mieux, il existe une composante réduite Y_i de Y qui est une F-variété rationnellement connexe.*

Dans cette direction, on a les résultats suivants.

Théorème 7.10 *(Kollár)[49] Soit F un corps de caractéristique zéro. Soit C une courbe lisse sur F, soit A l'anneau local de C en un point fermé de corps résiduel E, soit $\mathscr{X}$ un A-schéma régulier, propre et plat sur A, à fibre générique X lisse, à fibre spéciale un diviseur Y/E à croisements normaux stricts. Si X/K est une variété de Fano, alors il existe une composante réduite Y_i de Y qui est géométriquement irréductible sur E.*

Toute hypersurface est une dégénérescence d'une hypersurface lisse de même degré. Le résultat de Kollár établit ainsi la conjecture 5.6 (Ax) en caractéristique zéro : toute F-hypersurface de degré d dans $\mathbb{P}^n_F$ avec $n \geq d$ contient une sous-F-variété géométriquement intègre.

Corollaire 7.11 *Soit k un corps de caractéristique zéro. Il existe un corps L contenant k possédant les propriétés suivantes :*

(i) *Le corps k est algébriquement fermé dans L.*
(ii) *Le corps L est union de corps de fonctions de k-variétés géométriquement intègres.*
(iii) *Toute L-variété géométriquement intègre possède un point L-rationnel (le corps L est « pseudo-algébriquement clos »).*
(iv) *Le corps L est un corps C_1.*

Démonstration. La construction du paragraphe 2 donne un corps L satisfaisant les propriétés (i) à (iii). Le point (iv) est alors une application de la conjecture d'Ax.

Théorème 7.12 *(Starr)[66] Soit F un corps parfait contenant un corps algébriquement clos. Soit C une courbe lisse sur F, soit A l'anneau local de C en un point fermé de corps résiduel E, soit $\mathscr{X}$ un A-schéma régulier, propre et plat sur A, à fibre générique X/K lisse, à fibre spéciale un diviseur Y/E à croisements normaux stricts. Si X est une K-variété séparablement rationnellement connexe, alors il existe une composante réduite Y_i de Y qui est géométriquement irréductible sur E.*

Le théorème suivant implique en particulier que le théorème 7.10 vaut plus généralement lorsque la fibre générique est une variété rationnellement connexe.

Théorème 7.13 *(Hogadi et Xu)[40] Soient F un corps de caractéristique zéro, C une F-courbe lisse, A l'anneau local de C en un point fermé P, et E le corps résiduel en P. Soit $\mathscr{X}$ un A-schéma propre et plat sur A, de fibre générique X une K-variété rationnellement connexe. Alors*

(a) Il existe une E-variété rationnellement connexe Z et un E-morphisme de Z dans la fibre Y/E de $\mathscr{X} \to C$ en P.

(b) Si $\mathscr{X}$ est régulier, connexe, de dimension relative au plus 3, et si la fibre spéciale est un diviseur Y/E à croisements normaux stricts, alors il existe une composante réduite Y_i de Y qui est une E-variété rationnellement connexe.

Sous l'hypothèse supplémentaire que F contient un corps algébriquement clos, le résultat (a) avait été établi antérieurement par de Jong.

Corollaire 7.14 *Soit k un corps de caractéristique zéro. Il existe un corps L contenant k possédant les propriétés suivantes :*

(i) Le corps k est algébriquement fermé dans L.
(ii) Le corps L est union de corps de fonctions de k-variétés rationnellement connexes.
(iii) Toute L-variété rationnellement connexe possède un point L-rationnel.
(iv) Le corps L est un corps C_1.

Démonstration. On reprend la construction du paragraphe 2 mais à la place des F-variétés géométriquement intègres on utilise les F-variétés intègres F-birationnelles à une F-variété rationnellement connexe. La conditions (Stab) du paragraphe 2 est satisfaite grâce au théorème 7.4 (conséquence du théorème de Graber, Harris et Starr). Le corps L ainsi construit satisfait les propriétés (i) à (iii). Toute hypersurface est une dégénérescence d'une hypersurface lisse de même degré. Le théorème 7.13 implique donc que le corps L est un corps C_1 (voir [40], Cor. 1.5).

Une variante de la démonstration du théorème 7.5 permet de généraliser la partie (a) du théorème 7.13.

Théorème 7.15 *Soit A un anneau de valuation discrète, de corps des fractions K et de corps résiduel F de caractéristique nulle. Soit $\mathscr{X}$ un A-schéma projectif et plat sur A, de fibre générique une K-variété rationnellement connexe. Alors il existe une F-variété rationnellement connexe Z et un F-morphisme de Z dans la fibre spéciale $Y = \mathscr{X} \times_A F$.*

Démonstration. Pour établir le résultat, on peut remplacer A par son complété. Comme la caractéristique de F est nulle, ce complété est isomorphe à $F[[t]]$. On est donc réduit au cas $A = F[[t]]$. Soit R le hensélisé de $F[t]$ en $t = 0$. Soit L son corps des fractions. On a $\hat{R} = A$ et $\hat{L} = K$. La démonstration du théorème 7.5 montre qu'il existe un R-schéma projectif et plat $\mathscr{X}_1$ (non nécessairement régulier) de fibre générique une L-variété rationnellement connexe, tel que $\mathscr{X}_1 \times_R F \simeq \mathscr{X} \times_{F[[t]]} F = Y$, en d'autres termes, la fibre spéciale Y de $\mathscr{X}_1$ est F-isomorphe à la fibre spéciale de $\mathscr{X}$. D'après le théorème 7.13 (Hogadi et Xu), il existe une F-variété rationnellement connexe Z et un F-morphisme $Z \to \mathscr{X}_1 \times_R F$.

Remarque 7.16. (Wittenberg) Soit A un anneau de valuation discrète de corps des fractions K et de corps résiduel F de caractéristique zéro. Soit $\mathscr{X}$ un A-schéma régulier, propre et plat sur A, à fibre générique X lisse, à fibre spéciale un diviseur Y/F à croisements normaux stricts. Lorsque la fibre générique X est une K-compactification lisse d'un espace homogène d'un K-groupe algébrique linéaire connexe, on peut facilement établir le théorème 7.10 et l'énoncé (a) du théorème 7.13. On remplace A par $F[[t]]$. Comme rappelé au paragraphe 2, le corps F est algébriquement fermé dans un corps E de dimension cohomologique $cd(E) \leq 1$, corps qui est union de corps de fonctions de F-variétés d'un type spécial, en particulier rationnellement connexes. Le corps L limite inductive des corps $E((t^{1/n}))$ a le même groupe de Galois que E. Il est donc de dimension cohomologique 1, et X a un L-point. Ceci implique l'existence d'une F-application rationnelle d'une F-variété rationnellement connexe Z dans une composante réduite de la fibre spéciale, composante qui étant lisse doit en particulier être géométriquement intègre.

Remarque 7.17. De même que l'on ne peut espérer étendre le théorème 7.7 à d'autres classes de variétés que celle des variétés rationnellement connexes, de même il semble déraisonnable d'espérer une réponse positive à la suggestion 7.9 (a) pour d'autres classes que celle des variétés rationnellement connexes, par exemple pour les variétés projectives et lisses X telles que $H^i(X, O_X) = 0$ pour $i \geq 1$, ou telles que le groupe de Chow de X réduit des zéro-cycles sur tout corps algébriquement clos soit nul (voir le paragraphe suivant). Starr (communication privée) a donné un exemple de surface d'Enriques sur un corps $K(t)$ telle que pour tout modèle de type (R) de cette surface sur l'anneau local de $K[t]$ en $t = 0$, à croisements normaux stricts, aucune composante réduite ne soit géométriquement intègre.

8 Autour du théorème de Chevalley-Warning : variétés dont le groupe de Chow géométrique est trivial

Le théorème de Chevalley-Warning a fait l'objet de plusieurs généralisations (Ax, Katz, Esnault, voir [7]).

Théorème 8.1 *(Weil, 1954) Toute surface projective lisse géométriquement rationnelle sur un corps fini possède un point rationnel.*

Théorème 8.2 *(formule de Woods Hole 1964, de Lefschetz-Verdier, voir Grothendieck/Illusie SGA 5 III, Katz SGA7 XXII) Soit $\mathbb{F}$ un corps fini de caractéristique p. Soit $X/\mathbb{F}$ une variété propre. Si $H^0(X, O_X) = \mathbb{F}$ et si $H^r(X, O_X) = 0$ pour $r \geq 1$, alors le nombre de points rationnels de X est congru à 1 modulo p.*

En caractéristique nulle, les groupes $H^r(X, O_X)$ $(r \geq 1)$ s'annulent pour une variété de Fano, mais on ne sait pas le démontrer en caractéristique positive (sauf en dimension au plus 3, le cas de la dimension 3 étant dû à Shepherd-Barron).

H. Esnault a obtenu le résultat suivant.

Théorème 8.3 *(Esnault 2003) [25] Soit $\mathbb{F}$ un corps fini de cardinal q. Pour $X/\mathbb{F}$ lisse, projective, géométriquement intègre, et Ω un corps algébriquement clos contenant le corps $\mathbb{F}(X)$, si l'on a $A_0(X_\Omega) = 0$, alors le nombre de points $\mathbb{F}$-rationnels de X est congru à 1 modulo q.*

Soit l un nombre premier, $l \neq \operatorname{car}(\mathbb{F})$. Par un argument remontant à Spencer Bloch et développé par Bloch et Srinivas, l'hypothèse assure que la cohomologie l-adique de X est de coniveau 1, c'est-à-dire qu'elle satisfait $H^i_{\acute{e}t}(\overline{X}, \mathbb{Q}_l) = N^1 H^i_{\acute{e}t}(\overline{X}, \mathbb{Q}_l)$ pour tout $i \geq 1$ (toute classe de cohomologie s'annule sur un ouvert de Zariski non vide). Sous cette condition, H. Esnault utilise des résultats de Deligne pour établir la congruence annoncée.

Ce théorème s'applique pour les variétés rationnellement connexes par chaînes, et en particulier pour les variétés de Fano, à la différence du théorème 8.2.

A noter que le théorème s'applique aussi pour des variétés qui ne sont pas rationnellement connexes, comme les surfaces d'Enriques et aussi certaines surfaces de type général.

Comme dans l'énoncé initial de Chevalley-Warning et dans l'énoncé du théorème de Tsen, on dispose de versions portant sur les fibres spéciales, singulières, de telles variétés.

Théorème 8.4 *(N. Fakhruddin et C.S. Rajan, 2004) [30] Soit $f : X \to Y$ un morphisme propre dominant de variétés lisses et géométriquement irréductibles sur un corps fini $\mathbb{F}$ de cardinal q. Soit Z la fibre générique, supposée géométriquement intègre. Soit $\overline{\mathbb{F}(Y)}$ une clôture algébrique du corps des fonctions $\mathbb{F}(Y)$. Si l'on a $A_0(Z_{\overline{\mathbb{F}(Y)}}) = 0$, alors pour tout point $y \in Y(\mathbb{F})$, le cardinal de $X_y(\mathbb{F})$ est congru à 1 modulo q. Si l'hypothèse X lisse est omise mais si la fibre générique Z est lisse, on a $X_y(\mathbb{F}) \neq \emptyset$ pour tout $y \in Y(\mathbb{F})$.*

Donc sur toute dégénérescence de variété RCC (lisse) il y a un $\mathbb{F}$-point. Ceci vaut aussi sur une dégénérescence d'une surface d'Enriques ou de certaines surfaces de type général.

Théorème 8.5 *(Esnault [26, 27]; Esnault et Xu [28]) Soit A un anneau de valuation discrète complet de corps des fractions K et de corps résiduel $\mathbb{F}$ fini de cardinal q. Soit $\mathscr{X}$ un A-schéma intègre propre et plat. Soit l un nombre premier, $l \neq \mathrm{car}(\mathbb{F})$. Supposons la fibre générique géométriquement intègre, lisse et à cohomologie l-adique de coniveau 1. Soit $Y/\mathbb{F}$ la fibre spéciale. Alors*

(i) $Y(\mathbb{F}) \neq \emptyset$;
(ii) si de plus $\mathscr{X}$ est régulier, alors $\mathrm{card}(Y(\mathbb{F})) \equiv 1 \bmod q$.

L'hypothèse sur la cohomologie est satisfaite si $A_0(X \times_K \Omega) = 0$, où Ω est un corps algébriquement clos contenant $K(X)$, en particulier pour les variétés RCC mais aussi pour les surfaces d'Enriques et certaines surfaces de type général.

En particulier il y a un point rationnel sur la fibre spéciale. En particulier si toutes les composantes de la fibre spéciale sont lisses, alors l'une d'entre elles est géométriquement intègre sur $\mathbb{F}_q$.

Il y a des théorèmes de géométrie algébrique qui se démontrent par réduction au cas des corps finis. On part d'une variété sur un corps k. Une telle variété est obtenue par changement de base $A \to k$ à partir d'un A-schéma de type fini, pour une $\mathbb{Z}$-algèbre de type fini A convenable. On réduit ensuite aux points fermés de A (leurs corps résiduels sont finis) et on applique les résultats obtenus sur les corps finis.

Les théorèmes établis par H. Esnault sont de ce point de vue « trop bons » : la classe des K-variétés auxquelles ses résultats s'appliquent est plus large que celle des K-variétés rationnellement connexes. On ne peut donc espérer les utiliser pour établir des résultats comme le théorème 7.3 (Graber-Harris-Starr) ou le théorème 7.10 ci-dessus (Kollár) – pas plus d'ailleurs que l'on ne pouvait utiliser le théorème de Chevalley-Warning pour établir le théorème de Tsen ou la conjecture d'Ax. Un obstacle essentiel semble être le fait bien connu suivant : il existe des polynômes en une variable sur $\mathbb{Z}$ qui n'ont pas de zéro sur $\mathbb{Q}$ mais dont la réduction en tout premier p sauf un nombre fini a un zéro, par exemple $(x^2 - a)(x^2 - b)(x^2 - ab)$, avec $a, b \in \mathbb{Z}$ non carrés.

Les résultats sur les corps finis peuvent néanmoins en suggérer d'autres sur les corps de fonctions d'une variable. On en trouvera un exemple récent dans [44], §9.8, Remarque 3.

9 Approximation faible pour les variétés rationnellement connexes

Suggestion 9.1 *Soit K un corps de fonctions d'une variable sur un corps algébriquement clos. Pour toute variété rationnellement connexe X sur K,* *l'approximation faible vaut : pour tout ensemble fini I de places v de K, l'application diagonale*

$$X(K) \to \prod_{v \in I} X(K_v)$$

a une image dense. Ici K_v est le complété de K en v et $X(K_v)$ est muni de la topologie induite par la topologie de la valuation sur K_v.

Des arguments élémentaires ([11]) permettent d'établir l'approximation faible en tout ensemble fini de places pour les compactifications lisses d'espaces homogènes de groupes linéaires connexes, puis pour les variétés obtenues par fibrations en de telles variétés. On traite ainsi les intersections complètes lisses de deux quadriques dans $\mathbb{P}^n$ pour $n \geq 4$.

Théorème 9.2 *(Hassett-Tschinkel)[37] Soit K un corps de fonctions d'une variable sur un corps algébriquement clos de caractéristique zéro. Soit X/K une K-variété rationnellement connexe. Si I est un ensemble fini de places de K de bonne réduction pour X/K, alors l'approximation faible vaut pour X en ces places : l'application diagonale $X(K) \to \prod_{v \in I} X(K_v)$ a une image dense.*

Ceci généralise un résultat de Kollár, Miyaoka et Mori (cas où l'on demande une réduction fixée, sans obtenir d'approximations aux jets d'ordre supérieur).

Le cas particulier des surfaces cubiques lisses avait été traité par Madore [57].

Hassett et Tschinkel [38] ont aussi des résultats d'approximation en des places de mauvaise, mais pas trop mauvaise réduction. Mais comme ces auteurs le notent, le cas suivant est ouvert.

Question 9.3 *L'approximation faible en la place $\lambda = 0$ vaut-elle pour la surface cubique $x^3 + y^3 + z^3 + \lambda t^3 = 0$ sur le corps $K = \mathbb{C}(\lambda)$?*

Lorsque le nombre de variables est suffisamment grand par rapport au degré, on a pu établir l'approximation faible en toutes les places. Voir la section 12.5 ci-dessous.

10 *R*-équivalence sur les variétés rationnellement connexes

Soient k un corps non algébriquement clos et X une k-variété (séparablement) rationnellement connexe. Que sait-on sur l'ensemble $X(k)/R$?

Théorème 10.1 *(Kollár)[47] Soit K un corps local usuel (localement compact) et soit X une K-variété séparablement rationnellement connexe. Alors la R-équivalence sur $X(K)$ est une relation ouverte. L'ensemble $X(K)/R$ est fini. Dans le cas $K = \mathbb{R}$ les classes de R-équivalence coïncident avec les composantes connexes de $X(\mathbb{R})$.*

Ce résultat est une vaste généralisation de cas particuliers antérieurement connus (surfaces fibrées en coniques, compactifications de groupes algébriques linéaires connexes, hypersurfaces cubiques lisses, intersections lisses de deux quadriques dans $\mathbb{P}^n$ pour $n \geq 4$).

Suggestion 10.2 *(Kollár) Soient* $\mathbb{F}$ *un corps fini et X une* $\mathbb{F}$-*variété séparablement rationnellement connexe. Alors tous les points de* $X(\mathbb{F})$ *sont R-équivalents : l'ensemble* $X(\mathbb{F})/R$ *a un élément.*

Swinnerton-Dyer montra qu'il en est ainsi pour les surfaces cubiques lisses. Ce résultat a été récemment étendu par J. Kollár [50] à toutes les hypersurfaces cubiques lisses sur un corps fini de cardinal au moins 8.

Théorème 10.3 *(Kollár-Szabó)[52] Soient* $\mathbb{F}$ *un corps fini et X une* $\mathbb{F}$-*variété séparablement rationnellement connexe. Si l'ordre de* $\mathbb{F}$ *est plus grand qu'une certaine constante qui dépend seulement de la géométrie de X alors* $X(\mathbb{F})/R$ *est réduit à un point.*

Théorème 10.4 *(Kollár-Szabó)[52] Soit K un corps local non archimédien de corps résiduel le corps fini* $\mathbb{F}$. *Soit A l'anneau de la valuation. Soit* $\mathscr{X}$ *un A-schéma régulier, intègre, projectif et plat sur A, de fibre spéciale* $Y/\mathbb{F}$ *une* $\mathbb{F}$-*variété séparablement rationnellement connexe – ce qui implique que la fibre générique* $X = \mathscr{X} \times_A K$ *est SRC. Si l'ordre de* $\mathbb{F}$ *est plus grand qu'une certaine constante qui dépend seulement de la géométrie de X alors* $X(K)/R$ *est réduit à un point.*

Ici encore on se demande si la condition sur l'ordre du corps résiduel est nécessaire. A tout le moins, le résultat ci-dessus implique :

Théorème 10.5 *[52] Soient K un corps de nombres et* X/K *une K-variété rationnellement connexe. Alors pour presque toute place v de K, notant* K_v *le complété de K en v, on a* card $X(K_v)/R = 1$.

Soit A un anneau de valuation discrète de corps des fractions K et de corps résiduel F. Soit $\mathscr{X}$ un A-schéma intègre, propre et lisse. Soit $X = \mathscr{X} \times_A K$ la fibre générique et $Y = \mathscr{X} \times_A F$ la fibre spéciale.

La spécialisation $X(K) = \mathscr{X}(A) \to Y(F)$ passe au quotient par la R-équivalence (voir [55]). On a donc une application de spécialisation :

$$X(K)/R \to Y(F)/R.$$

Théorème 10.6 *(Kollár) [48] Dans la situation ci-dessus, si* Y/F *est SRC, et si A est hensélien, alors l'application de spécialisation* $X(K)/R \to Y(F)/R$ *est une bijection.*

On dit qu'un corps K est fertile (les anglo-saxons disent « large field ») si sur toute K-variété lisse intègre avec un K-point les K-points sont denses pour la topologie de Zariski.

Exemples :

(a) Une extension algébrique infinie d'un corps fini (estimations de Lang–Weil).

(b) Un corps local usuel (non archimédien, à corps résiduel fini), plus généralement le corps des fractions d'un anneau de valuation discrète hensélien de corps résiduel quelconque.

(c) Le corps $\mathbb{R}$ des réels, plus généralement un corps réel clos, plus généralement un corps dont le groupe de Galois absolu est un pro-p-groupe (p étant un nombre premier).

(d) Un corps pseudo-algébriquement clos.

Théorème 10.7 *(Kollár) Soient K un corps fertile et X une K-variété séparablement rationnellement connexe.*

(1) [47] Pour tout point $M \in X(K)$, il existe un K-morphisme très libre $f : \mathbb{P}^1_K \to X$ tel que M appartienne à $f(\mathbb{P}^1(K))$.

(2) [48] Si deux K-points sont R-équivalents, alors il existe un K-morphisme $\mathbb{P}^1_K \to X$ tel que ces deux points soient dans l'image de $\mathbb{P}^1(K)$.

Corollaire 10.8 *[48] Pour K corps fertile et X/K comme ci-dessus, pour tout ouvert de Zariski non vide $U \subset X$, l'application $U(K)/R \to X(K)/R$ est bijective.*

On ne sait pas si les deux énoncés précédents valent sur un corps K infini quelconque.

Théorème 10.9 *(Kollár) [48] Soit K un corps local usuel, soit $f : X \to Y$ un K-morphisme projectif et lisse de K-variétés lisses, dont les fibres géométriques sont des variétés SRC. L'application $Y(K) \to \mathbb{N}$ qui à un point $y \in Y(K)$ associe le cardinal de $X_y(K)/R$ est semi-continue supérieurement : tout point de $Y(K)$ admet un voisinage (pour la topologie sur $Y(K)$ définie par celle du corps local K) tel que pour z dans ce voisinage le cardinal de $X_z(K)/R$ soit au plus égal à celui de $X_y(K)/R$.*

Question 10.10 *[48] Le cardinal de $X_y(K)/R$ est-il localement constant quand y varie dans $Y(K)$?*

Question 10.11 *Soient k un corps et X une k-variété séparablement rationnellement connexe. Dans chacun des cas suivants :*

(a) $k = \mathbb{C}(C)$ est un corps de fonctions d'une variable sur les complexes,
(b) $k = \mathbb{C}((t))$ est un corps de séries formelles en une variable,
(c) k est un corps C_1,
(d) k est un corps parfait de dimension cohomologique $cd(k) \leq 1$,

l'ensemble $X(k)/R$ a-t-il au plus un élément ?

On ne s'attend pas à une réponse positive. Cependant, pour k de caractéristique nulle, sous la simple hypothèse $cd(k) \leq 1$, c'est connu dans les cas suivants :

(i) X est une compactification lisse d'un groupe linéaire connexe [16].
(ii) X est une surface fibrée en coniques de degré 4 sur la droite projective [18].
(iii) X est une intersection lisse de deux quadriques dans $\mathbb{P}^n_k$ et $n \geq 5$ ([17], Thm. 3.27 (ii)).
(iv) Le corps k est C_1, la variété X est une hypersurface cubique lisse dans $\mathbb{P}^n_k$ avec $n \geq 5$ [58].

On a aussi le résultat suivant, portant sur des variétés singulières :

(v) Soit k un corps de caractéristique nulle tel que toute forme quadratique sur k en 3 variables ait un zéro non trivial. Alors pour *toute* surface cubique *singulière* $X \subset \mathbb{P}_k^3$ l'ensemble $X(k)/R$ a au plus un élément.

Lorsque X possède un point singulier k-rationnel, ceci est établi dans [58], §1. Dans le cas général, on établit ce résultat en utilisant la classification des surfaces cubiques singulières. Le seul cas non couvert par les arguments donnés au §5 de [56] (voir aussi [58], Remarque 1) est le cas des surfaces de Châtelet (cas 7 p. 182 de [56]). Le résultat dans ce cas s'obtient en combinant le Théorème 8.6 (d) de [17] et les résultats de [20].

Une réponse positive à la question 10.11 pour les surfaces (projectives et lisses) géométriquement rationnelles définies sur $\mathbb{C}(t)$ impliquerait l'unirationalité des variétés de dimension 3 sur $\mathbb{C}$ qui admettent une fibration en coniques sur le plan projectif. Il s'agit là d'une question largement ouverte.

Question 10.12 *Soient K un corps de nombres et X une K-variété rationnellement connexe. Le quotient $X(K)/R$ est-il fini ?*

C'est connu dans les cas suivants :

(i) La variété X est une compactification lisse d'un groupe linéaire connexe G. L'immersion ouverte $G \subset X$ induit une bijection $G(k)/R \simeq X(k)/R$ ([32]). La finitude dans le cas général est due à Gille [31], elle s'appuie sur des résultats antérieurs de Margulis (groupes semi-simples simplement connexes) et CT-Sansuc ([16], cas des tores algébriques).

(ii) La variété X est une surface fibrée en coniques de degré 4 sur la droite projective (CT-Sansuc, cf. [18]).

(iii) La variété X est une intersection lisse de deux quadriques dans $\mathbb{P}_K^n$ et $n \geq 6$ [17].

La question de la finitude de $X(K)/R$ sur K un corps de nombres est ouverte pour les compactifications lisses d'espaces homogènes de groupes linéaires connexes, même en supposant les groupes d'isotropie géométrique connexes.

On pourrait se poser la question de la finitude de $X(K)/R$ pour X/K rationnellement connexe et K de type fini sur l'un quelconque des corps suivants : un corps fini, $\mathbb{Q}$, $\mathbb{C}$, $\mathbb{R}$, $\mathbb{Q}_p$.

On a par exemple la finitude dans ce cadre dans le cas (ii) ci-dessus [18], et c'est une question ouverte lorsque X est de dimension 2, i.e. est une surface géométriquement rationnelle. Dans le cas (i), on a la finitude lorsque G est un tore [16]. C'est une question largement ouverte pour G un groupe linéaire quelconque.

Mais, sur chacun des corps $\mathbb{Q}(t)$, $\mathbb{R}(t)$, $\mathbb{R}((t))$, la réunion pour tout $n \geq 1$ des $\mathbb{R}((t^{1/n}))$ (qui est un corps réel clos non archimédien), Kollár [48] a construit des exemples d'hypersurfaces lisses X de degré 4 dans $\mathbb{P}_K^n$, avec n arbitrairement grand, telles que $X(K)/R$ soit infini.

11 Équivalence rationnelle sur les zéro-cycles des variétés rationnellement connexes

Proposition 11.1 *[9] Soient k un corps et X une k-variété RCC. Il existe un entier $N = N(X) > 0$ tel que pour toute extension de corps L/k on ait $NA_0(X \times_k L) = 0$.*

Soient $\mathbb{F}$ un corps fini et X une $\mathbb{F}$-variété séparablement rationnellement connexe. Du théorème de Kollár et Szabó [52] il résulte que l'on a $A_0(X) = 0$. Mais ceci n'est qu'un cas particulier d'un théorème général en théorie du corps de classes supérieur :

Théorème 11.2 *(K. Kato et S. Saito, 1983) Soient $\mathbb{F}$ un corps fini et X une $\mathbb{F}$-variété projective et lisse géométriquement intègre. Soit Alb_X la variété d'Albanese de X (c'est une variété abélienne) et μ le $\mathbb{F}$-groupe fini commutatif dual de la torsion du groupe de Néron-Severi géométrique de X. Le groupe $A_0(X)$ est fini, et l'on a une suite exacte*

$$0 \to H^1(\mathbb{F}, \mu) \to A_0(X) \to Alb_X(\mathbb{F}) \to 0.$$

Question 11.3 *Soient k un corps et X une k-variété séparablement rationnellement connexe. Dans chacun des cas suivants :*

(a) $k = \mathbb{C}(C)$ est un corps de fonctions d'une variable sur les complexes,
(b) $k = \mathbb{C}((t))$ est un corps de séries formelles en une variable,
(c) k est un corps parfait de dimension cohomologique 1,

a-t-on $A_0(X) = 0$?

On ne s'attend pas à une réponse positive. Cependant, pour k de caractéristique nulle, sous la simple hypothèse $cd(k) \leq 1$, il en est ainsi dans chacun des cas suivants :

(i) Compactification lisse d'espace homogène principal de groupe algébrique linéaire connexe [16].
(ii) Surface SRC, i.e. surface géométriquement rationnelle. La situation est ici bien meilleure que pour la R-équivalence (voir la question 10.11). On établit $A_0(X) = 0$ par des méthodes de K-théorie algébrique [8].
(iii) Hypersurface cubique lisse dans $\mathbb{P}_k^n$ ($n \geq 3$) avec un k-point, pour $n \geq 3$. Soit $P \in X(k)$. Pour établir (iii), il suffit de montrer que tout k-point M est rationnellement équivalent au point P (on applique ensuite cet énoncé sur toute extension finie de k.)

Soit $L \subset \mathbb{P}_k^n$ un espace linéaire de dimension 3 contenant P et M. Soit $Y = X \cap L \subset L \simeq \mathbb{P}_k^3$ la surface cubique découpée par L. Si Y est singulière, alors P et M sont R-équivalents sur Y, donc sur X : voir l'énoncé (v) après la question 10.11. Si Y est non singulière, on a $A_0(Y) = 0$ d'après le point (ii) ci-dessus. Dans tous les cas on voit que P et M sont rationnellement équivalents.

(iv) Intersection lisse de deux quadriques dans $\mathbb{P}_k^n$ avec un k-point, pour $n \geq 5$. Ceci résulte de l'énoncé (iii) suivant la question 10.11. Une adaptation de l'argument donné ci-dessus pour les hypersurfaces cubiques devrait donner le résultat pour $n \geq 4$.

Théorème 11.4 *(CT-Ischebeck 1981) Soit X une $\mathbb{R}$-variété projective et lisse géométriquement intègre avec $X(\mathbb{R}) \neq \emptyset$. Soit s le nombre de composantes connexes de $X(\mathbb{R})$.*

Le sous-groupe $2A_0(X)$ est le sous-groupe divisible maximal de $A_0(X)$ et le quotient $A_0(X)/2A_0(X) = (\mathbb{Z}/2)^{s-1}$.

En particulier si X est rationnellement connexe et $X(\mathbb{R}) \neq \emptyset$, alors $A_0(X)$ est fini et $A_0(X) = (\mathbb{Z}/2)^{s-1}$.

Soit R un corps réel clos. Knebusch et Delfs ont montré comment l'on peut, pour toute R-variété algébrique X, donner une définition adéquate des « composantes connexes » de $X(R)$. Celles-ci sont en nombre fini. Le théorème ci-dessus vaut dans ce cadre plus large. On comparera ceci avec la remarque finale de la section 10.

Question 11.5 *Soient K un corps p-adique (extension finie de $\mathbb{Q}_p$) et X une K-variété rationnellement connexe. Le groupe $A_0(X)$ est-il fini ?*

Soit A l'anneau de la valuation du corps local K, soit $\mathbb{F}$ le corps fini résiduel. Voici des résultats obtenus dans cette direction.

(i) Si $\dim(X) = 2$, le groupe $A_0(X)$ est fini [8].

(ii) Si X est une intersection lisse de deux quadriques dans $\mathbb{P}_K^n, n \geq 4$ et $X(K) \neq \emptyset$, le groupe $A_0(X)$ est fini ([17, 19] et [62]).

(iii) Si X est un fibré en quadriques de dimension relative au moins 1 sur la droite projective, le groupe $A_0(X)$ est fini [19, 62].

(iv) Si X est une K-compactification lisse d'un K-groupe linéaire connexe, alors $A_0(X)$ est somme d'un groupe fini et d'un groupe de torsion p-primaire (d'exposant fini) [9].

(v) (Kollár-Szabó) [52] Si X a bonne réduction SRC, i.e. s'il existe un A-schéma $\mathscr{X}$ régulier, intègre, propre et lisse de fibre spéciale $Y/\mathbb{F}$ SRC, alors $A_0(X) = 0$.

(vi) (S. Saito et K. Sato) [64] Soit X une K-variété projective, lisse, géométriquement connexe. Supposons que X/K possède un modèle $\mathscr{X}/A$ régulier intègre, propre et plat, de fibre spéciale réduite $Y_{red}/\mathbb{F}$ à croisements normaux stricts. Alors le groupe $A_0(X)$ est somme directe d'un groupe fini et d'un groupe divisible par tout entier premier à p. Si en outre X est une variété rationnellement connexe, alors $A_0(X)$ est somme d'un groupe fini et d'un groupe de torsion p-primaire d'exposant fini.

On s'est longtemps posé la question de savoir si pour toute variété projective lisse X sur un corps p-adique le sous-groupe de torsion de $A_0(X)$ est fini. M. Asakura et S. Saito ont montré récemment qu'il n'en est rien (exemples : surfaces de degré $d \geq 5$ suffisamment générales dans $\mathbb{P}^3$).

Question 11.6 *Soient K un corps de type fini sur le corps premier et X une K-variété rationnellement connexe. Le groupe $A_0(X)$ est-il fini ?*

C'est connu lorsque $\dim(X) = 2$ et $X(K) \neq \emptyset$ [8], et lorsque X est une compactification lisse d'un K-tore de dimension 3 (Merkur'ev [60]).

Mais le cas général des compactifications lisses de tores sur un corps de nombres est ouvert.

De façon générale, on se demande si pour toute variété X connexe, projective et lisse sur un corps K de type fini sur le corps premier, le groupe $A_0(X)$ est un groupe de type fini.

12   Vers les variétés supérieurement rationnellement connexes

12.1   Deux exemples

12.1.1   Formes tordues d'hyperquadriques

D. Tao [68] a obtenu les résultats suivants. Soit K un corps possédant une algèbre simple centrale A de degré $2n \geq 6$ dont la classe $[A]$ dans le groupe de Brauer de K est non nulle et d'exposant 2. La condition $2.[A] = 0$ assure l'existence sur A d'une involution de première espèce σ qu'on peut choisir orthogonale. A une telle situation est alors associée une K-variété X qui est une forme tordue d'une quadrique lisse dans $\mathbb{P}^{2n-1}$ et pour laquelle

$$\mathrm{Ker}(\mathrm{Br}\,K \to \mathrm{Br}\,K(X)) = \mathbb{Z}/2 = \mathbb{Z}.[A] \subset \mathrm{Br}\,K.$$

Il y a donc une obstruction élémentaire à l'existence d'un K-point, en particulier $X(K) = \emptyset$.

On peut trouver des algèbres A du type requis sur l'un quelconque des corps suivants : un corps p-adique, le corps des séries formelles itérées $\mathbb{C}((u))((v))$, un corps de fonctions de deux variables sur $\mathbb{C}$.

Sur K l'un quelconque de ces corps, pour $m \geq 4$, les quadriques dans $\mathbb{P}^m_K$ ont un point rationnel. Mais pour m impair les formes tordues obtenues n'ont pas de point K-rationnel.

Si l'on considère une telle forme tordue X sur le corps $K = \mathbb{C}((u))((v))$, pour laquelle l'application $\mathrm{Br}\,K \to \mathrm{Br}\,K(X)$ n'est pas injective, il résulte de la proposition 4.1 que la fibre spéciale sur $F = \mathbb{C}((u))$ d'un modèle propre, plat, régulier de X sur l'anneau de valuation discrète $A = \mathbb{C}((u))[[v]]$ n'a aucune composante géométriquement intègre de multiplicité 1.

12.1.2   Une hypersurface cubique

L'hypersurface cubique diagonale $X \subset \mathbb{P}^8_K$ de coefficients

$$(1, u, u^2, v, vu, vu^2, v^2, v^2 u, v^2 u^2)$$

sur le corps $K = \mathbb{C}((u))((v))$ n'a pas de point rationnel.

La condition d'injectivité sur le groupe de Brauer $\mathrm{Br}\,K \hookrightarrow \mathrm{Br}\,K(X)$ est ici satisfaite, plus généralement, l'obstruction élémentaire s'annule : il en est ainsi pour toute hypersurface lisse de dimension au moins 3 (cf. [4]).

L'anneau de valuation discrète $A = \mathbb{C}((u))[[v]]$ a pour corps des fractions K. La K-hypersurface cubique X admet un modèle régulier $\mathscr{X}$ projectif sur A dont une composante réduite de la fibre sur $F = \mathbb{C}((u))$ est géométriquement intègre et de multiplicité 1 : un ouvert est donné par un ouvert du cône de $\mathbb{P}_F^8$ d'équation homogène $x^3 + uy^3 + u^2z^3 = 0$. Mais cette composante n'est pas rationnellement connexe, elle ne possède même pas de $\mathbb{C}((u))$-point lisse. De fait, la fibre spéciale Y de $\mathscr{X}$ ne saurait posséder une composante géométriquement intègre rationnellement connexe de multiplicité 1 : d'après le théorème 7.5 toute telle composante posséderait des points lisses sur $\mathbb{C}((u))$, points qui seraient Zariski-denses car $\mathbb{C}((u))$ est fertile, et l'on pourrait relever un $\mathbb{C}((u))$-point non situé sur les autres composantes en un K-point de X. Le même argument montre qu'aucune composante de multiplicité 1 de la fibre spéciale d'un modèle $\mathscr{X}$ de type (R) de X, à croisements normaux stricts, n'est le but d'une application rationnelle depuis une $\mathbb{C}((u))$-variété rationnellement connexe.

12.2 Fibres spéciales avec une composante géométriquement intègre de multiplicité 1

Soit A un anneau de valuation discrète, K son corps des fractions, F son corps résiduel. Soit π une uniformisante de A. Soit $\mathscr{X}/A$ un A-schéma de type (R) (voir le paragraphe 2), X/K sa fibre générique, Y/F sa fibre spéciale. Si l'on a $X(K) \neq \emptyset$ alors l'on a $\mathscr{X}(A) \neq \emptyset$. Comme $\mathscr{X}$ est régulier, une A-section de X/A rencontre Y en un F-point M possédant les propriétés suivantes : il est sur une unique composante réduite de Y, il est lisse sur cette composante, cette composante est de multiplicité 1 et géométriquement intègre. Inversement, si A est hensélien, un tel point M se relève en un K-point de X.

On voit donc qu'une condition nécessaire pour l'existence d'un K-point sur X est l'existence d'une composante géométriquement intègre de multiplicité 1 de la fibre spéciale Y. Au paragraphe 3.4 on a discuté cette propriété. Par analogie avec les suggestions 7.9 on est amené ici à s'intéresser aux propriétés suivantes d'un A-schéma $\mathscr{X}$ de type (R).

(i) *La fibre spéciale Y/F contient une composante géométriquement intègre de multiplicité 1.*

(ii) *La fibre spéciale Y/F contient une composante géométriquement intègre de multiplicité 1 qui admet un F-morphisme depuis une F-variété séparablement rationnellement connexe.*

(iii) *La fibre spéciale Y/F contient une composante géométriquement intègre de multiplicité 1 qui est une F-variété séparablement rationnellement connexe.*

On laisse ici au lecteur le soin de vérifier que la propriété (ii) satisfait la même propriété d'invariance K-birationnelle que la propriété (i) (cf. §3.4).

Théorème 12.1 (*CT-Kunyavskiĭ 2006*) *[13] Soit A un anneau de valuation discrète, de corps des fractions K, de corps résiduel F de caractéristique zéro. Soit $\mathscr{X}$ un A-schéma régulier propre intègre de fibre générique $\mathscr{X}_K$ une compactification lisse d'un espace homogène principal d'un groupe semi-simple simplement connexe, à fibre spéciale un diviseur à croisements normaux stricts. Il existe alors une composante de la fibre spéciale qui est géométriquement intègre et de multiplicité 1, et qui de plus admet un F-morphisme depuis une F-variété rationnellement connexe.*

Démonstration. Comme rappelé au paragraphe 2, on peut suivant Ducros [23] plonger F dans un corps L satisfaisant :

(i) Le corps F est algébriquement fermé dans L.

(ii) Le corps L est un corps de dimension cohomologique 1.

(iii) Le corps L est limite inductive de corps de fonctions de F-variétés admettant des fibrations successives (par applications rationnelles) en variétés qui sont des restrictions à la Weil de variétés de Severi-Brauer. On voit aisément que de telles variétés sont birationnelles à des variétés rationnellement connexes (on n'a pas ici besoin d'invoquer le théorème 7.4).

D'après Bruhat et Tits, tout espace homogène principal sous un groupe semi-simple simplement connexe sur le corps local $L((t))$, dont le corps résiduel est parfait et de dimension cohomologique 1, est trivial, i.e. possède un point $L((t))$-rationnel. Je renvoie ici le lecteur à [13] pour l'algèbre commutative utilisée pour terminer la démonstration.

Remarque 12.2. L'assertion sur l'existence d'un F-morphisme depuis une F-variété rationnellement connexe ne figurait pas dans [13].

Théorème 12.3 *[10] Soit A un anneau de valuation discrète de corps des fractions K, de corps résiduel F de caractéristique zéro. Soit $\Phi \in A[x_0,\ldots,x_n]$ une forme homogène de degré d en $n+1 > d^2$ variables. Supposons que l'hypersurface X/K définie par $\Phi = 0$ dans $\mathbb{P}^n_K$ est lisse. Soit $\mathscr{X}/A$ un modèle régulier de cette hypersurface, propre et plat sur A, à fibre spéciale à croisements normaux stricts. Il existe alors une composante de la fibre spéciale qui est géométriquement intègre et de multiplicité 1, et qui de plus admet un F-morphisme depuis une F-variété rationnellement connexe.*

Démonstration. Pour établir le résultat on peut supposer $A = F[[t]]$. D'après le théorème 7.14, on peut plonger F dans un corps L qui est union de corps de fonctions de F-variétés rationnellement connexes, et qui est un corps C_1. On remplace $F[[t]]$ par $L[[t]]$ et on utilise le fait que $L((t))$ est un corps C_2 puisque L est un corps C_1 (théorème 5.4). On a donc $X(L((t)) \neq \emptyset$ et donc $\mathscr{X}(L[[t]]) \neq \emptyset$. De ceci on déduit que la fibre spéciale contient une composante géométriquement intègre de multiplicité 1. On termine alors la démonstration comme dans le théorème 12.1 ci-dessus.

Remarque 12.4. L'assertion sur l'existence d'un F-morphisme depuis une F-variété rationnellement connexe ne figurait pas dans [10]. C'est l'utilisation du théorème 7.13 (Hogadi et Xu) au lieu du théorème 7.10 (Kollár) qui permet ici de l'obtenir.

Remarque 12.5. Soit A l'anneau des entiers d'un corps p-adique, $\mathbb{F}$ son corps résiduel. Soit Φ, n, d et $\mathcal{X}/A$ comme dans l'énoncé du théorème 12.3, en particulier on suppose donné un modèle à fibre spéciale à croisements normaux stricts. Supposons que le théorème 12.3 vaille encore dans ce cas d'inégale caractéristique.

On dispose alors d'une composante de Y qui est de multiplicité 1 et est géométriquement intègre sur le corps fini $\mathbb{F}$. Par les estimations de Lang-Weil, il existe un zéro-cycle de degré 1 (par rapport au corps $\mathbb{F}$) de support dans le lieu lisse de cette composante et non situé sur les autres composantes. Par le lemme de Hensel, on peut relever ce zéro-cycle sur K et l'on obtient que X/K possède un zéro-cycle de degré 1. Pour K un corps p-adique, l'existence d'un zéro-cycle de degré 1 sur toute hypersurface lisse de degré d dans $\mathbb{P}^n_K$, avec $n \geq d^2$, est une conjecture de Kato et Kuzumaki [45], établie par ces auteurs lorsque d est un nombre premier.

On dispose plus précisément d'une $\mathbb{F}$-application rationnelle d'une $\mathbb{F}$-variété séparablement rationnellement connexe sur un corps fini vers une composante lisse de multiplicité 1 de la fibre spéciale. Le théorème 8.3 (Esnault) assure l'existence d'un $\mathbb{F}$-point sur toute $\mathbb{F}$-variété séparablement rationnellement connexe, donc par le lemme 2.1 sur la composante lisse. Mais tout tel $\mathbb{F}$-point peut se trouver aussi sur une autre composante, donc ne pas être lisse sur Y, ce qui empêche de le relever en un K-point de X. C'est heureux. Sinon (modulo l'existence de bons modèles) $\mathbb{Q}_p$ serait un corps C_2 (ex-conjecture d'E. Artin). Mais les exemples fameux de Terjanian et de ses successeurs montrent que $\mathbb{Q}_p$ n'est pas C_2.

Il vaudrait d'ailleurs la peine de regarder les nombreux contre-exemples à la conjecture d'Artin qui ont été construits et de vérifier qu'il existe toujours dans ces cas un zéro-cycle de degré 1. Il en est ainsi pour l'exemple initial de Terjanian sur $\mathbb{Q}_2$.

12.3 *Variétés rationnellement simplement connexes*

Les variétés rationnellement connexes sont un analogue algébrique des espaces topologiques connexes par arcs. B. Mazur a demandé s'il y a un analogue en géométrie algébrique des espaces simplement connexes. En topologie, on demande que l'espace des lacets pointés soit connexe par arcs. A la suite d'une suggestion de Mazur, de Jong et Starr [43] proposent les définitions suivantes. Dans l'état actuel des recherches, il faut considérer ces définitions comme provisoires.

Soit X une variété projective et lisse sur $\mathbb{C}$, équipée d'un fibré ample H. Soit $\overline{M}_{0,2}(X, e)$ l'espace de Kontsevich paramétrisant les données suivantes : une courbe C propre, réduite, connexe, à croisements normaux, de genre arithmétique 0, un couple ordonné (p, q) de points lisses de C, un morphisme $h : C \to X$ de cycle image de degré e, tels que de plus la situation n'ait qu'un nombre fini d'automorphismes.

On dispose alors d'un morphisme d'évaluation

$$\overline{M}_{0,2}(X,e) \to X \times X.$$

La fibre générale de ce morphisme est un analogue de l'espace des chemins à points base en topologie.

La variété (projective et lisse) X est dite *rationnellement simplement connexe* si pour $e \geq 1$ suffisamment grand il existe une composante M de $\overline{M}_{0,2}(X,e)$ dominant $X \times X$ telle que la fibre générique de $M \to X \times X$ soit une variété rationnellement connexe.

De Jong et Starr [43] considèrent aussi l'espace

$$\overline{M}_{0,m}(X,e)$$

où cette fois-ci l'on fixe $m \geq 2$ points lisses ordonnés sur la courbe de genre arithmétique zéro, et l'évaluation

$$\overline{M}_{0,m}(X,e) \to X^m.$$

Ils appellent X *fortement rationnellement simplement connexe* si pour tout $m \geq 2$ et tout entier e suffisamment grand (fonction de m) il existe une composante M de $\overline{M}_{0,m}(X,e)$ dominant X^m telle que la fibre générique de $M \to X^m$ soit une variété rationnellement connexe.

De Jong et Starr (travaux en cours) ont obtenu une série de résultats sur les intersections complètes lisses dans l'espace projectif. Pour simplifier, je cite leurs résultats pour les hypersurfaces.

Théorème 12.6 *(de Jong-Starr) [43] Une hypersurface lisse de degré $d \geq 2$ dans* $\mathbb{P}^n_{\mathbb{C}}$ *avec*

$$n \geq d^2 - 1$$

est rationnellement simplement connexe, à l'exception des quadriques dans $\mathbb{P}^3_{\mathbb{C}}$.

Théorème 12.7 *(de Jong-Starr) [43] Une hypersurface lisse de degré $d \geq 2$ dans* $\mathbb{P}^n_{\mathbb{C}}$ *avec*

$$n \geq 2d^2 - d - 1$$

est fortement rationnellement simplement connexe.

Dans la définition ci-dessus on peut prendre $e \geq 4m - 6$.

Théorème 12.8 *(de Jong-Starr) [43] Pour $n \geq d^2$, il existe un ouvert de Zariski non vide de l'espace des hypersurfaces de degré d dans* $\mathbb{P}^n_{\mathbb{C}}$ *tel que toute hypersurface paramétrée par un point de cet espace est fortement rationnellement simplement connexe.*

La suggestion suivante est une version locale d'une suggestion globale de de Jong (12.11 ci-après).

Suggestion 12.9 *Soit A un anneau de valuation discrète de corps des fractions $K \subset \mathbb{C}$ et soit F son corps résiduel, supposé de caractéristique zéro. Soit $\mathscr{X}$ un A-schéma de type (R), X/K sa fibre générique, Y/F sa fibre spéciale. Si les conditions suivantes sont satisfaites :*

(i) la $\mathbb{C}$-variété $X \times_K \mathbb{C}$ est fortement rationnellement simplement connexe,
(ii) l'obstruction élémentaire pour X/K s'annule,

alors la fibre spéciale Y/F contient une composante géométriquement intègre de multiplicité 1 qui admet un F-morphisme depuis une F-variété rationnellement connexe.

Remarque 12.10. Dans (i), l'exemple 12.1.2 et le théorème 12.6 justifient la restriction aux variétés fortement rationnellement simplement connexes, plutôt qu'aux variétés rationnellement simplement connexes. L'exemple 12.1.1 justifie la condition (ii).

12.4   Existence d'un point rationnel sur un corps de fonctions de deux variables

Sur K un corps de fonctions d'une variable sur $\mathbb{C}$, le théorème de Graber-Harris-Starr dit que les K-variétés rationnellement connexes ont automatiquement un point K-rationnel (et le théorème de Graber-Harris-Starr-Mazur dit que ce sont essentiellement les seules).

On peut se demander s'il existe une classe de variétés qui ont la propriété que lorsqu'elles sont définies sur un corps K de fonctions de deux variables sur $\mathbb{C}$, elles ont automatiquement un K-point.

Voici deux familles de variétés pour lesquelles ceci est connu.

Le théorème de Tsen-Lang implique que toute hypersurface de degré d dans $\mathbb{P}^n_K$ avec $n \geq d^2$ possède un K-point.

Soit G un K-groupe semi-simple simplement connexe, E un espace homogène principal de G et X une K-compactification lisse de E. La conjecture II de Serre pour le corps K affirme que E et donc aussi X ont un K-point. Ceci est connu lorsque G n'a pas de facteur de type E_8 (Merkur'ev-Suslin, Suslin, Bayer-Parimala, P. Gille). Pour avoir l'énoncé dans tous les cas il reste à traiter le cas E_8 déployé. La résolution de ce dernier cas a été récemment annoncée par de Jong et Starr, leur démonstration utilise les techniques de variétés rationnellement simplement connexes.

Suggestion 12.11 *(de Jong) Soit $K = \mathbb{C}(S)$ le corps de fonctions d'une surface sur le corps des complexes. Soit X une K-variété fortement rationnellement simplement connexe. Supposons l'application de restriction $\mathrm{Br}\, K \to \mathrm{Br}\, K(X)$ injective. Alors X possède un point K-rationnel.*

Remarque 12.12. L'exemple 12.1.1 montre la nécessité de la condition non géométrique portant sur le groupe de Brauer. Des conditions supplémentaires de même nature pourraient être nécessaires. Par exemple on peut demander que pour toute extension finie (ou non) de corps L/K l'application $\operatorname{Br} L \to \operatorname{Br} L(X)$ soit injective. De façon encore plus générale, on peut demander que pour toute extension finie (ou non) de corps L/K il n'y ait pas d'obstruction élémentaire à l'existence d'un L-point sur $X \times_K L$ (voir [4]).

Pour une K-variété X intersection complète lisse de dimension au moins 3 dans un espace projectif $\mathbb{P}^n_K$, l'obstruction élémentaire s'annule. Il en est de même pour une K-variété projective et lisse géométriquement connexe qui contient un ouvert U qui est un espace homogène principal d'un groupe semi-simple simplement connexe. Pour ces résultats, voir [4].

Remarque 12.13. Dans [4] on s'intéresse aux compactifications lisses d'espaces homogènes de groupes linéaires connexes sur $K = \mathbb{C}(S)$ un corps de fonctions de deux variables, lorsque les stabilisateurs géométriques sont connexes (et qu'il n'y a pas de facteur E_8). On montre que dans ce cas l'obstruction élémentaire à l'existence d'un point rationnel est la seule obstruction.

Pour une hypersurface cubique lisse de dimension au moins 3 sur un corps K, l'obstruction élémentaire s'annule. Sur l'exemple 12.1.2 on voit donc que l'obstruction élémentaire est loin de contrôler l'existence d'un point rationnel pour les variétés rationnellement connexes sur un corps de fonctions de deux variables sur les complexes.

Remarque 12.14. De Jong et Starr ont un travail en préparation sur les variétés rationnellement simplement connexes où ils montrent que certains espaces homogènes projectifs sur un corps de fonctions de deux variables sur $\mathbb{C}$ ont automatiquement un point rationnel. Cela leur permet de donner une nouvelle démonstration (la troisième !) du théorème de de Jong [41] qu'indice et exposant coïncident pour les algèbres simples centrales sur un tel corps.

12.5 Approximation faible en toutes les places d'un corps de fonctions d'une variable

Rappelons que c'est une question ouverte (9.1) de savoir si toute variété rationnellement connexe sur un corps de fonctions d'une variable satisfait l'approximation faible en toute place.

Théorème 12.15 (*Hassett-Tschinkel*)*[39] Soit K un corps de fonctions d'une variable sur un corps algébriquement clos de caractéristique zéro. Il existe une fonction $\varphi : \mathbb{N} \to \mathbb{N}$ satisfaisant la propriété suivante. Pour toute hypersurface lisse de degré d dans $\mathbb{P}^n$ avec $n \geq \varphi(d)$, l'approximation faible vaut en tout ensemble fini de places de K.*

Pour $d = 3$, $\varphi(3) = 6$ convient.

Un travail en cours sur les variétés rationnellement simplement connexes (de Jong-Starr [43], appendice de Hassett) donne $\varphi(d) \leq 2d^2 - d - 1$ et, si l'hypersurface est « générale », $\varphi(d) \leq d^2$.

Théorème 12.16 *(Hassett) Soit $K = \mathbb{C}(C)$ le corps des fonctions d'une courbe. Si X/K est une variété fortement rationnellement simplement connexe, alors elle satisfait l'approximation faible par rapport à tout ensemble fini de places de K.*

12.6   R-équivalence et équivalence rationnelle

Dans la recherche de la bonne définition de variétés « supérieurement » rationnellement connexes, on peut aussi penser à des conditions de trivialité de $X(k)/R$ et de $A_0(X)$ sur les corps « de dimension 2 », comme les corps p-adiques, les corps de fonctions de deux variables sur les complexes, les corps de séries formelles itérées $\mathbb{C}((a))((b))$.

12.6.1   Groupes semi-simples simplement connexes

Si K est un corps p-adique, ou si K est un corps de fonctions de deux variables sur les complexes, ou si $K = \mathbb{C}((a))((b))$, et si G/K est un groupe semi-simple simplement connexe sans facteur de type E_8, on sait établir $G(K)/R = 1$ et $X(K)/R = 1$ (voir [12]). Pour X une compactification lisse d'un tel G, ceci implique $A_0(X) = 0$.

12.6.2   Hypersurfaces cubiques lisses

Proposition 12.17 *(Madore) [58] Soit K un corps p-adique ou un corps C_2. Soit $X \subset \mathbb{P}^n_K$ une hypersurface cubique lisse. Pour $n \geq 11$, on a card $X(K)/R = 1$ et $A_0(X) = 0$.*

Soit K un corps p-adique. Pour $n = 3$, on sait donner des exemples avec $X(K)/R$ et $A_0(X)$ d'ordre plus grand que 1. On ignore ce qui se passe pour $4 \leq n \leq 10$.

Par exemple, qu'en est-il pour l'hypersurface cubique d'équation :

$$x^3 + y^3 + z^3 + pu^3 + p^2v^3 = 0$$

dans $\mathbb{P}^4_{\mathbb{Q}_p}$?

Supposons $p \equiv 1 \bmod 3$, et soit $a \in \mathbb{Z}_p^\times$ non cube. Qu'en est-il pour l'hypersurface

$$x^3 + y^3 + z^3 + p(u_1^3 + au_2^3) + p^2(v_1^3 + av_2^3) = 0$$

dans $\mathbb{P}^6_{\mathbb{Q}_p}$?

Sur le corps $K = \mathbb{C}((a))((b))$, en utilisant la théorie de l'intersection sur un modèle au-dessus de $\mathbb{C}((a))[[b]]$, Madore [58] a montré que pour l'hypersurface cubique lisse $X \subset \mathbb{P}_K^4$ d'équation

$$x^3 + y^3 + az^3 + bu^3 + abv^3 = 0,$$

on a $A_0(X) \neq 0$.

12.6.3 Intersections lisses de deux quadriques

Soit K un corps p-adique, et soit $X \subset \mathbb{P}_K^n$, avec $n \geq 4$, une intersection complète lisse de deux quadriques possédant un K-point. Si $n \geq 7$, alors card $X(K)/R = 1$ [17] et donc $A_0(X) = 0$ (ceci vaut aussi pour un corps C_2). L'ensemble fini $X(K)/R$ peut être non trivial pour $n = 4$. Les cas $n = 5$ et $n = 6$ sont ouverts. Le groupe $A_0(X)$ est nul si $n = 6$ et k est non dyadique [62]. Le groupe fini $A_0(X)$ peut être non trivial pour $n = 4$. Les cas $n = 5$ et $n = 6$ (k dyadique) sont ouverts.

Soient $a_i, i = 1, \ldots, 3, b_i, i = 1, \ldots, 3$ dans $\mathbb{Z}_p$ satisfaisant $a_i \neq b_i$ et $a_i b_j - a_j b_i \in \mathbb{Z}_p^\times$ pour $i \neq j$. Soit $X \subset \mathbb{P}_{\mathbb{Q}_p}^5$ l'intersection complète lisse de deux quadriques donnée par le système

$$\sum_{i=0}^3 a_i X_i^2 + p X_4^2 = 0, \quad \sum_{i=0}^3 b_i X_i^2 + p X_5^2 = 0.$$

Que valent $X(\mathbb{Q}_p)/R$ et $A_0(X)$?

12.6.4 Fibrés en quadriques sur la droite projective

Soit K un corps p-adique, et soit X une K-variété géométriquement intègre, projective et lisse sur K, fibrée en quadriques de dimension $d \geq 1$ sur la droite projective $\mathbb{P}_K^1$. Si $p \neq 2$ et $d \geq 3$, alors $A_0(X) = 0$ [62]. Dans le cas $d = 2$, Parimala et Suresh [62] ont un exemple intéressant avec $A_0(X) \neq 0$. Dans cet exemple, un élément non nul de $A_0(X)$ est détecté par la mauvaise réduction de X de façon subtile, le groupe de Brauer de X ne permet pas de détecter cet élément.

13 Surjectivité arithmétique et surjectivité géométrique

Mis à part bien sûr le théorème d'Ax et Kochen, les énoncés de ce paragraphe sont établis dans des notes non publiées de l'auteur. Le lecteur ne devrait pas avoir de difficulté à reconstituer les démonstrations.

13.1   Morphismes définis sur un corps de nombres et applications induites sur les points locaux

On s'intéresse dans la suite à la situation suivante.

(*) On est sur un corps de nombres k, X et Y sont deux k-variétés lisses géométriquement intègres, la k-variété Y est projective, on a un k-morphisme projectif $f : X \to Y$ de fibre générique géométriquement intègre. On note $U \subset X$ l'ouvert de lissité du morphisme f.

On demande quels sont les liens entre la géométrie du morphisme f et les propriétés de surjectivité des applications induites $X(k_v) \to Y(k_v)$ pour presque toute place v, ou déjà pour une infinité de places v du corps de nombres k.

Les théorèmes 13.1 et 13.3 ci-après jouent un rôle central dans l'étude du principe de Hasse pour les variétés algébriques sur un corps de nombres.

Théorème 13.1 *Sous les hypothèses (*), si Y est une courbe et si l'application induite $U \to Y$ est surjective (ce qui équivaut à : $f : X \to Y$ est localement scindé pour la topologie étale sur Y), alors il existe une infinité de places v de k pour lesquelles l'application induite $X(k_v) \to Y(k_v)$ est surjective.*

Démonstration. Pour chaque point fermé P de Y à fibre $X_P = f^{-1}(P)$ non lisse on choisit une composante Z_P de multiplicité 1 de X_P. L'existence d'une telle composante est garantie par l'hypothèse que la fibration est localement scindée pour la topologie étale sur Y. Soit k_P le corps résiduel de Y en P. Soit K_P la clôture intégrale de k_P dans le corps des fonctions de Z_P. Soit K/k une extension finie galoisienne de k dans laquelle se plongent toutes les extensions K_P/k. La fibration $f_K : X_K \to Y_K$ satisfait alors les hypothèses du théorème 13.3 ci-après. En combinant ce théorème et le théorème de Tchebotarev, qui garantit l'existence d'une infinité de places v de k décomposées dans K, on conclut. Cette démonstration montre que l'ensemble infini de places cherché contient un ensemble de places de k de densité positive.

Remarques 13.2. (1) L'hypothèse que l'application $f : X \to Y$ est localement scindée pour la topologie étale sur Y est en particulier satisfaite si après extension finie convenable de k la fibration f admet une section. D'après le théorème 7.3 (Graber, Harris et Starr), c'est le cas si la fibre générique est une variété rationnellement connexe.

(2) Le théorème ne s'étend pas à Y de dimension supérieure, comme l'on voit en considérant une fibration en coniques sur $\mathbb{P}^2_{\mathbb{Q}}$ dont le lieu de ramification est une courbe C lisse et dont le revêtement double $D \to C$ associé est donné par une courbe $D/\mathbb{Q}$ géométriquement intègre. On peut par exemple prendre pour $C \subset \mathbb{P}^2_{\mathbb{Q}}$ une courbe elliptique d'équation affine $v^2 = u(u-a)(u-b)$ et une famille de coniques d'équation générique

$$X^2 - uY^2 - (v^2 - u(u-a)(u-b))T^2 = 0.$$

En utilisant le théorème de Lang-Weil, on établit le théorème suivant.

Théorème 13.3 *Sous les hypothèses (*), s'il existe un ouvert $V \subset X$ tel que le morphisme induit $V \to Y$ soit lisse, surjectif, à fibres géométriquement intègres, alors pour presque toute place v de k, l'application induite $X(k_v) \to Y(k_v)$ est surjective.*

Remarque 13.4. Il ne suffit pas d'avoir la propriété en codimension 1 sur Y, comme le montre l'exemple suivant. Prendre $a \in \mathbb{Q}$ non carré et $X \subset \mathbb{P}^3 \times_{\mathbb{Q}} \mathbb{P}^2$ donnée par

$$uX_0^2 - avX_1^2 + wX_2^2 - a(u+v+w)X_3^2 = 0.$$

Pour une infinité de p, la flèche $X(\mathbb{Q}_p) \to \mathbb{P}^2(\mathbb{Q}_p)$ n'est pas surjective : pour $a \in \mathbb{Z}_p$, a non carré dans $\mathbb{Z}_p$, et M un point $(p^{2n+1}\alpha, p^{2m+1}\beta, 1)$ avec $n, m \geq 0$, α et β in $\mathbb{Z}_p^*$ et $\alpha.\beta$ carré dans $\mathbb{Z}_p$, la fibre en M n'a pas de $\mathbb{Q}_p$-point.

Le célèbre théorème d'Ax et Kochen [3] peut se formuler de la façon suivante.

Théorème 13.5 *(Ax et Kochen) Fixons des entiers d et $n \geq d^2$. Soit $N + 1$ la dimension de l'espace des formes homogènes de degré d en $n + 1$ variables. Soit $F(x_0, \ldots, x_N; y_0, \ldots, y_n)$ la forme universelle de degré d en $n + 1$-variables. Soit $Z \subset \mathbb{P}^N \times_{\mathbb{Q}} \mathbb{P}^n$ le fermé défini par l'annulation de cette forme. Soit $\pi : Z \to \mathbb{P}^N$ la projection sur le premier facteur. Sur tout corps de nombres k, pour presque toute place v de k, la projection induite $Z(k_v) \to \mathbb{P}^N(k_v)$ est surjective.*

Remarque 13.6. En combinant le théorème 13.3 et le théorème 12.3, on établit un énoncé du type Ax-Kochen pour la restriction de $Z \to \mathbb{P}^N$ au-dessus d'une droite de $\mathbb{P}^N$ (passant par un point à fibre lisse).

Si l'on pouvait répondre par l'affirmative à la question 3.10, la combinaison du théorème 12.3 et du théorème 13.3 donnerait une nouvelle démonstration du théorème d'Ax et Kochen.

Sans répondre à la question 3.10, Jan Denef a tout récemment (juin 2008) obtenu une nouvelle démonstration du théorème d'Ax et Kochen, en établissant une conjecture générale de [10].

On s'intéresse aux réciproques des énoncés ci-dessus.

Théorème 13.7 *Plaçons-nous sous les hypothèses (*), avec Y une courbe.*

(i) *Si pour une infinité de places v l'application $X(k_v) \to Y(k_v)$ est surjective, alors pour tout point $P \in Y(k)$ il existe une composante de multiplicité 1 de $f^{-1}(P)$.*

(ii) *Si pour toute extension finie K/k, pour une infinité de places w de K, l'application $X(K_w) \to Y(K_w)$ est surjective, alors l'application induite $U \to Y$ est surjective : le morphisme $X \to Y$ est localement scindé pour la topologie étale sur Y.*

Théorème 13.8 *Plaçons-nous sous les hypothèses (*), avec Y une courbe. Supposons que pour presque toute place v de k l'application $X(k_v) \to Y(k_v)$ est surjective. Alors :*

(a) *L'application induite $U \to Y$ est surjective : le morphisme $X \to Y$ est localement scindé pour la topologie étale sur Y.*

(b) *Toute fibre connexe de $U \to Y$ est géométriquement connexe.*

(c) *Si P est un point fermé de Y, de corps résiduel κ et la fibre $f^{-1}(P)$ s'écrit $\sum_i e_i D_i$ avec chaque D_i diviseur intègre, de corps des fonctions κ_i, la flèche*

$$H^1(\kappa, \mathbb{Q}/\mathbb{Z}) \to \oplus_i H^1(\kappa_i, \mathbb{Q}/\mathbb{Z})$$

obtenue par somme des applications $e_i.\mathrm{Res}_{\kappa_i/\kappa}$ est injective.

Remarque 13.9. Le théorème 13.8 est le meilleur possible, comme le montre l'exemple suivant. Soit k un corps de nombres, $a, b \in k^*$ avec $a, b, ab \notin k^{*2}$. Soit $f : X \to \mathbb{P}^1_k$ un modèle projectif de la situation affine suivante :

$$(x^2 - ay^2)(u^2 - bv^2)(z^2 - abw^2) = t,$$

la flèche de projection sur $\mathbb{A}^1_k$ étant donnée par la coordonnée t. Alors

a) La fibre de f en 0 ne contient aucune composante géométriquement intègre de multiplicité 1.

b) Pour toute extension finie K/k, pour presque toute place w de K, l'application $X(K_w) \to Y(K_w)$ est surjective.

Tout le problème est qu'un polynôme en une variable sur un corps de nombres peut avoir une solution partout localement sans en avoir sur le corps de nombres, dès qu'il est réductible. Ainsi on ne peut pas partir du théorème d'Ax et Kochen pour en déduire le théorème 12.3 sur la réduction des formes lisses de degré d en $n > d^2$ variables.

13.2 Quelques autres questions

Diverses questions connexes ont été discutées dans la littérature.

Soient k un corps de nombres et $f : X \to Y$ un k-morphisme propre de k-variétés lisses géométriquement intègres, Y étant une courbe.

Question 1. Si sur toute extension finie K/k l'application $X(K) \to Y(K)$ est surjective (à un nombre fini de points près), le morphisme admet-il une section ?

Question 2. Si sur tout complété k_v le k_v-morphisme $X_{k_v} \to Y_{k_v}$ a une section, le morphisme f admet-il une section ?

En dimension relative 1, pour les courbes relatives de genre zéro, la réponse à ces deux questions est oui pour $Y = \mathbb{P}^1$ (Schinzel, Salberger, Serre). Ceci utilise l'injection $\mathrm{Br}\, k(t) \hookrightarrow \prod_v \mathrm{Br}\, k(t)$ qui s'établit en considérant la suite exacte de localisation pour le groupe de Brauer sur la droite projective. La réponse est non pour Y une courbe de genre 1 : on utilise une courbe elliptique Y avec un élément de 2-torsion dans son groupe de Tate-Shafarevich représenté par une algèbre de quaternions sur le corps de fonctions $k(Y)$ (voir [61]).

En dimension relative 1, pour les courbes relatives de genre 1 et $Y = \mathbb{P}^1$, la question 1 est ouverte. La question 2 a une réponse négative (prendre $X = C \times_k \mathbb{P}^1$ avec C une courbe de genre 1 qui est un contre-exemple au principe de Hasse).

Pour les familles de quadriques de dimension relative $d \geq 2$ au-dessus de $Y = \mathbb{P}^1_{\mathbb{Q}}$ la réponse aux deux questions ci-dessus est négative, et ce pour tout tel d.

Soit k un corps de nombres totalement imaginaire. Pour les familles de quadriques de dimension relative $d \geq 2$ au-dessus de $Y = \mathbb{P}^1_k$, la réponse aux deux questions est négative pour $2 \leq d \leq 6$. (Pour $d \geq 7$, on conjecture qu'il y a toujours une section.)

Pour justifier ces réponses négatives, partons d'un couple de formes quadratiques $f(x_1, \ldots, x_n), g(y_1, \ldots, y_m)$ sur le corps de nombres k tel que sur tout complété de k l'une des deux formes ait un zéro non trivial (donc n et m sont au moins égaux à 2) mais que pour chacune de ces formes il existe un complété k_v sur lequel la forme n'a pas de zéro non trivial.

Un théorème d'Amer et de Brumer (voir les références dans [17]) garantit que sur toute extension F de k, la forme quadratique $f(x_1, \ldots, x_n) + tg(y_1, \ldots, y_m)$ sur le corps $F(t)$ admet un zéro non trivial sur $F(t)$ si et seulement si le système

$$f(x_1, \ldots, x_n) = 0, \; g(y_1, \ldots, y_m) = 0$$

admet un zéro non trivial dans F^{n+m}.

La forme $f(x_1, \ldots, x_n) + tg(y_1, \ldots, y_m)$ sur le corps $k(t)$ a alors un zéro sur chaque $k_v(t)$ mais n'en a pas sur $k(t)$. Ceci donne les réponses négatives à la question 2, et les réponses négatives à la question 1 résultent du principe de Hasse pour les formes quadratiques sur un corps de nombres.

Bibliographie

1. Araujo, C., Kollár, J. : Rational curves on varieties. In : Higher Dimensional Varieties and Rational Points. Bolyai Society Mathematics Studies, vol. 12, pp. 13–92. Springer, Heidelberg (2003)
2. Ax, J. : The elementary theory of finite fields. Ann. Math. **88**(2), 239–271 (1968)
3. Ax, J., Kochen, S. : Diophantine problems over local fields, I. Amer. J. Math. **87**, 605–631 (1965)
4. Borovoi, M., Colliot-Thélène, J.-L., Skorobogatov, A.N. : The elementary obstruction and homogeneous spaces. Duke Math. J. **141**, 321–364 (2008)
5. Bosch, S., Lütkebohmert, W., Raynaud, M. : Néron Models, Ergebnisse der Math. und ihrer Grenzg. 3. Folge Band, vol. 21. Springer, Heidelberg
6. Campana, F., Connexité rationnelle des variétés de Fano. Ann. Scient. École Norm. Sup. **25**, 539–545 (1992)
7. Chambert-Loir, A. : Points rationnels et groupes fondamentaux : applications de la cohomologie p-adique (d'après P. Berthelot, T. Ekedahl, H. Esnault, etc.), Séminaire Bourbaki 2002–2003, exposé 914. Astérisque **294**, 125–146 (2004)
8. Colliot-Thélène, J.-L. : Hilbert's theorem 90 for K_2, with application to the Chow groups of rational surfaces. Invent. math. **71**, 1–20 (1983)
9. Colliot-Thélène, J.-L. : Un théorème de finitude pour le groupe de Chow des zéro-cycles d'un groupe algébrique linéaire sur un corps p-adique. Invent. math **159**, 589–606 (2005)

10. Colliot-Thélène, J.-L. : Fibres spéciales des hypersurfaces de petit degré. C. R. Acad. Sc. Paris **346**, 63–65 (2008)

11. Colliot-Thélène, J.-L., Gille, P. : Remarques sur l'approximation faible sur un corps de fonctions d'une variable. In : Poonen, B., Tschinkel, Yu. (eds.) Arithmetic of Higher Dimensional Arithmetic Varieties, Progress in Mathematics, vol. 226, pp. 121–133. Birkhäuser, Boston (2003)

12. Colliot-Thélène, J.-L., Gille, P., Parimala, R. : Arithmetic of linear algebraic groups over two-dimensional fields. Duke Math. J. **121**, 285–341 (2004)

13. Colliot-Thélène, J.-L., Kunyavskiĭ, B. : Groupe de Picard et groupe de Brauer des compactifications lisses d'espaces homogènes. J. Algebraic Geom. **15**, 733–752 (2006)

14. Colliot-Thélène, J.-L., Madore, D. : Surfaces de Del Pezzo sans point rationnel sur un corps de dimension cohomologique un. Journal de l'Institut Mathématique de Jussieu **3**, 1–16 (2004)

15. Colliot-Thélène, J.-L., Saito, S. : Zéro-cycles sur les variétés p-adiques et groupe de Brauer. IMRN **4**, 151–160 (1996)

16. Colliot-Thélène, J.-L., Sansuc, J.-J. : La R-équivalence sur les tores. Ann. Scient. École Norm. Sup. **10**, 175–229 (1977)

17. Colliot-Thélène, J.-L., Sansuc, J.-J., Sir Peter Swinnerton-Dyer : Intersections of two quadrics and Châtelet surfaces. I, J. für die reine und angew. Math. (Crelle) **373**, 37–107 (1987) ; II, **374**, 72–168 (1987)

18. Colliot-Thélène, J.-L., Skorobogatov, A.N. : R-equivalence on conic bundles of degree 4. Duke Math. J. **54**, 671–677 (1987)

19. Colliot-Thélène, J.-L., Skorobogatov, A.N. : Groupes de Chow des zéro-cycles des fibrés en quadriques. K-Theor. **7**, 477–500 (1993)

20. Coray, D.F., Tsfasman, M.A. : Arithmetic on singular Del Pezzo surfaces. Proc. Lond. Math. Soc. (3) **57**(1), 25–87 (1988)

21. Debarre, O. : Higher-dimensional algebraic geometry, Universitext. Springer, Heidelberg (2001)

22. Denef, J., Jarden, M., Lewis, D.J. : On Ax-fields which are C_i. Quart. J. Math. Oxford Ser. (2) **34**(133), 21–36 (1983)

23. Ducros, A. : Dimension cohomologique et points rationnels sur les courbes. J. Algebra **203**, 349–354 (1998)

24. Ducros, A. : Points rationnels sur la fibre spéciale d'un schéma au-dessus d'un anneau de valuation. Math. Z. **238**, 177–185 (2001)

25. Esnault, H. : Varieties over a finite field with trivial Chow group of 0-cycles have a rational point. Invent. math. **151**(1), 187–191 (2003)

26. Esnault, H. : Deligne's integrality theorem in unequal characteristic and rational points over finite fields (with an appendix by H. Esnault and P. Deligne). Ann. Math. **164**, 715–730 (2006)

27. Esnault, H. : Coniveau over p-adic fields and points over finite fields. C. R. Acad. Sc. Paris Sér. I **345**, 73–76 (2007)

28. Esnault, H., Xu, C. Congruence for rational points over finite fields and coniveau over local fields. Trans. Amer. Math. Soc. **361**(5), 2679–2688 (2009)

29. Euler, L. : Demonstratio theorematis Fermatiani omnem numerum sive integrum sive fractum esse summam quatuor pauciorumve quadratorum. N. Comm. Ac. Petrop. 5 (1754/5), 1760, pp. 13–58. E.242, Opera omnia. vol. I.2, pp. 338–372. Birkhäuser, Boston (2003)

30. Fakhruddin, N., Rajan, C.S. : Congruences for rational points on varieties over finite fields. Math. Ann. **333**, 797–809 (2005)

31. Gille, P. : La R-équivalence sur les groupes algébriques réductifs définis sur un corps global. Inst. Hautes Études Sci. Publ. Math. **86**, 199–235 (1997)

32. Gille, P. : Spécialisation de la R-équivalence pour les groupes réductifs. Trans. Amer. Math. Soc. **356**, 4465–4474 (2004)

33. Graber, T., Harris, J., Mazur, B., Starr, J. : Rational connectivity and sections of families over curves. Ann. Scient. École Norm. Sup. **38**(4), 671–692 (2005)

34. T. Graber, J. Harris, et J. Starr, Families of rationally connected varieties, J. Amer. Math. Soc. **16** (2003) 57–67

35. Greenberg, M.J. : Rational points in Henselian discrete valuation rings. Inst. Hautes Études Sci. Publ. Math. **31**, 59–64 (1966)
36. Grothendieck, A. : Le groupe de Brauer I, II, III. In : *Dix exposés sur la cohomologie des schémas*. North-Holland, Amsterdam (1968)
37. Hassett, B., Tschinkel, Yu. : Weak approximation over function fields. Invent. math. **163**(1), 171–190 (2006)
38. Hassett, B., Tschinkel, Yu. : Approximation at places of bad reduction for rationally connected varieties. Pure and Applied Math Quarterly, Bogomolov Festschrift **4**(3), 1–24 (2008)
39. Hassett, B., Tschinkel, Yu. : Weak approximation for hypersurfaces of low degree. Algebraic geometry – Seattle 2005. Part 2, 937–955. Proc. Sympos. Pure Math. **80**, Part 2, Amer. Math. Soc. Providence, RI (2009)
40. Hogadi, A., Xu, C. : Degenerations of rationally connected varieties. Trans. Amer. Math. Soc. **361**(7), 3931–3949 (2009)
41. de Jong, A.J. : The period-index problem for the Brauer group of an algebraic surface. Duke Math. J. **123**, 71–94 (2004)
42. de Jong, A.J., Starr, J. : Every rationally connected variety over the function field of a curve has a rational point. Amer. J. Math. **125**, 567–580 (2003)
43. de Jong, A.J., Starr, J. : Low degree complete intersections are rationally simply connected (prépublication)
44. Kahn, B. : Zeta functions and motives. Pure Appl. Math. Quart. **5**(1), à paraître
45. Kato, K., Kuzumaki, T. : The dimension of fields and algebraic K-theory. J. Number Theor. **24**, 229–244 (1986)
46. Kollár, J. : Rational curves on algebraic varieties, Ergebnisse der Mathematik und ihrer Grenzgebiete, 3. Folge, Band, vol. 32. Springer, Heidelberg (1996, réédition avec corrections 1999)
47. Kollár, J. : Rationally connected varieties over local fields. Ann. Math. (2) **150**(1), 357–367 (1999)
48. Kollár, J. : Specialization of zero cycles. Publ. Res. Inst. Math. Sci. **40**(3), 689–708 (2004)
49. Kollár, J. : A conjecture of Ax and degenerations of Fano varieties. Israel J. Math. **162**, 235–252 (2007)
50. Kollár, J. : Looking for rational curves on cubic hypersurfaces. In : Kaledin, D., Tschinkel, Y. (eds.) Higher-Dimensional Geometry over Finite Fields, vol. 16 NATO Science for Peace and Security Series : Information and Communication Security, (2008). Voir aussi : U. Derenthal and J. Kollár, Looking for rational curves on cubic hypersurfaces, matharXivv :0710.5516
51. Kollár, J., Miyaoka, Y., Mori, S. : Rational connectedness and boundedness of Fano manifolds. J. Diff. Geom. **36**(3), 765–779 (1992)
52. Kollár, J., Szabó, E. : Rationally connected varieties over finite fields. Duke Math. J. **120**(2), 251–267 (2003)
53. Lafon, G. : Une surface d'Enriques sans point sur $\mathbb{C}((t))$. C.R. Math. Acad. Sci. Paris **138**(1), 51–54 (2004)
54. Lang, S. : On quasi algebraic closure. Ann. Math. **55**(2), 373–390 (1952)
55. Madore, D. : Sur la spécialisation de la R-équivalence. In : Hypersurfaces cubiques, R-équivalence et approximation faible, thèse de doctorat. Université Paris-Sud (2005)
56. Madore, D. : Équivalence rationnelle sur les hypersurfaces cubiques sur les corps p-adiques. Manuscripta Math. **110**, 171–185 (2003)
57. Madore, D. : Approximation faible aux places de bonne réduction sur les surfaces cubiques sur les corps de fonctions. Bull. Soc. Math. France **134**(4), 475–485 (2006)
58. Madore, D. : Équivalence rationnelle sur les hypersurfaces cubiques de mauvaise réduction. J. Number Theor. **128**, 926–944 (2008)
59. Manin, Yu.I. : Cubic forms. Algebra, geometry, arithmetic. Translated from the Russian by M. Hazewinkel, 2nd edn. North-Holland Mathematical Library, 4. North-Holland, Amsterdam (1986)
60. Merkur'ev, A.S. : R-equivalence on three-dimensional tori and zero cycles. Algebra Number Theor. **2**, 69–89 (2008)

61. Parimala, R. Sujatha, R. : Hasse principle for Witt groups of function fields with special re-
    ference to elliptic curves. With an appendix by Colliot-Thélène, J.-L. Duke Math. J. **85**(3),
    555–582 (1996)
62. Parimala, R., Suresh, V. : Zero-cycles on quadric fibrations : Finiteness theorems and the cycle
    map. Invent. math. **122**, 83–117 (1995)
63. Pfister, A. : Quadratic forms with applications to algebraic geometry and topology, London
    Mathematical Society Lecture Note Series, vol. 217. Cambridge University Press, Cambridge
    (1995)
64. Saito, S., Sato, K. : A finiteness theorem for zero-cycles over p-adic fields. à paraître dans Ann.
    Math
65. Serre, J.-P. : Cohomologie galoisienne, Cinquième édition, révisée et complétée. Springer Lec-
    ture Notes in Mathematics, vol. 5 (1994)
66. Starr, J. : Degenerations of rationally connected varieties and PAC fields, arXiv preprint
67. Starr, J.M. : Arithmetic over function fields. Arithmetic geometry, 375–418, Clay Math. Proc.
    8, Amer. Math. Soc., Providence, RI (2009)
68. Tao, D. : A variety associated to an algebra with involution. J. Algebra **168**(2), 479–520 (1994)
69. Wittenberg, O. : La connexité rationnelle en arithmétique, notes pour un mini-cours, Session
    SMF États de la Recherche « Variétés rationnellement connexes », Strasbourg, mai (2008)

Texte reçu le 4 juin 2008, révisé le 11 août 2008.

Topics in Diophantine Equations

Sir Peter Swinnerton-Dyer

1 Introduction

These notes fall into two parts. The first part, which goes up to the end of Sect. 5, is a general survey of some of the topics in the theory of Diophantine equations which interest me and on which I hope to see progress within the next 10 years. Because of the second condition, I have for example not covered the Riemann Hypothesis or the Birch/Swinnerton-Dyer conjectures, both of which at the moment appear intractable. Another such survey can be found in Silverberg [1]; it has little overlap with this one but should appeal to the same readers. In the second part of these notes, I go into more detail on some particular topics than there was time for in the lectures.

A *Diophantine problem* over $\mathbf{Q}$ or $\mathbf{Z}$ is concerned with the solutions either in $\mathbf{Q}$ or in $\mathbf{Z}$ of a finite system of polynomial equations

$$F_i(X_1,\ldots,X_n) = 0 \quad (1 \le i \le m) \tag{1}$$

with coefficients in $\mathbf{Q}$. Without loss of generality we can obviously require the coefficients to be in $\mathbf{Z}$. A system (1) is also called a system of *Diophantine equations*. Often one will be interested in a family of such problems rather than a single one; in this case one requires the coefficients of the F_i to lie in some $\mathbf{Q}(c_1,\ldots,c_r)$, and one obtains an individual problem by giving the c_j values in $\mathbf{Q}$. Again one can get rid of denominators. Some of the most obvious questions to ask about such a family are:

(A) Is there an algorithm which will determine, for each assigned set of values of the c_j, whether the corresponding Diophantine problem has solutions, either in $\mathbf{Z}$ or in $\mathbf{Q}$?

(B) When the answer to (A) is positive, is there for values of the c_j for which the system is soluble an algorithm for exhibiting a solution? For example, is there

Sir P. Swinnerton-Dyer (✉)
Department of Pure Mathematics and Mathematical Statistics, Centre for Mathematical Sciences, University of Cambridge, Wilberforce Road, Cambridge, CB3 0WB, UK
e-mail: H.P.F.Swinnerton-Dyer@dpmms.cam.ac.uk

P. Corvaja and C. Gasbarri (eds.), *Arithmetic Geometry*, Lecture Notes in Mathematics 2009, DOI 10.1007/978-3-642-15945-9_2,
© Springer-Verlag Berlin Heidelberg 2011

an upper bound for the height of the smallest solution in terms of the heights of the coefficients of (1)?

For individual members of such a family, it is also natural to ask:

(C)   Can we describe the set of all solutions, or even its structure?
(D)   Is the phrase "density of solutions" meaningful, and if so, what can we say about it?

The attempts to answer these questions have led to the introduction of new ideas and these have generated new questions. Progress in mathematics usually comes by proving results; but sometimes a well justified conjecture throws new light on the structure of the subject. (For similar reasons, well motivated computations can be helpful; but computations not based on a feeling for the structure of the subject have generally turned out to be a waste of time.)

Though the problems associated with solutions in $\mathbf{Z}$ and in $\mathbf{Q}$ may look very similar (and indeed were believed for a long time to be so), it now appears that the methods which are useful are actually very different; and currently the theory for solutions in $\mathbf{Q}$ has much more structure than that for solutions in $\mathbf{Z}$. The main reason for this seems to be that in the rational case the system (1) defines a variety in the sense of algebraic geometry, and many of the tools of that discipline can be used. Despite the advent of Arakelov geometry, this is much less true of integral problems. However, for most families of varieties of degree greater than 2 it is only in low dimension that we yet know enough of the geometry for it to be useful. Uniquely, the Hardy-Littlewood method is useful both for integral and for rational problems; it was designed for integral problems but it can also be applied to rational problems by making the equations homogeneous. There is a brief discussion of this method in Sect. 5 and a comprehensive survey in [2].

Denote by V the variety defined by (1) and let V' be any variety birationally equivalent to V over $\mathbf{Q}$. Rational solutions of (1) in $\mathbf{Q}$ are just rational points on V, and finding them is almost the same as finding rational points on V'. Hence (except for Question (D) above) one expects the properties of the rational solutions of (1) to be essentially determined by the birational equivalence class of V over $\mathbf{Q}$. Classifying Diophantine problems over $\mathbf{Q}$ therefore corresponds to classifying birational equivalence classes of varieties over $\mathbf{Q}$. A first crude approximation to this is to classify them over $\mathbf{C}$. So number theorists would be helped if geometers could develop an adequate classification of varieties. At the moment, such a classification is reasonably complete for curves and surfaces, but it is still fragmentary even in dimension 3; so for those number theorists who use geometric methods it is natural to concentrate on curves and surfaces and on certain particularly simple kinds of variety of higher dimension.

The definitions and the questions above can be generalized to an arbitrary algebraic number field and the ring of integers in it; the answers are usually known or conjectured to be similar to those over $\mathbf{Q}$ or $\mathbf{Z}$, though the proofs can be very much harder. (But there are exceptions; for example, the modularity of elliptic curves only holds over $\mathbf{Q}$.) Some of the questions above can also be posed for other fields of number-theoretic interest – in particular for finite fields and for completions of algebraic number fields – and when one studies Diophantine problems it is often essential to consider these fields also. If V is defined over a field K, the set of points on V defined over K is denoted by $V(K)$. If $V(K)$ is not empty we say that V is *soluble in K*. In the special case where $K = k_v$, the completion of an algebraic number field k at the place v, we also say that V is *locally soluble at v*. From now on we denote by $\mathbf{Q}_v$ any completion of $\mathbf{Q}$; thus $\mathbf{Q}_v$ means $\mathbf{R}$ or some $\mathbf{Q}_p$.

One major reason for considering solubility in complete fields and in finite fields is that a necessary condition for (1) to be soluble in $\mathbf{Q}$ is that it is soluble in every $\mathbf{Q}_v$. The condition of solubility in every $\mathbf{Q}_v$ is computationally decidable; see Sect. 2. Moreover the first step in deciding solubility in $\mathbf{Q}_p$ is to study the solutions of the system reduced mod p in the finite field $\mathbf{F}_p$ of p elements.

Diophantine problems were first introduced by Diophantus of Alexandria, the last of the great Greek mathematicians, who lived at some time between 300 BC and 300 AD; but he was handicapped by having only one letter available to represent variables, all the others being used in the classical world to represent specific numbers. Individual Diophantine problems were studied by such great mathematicians as Fermat, Euler and Gauss. But it was Hilbert's address to the International Congress in 1900 which started the development of a systematic theory. His tenth problem asked:

> Given a Diophantine equation with any number of unknown quantities and with rational integral numerical coefficients: to devise a process according to which it can be determined by a finite number of operations whether the equation is soluble in rational integers.

Most of the early work on Diophantine equations was concerned with rational rather than integral solutions; presumably Hilbert posed this problem in terms of integral solutions because such a process for integral solutions would automatically provide the corresponding process for rational solutions also. In the confident days before the First World War, it was assumed that such a process must exist; but in 1970 Matijasevič showed that this was impossible. He exhibited a polynomial $F(c; x_1, \ldots, x_n)$ such that there cannot exist an algorithm which will decide for every given integer c whether $F = 0$ is soluble in integers. His proof is part of the great program on decidability initiated by Gödel; good accounts of it can be found in [3], pp 323–378 or [4]. The corresponding question for rational solutions is still open; I am among the few who believe that it may have a positive answer. Certainly it is important to ask for which families of varieties such a process exists, and to find such a process when it does exist.

2   The Hasse Principle and the Brauer-Manin Obstruction

Let V be a variety defined over $\mathbf{Q}$. If V is locally soluble at every place of $\mathbf{Q}$, we say that it satisfies the *Hasse condition*. If $V(\mathbf{Q})$ is not empty then V certainly satisfies the Hasse condition, so the latter is necessary for solubility. What makes this remark valuable is that the Hasse condition is computable – that is, one can decide in finitely many steps whether a given V satisfies the Hasse condition. This follows from the next two lemmas.

Lemma 2.1. *Let W be an absolutely irreducible variety of dimension n defined over the finite field $k = \mathbf{F}_p$. Then $N(p)$, the number of points on W defined over k, satisfies*

$$|N(p) - p^n| < Cp^{n-1/2}$$

where the constant C depends only on the degree and dimension of W and is computable.

This follows from the Weil conjectures, for which see Sect. 3; but weaker results which are adequate for the proof that the Hasse condition is computable were known much earlier. Since the singular points of W lie on a proper subvariety, there are at most $C_1 p^{n-1}$ of them, where C_1 is also computable. It follows that if p exceeds a computable bound depending only on the degree and dimension of W then W contains a nonsingular point defined over $\mathbf{F}_p$.

Let V be an absolutely irreducible nonsingular variety defined over $\mathbf{Q}$, embedded in affine or projective space. We obtain $\tilde{V}_p$, its reduction mod p, by taking all the equations for V with coefficients in $\mathbf{Z}$ and mapping the coefficients into $\mathbf{F}_p$. If $\tilde{V}_p$ is nonsingular and has the same dimension as V, then V is said to have *good reduction* at p; this happens for all but a finite computable set of primes p. If p is large enough, it follows from the remarks above that $\tilde{V}_p$ contains a nonsingular point Q_p defined over $\mathbf{F}_p$. The result which follows, which is known as Hensel's Lemma though the idea of the proof goes back to Newton, now shows that V contains a point P_p defined over $\mathbf{Q}_p$.

Lemma 2.2. *Let V be an absolutely irreducible variety defined over $\mathbf{Q}$ which has a good reduction $\tilde{V}_p$ mod p. If $\tilde{V}_p$ contains a nonsingular point Q_p defined over $\mathbf{F}_p$ then V contains a nonsingular point P_p defined over $\mathbf{Q}_p$ whose reduction mod p is Q_p.*

In view of this, to decide whether V satisfies the Hasse condition one only has to check solubility in $\mathbf{R}$ and in finitely many $\mathbf{Q}_p$. Each of these checks can be shown to be a finite process.

A family $\mathscr{F}$ of varieties is said to satisfy the *Hasse Principle* if every V contained in $\mathscr{F}$ and defined over $\mathbf{Q}$ which satisfies the Hasse condition actually contains at least one point defined over $\mathbf{Q}$. Again, a family $\mathscr{F}$ is said to admit *weak approximation* if every V contained in $\mathscr{F}$ and defined over $\mathbf{Q}$, and such that $V(\mathbf{Q})$ is not empty, has the following property: given any finite set of places v and corresponding non-empty sets $\mathscr{N}_v \subset V(\mathbf{Q}_v)$ open in the v-adic topology, there is a point P in $V(\mathbf{Q})$ which lies in each of the $\mathscr{N}_v$. In the special case when $\mathscr{F}$ consists of a single variety V, and $V(\mathbf{Q})$ is not empty, we simply say that V admits weak approximation. Whether V admits weak approximation appears not to be computable in general; for a case where it is, see [5]. All this generalizes effortlessly to an arbitrary algebraic number field.

The most important families which are known to have either of these properties (and which actually have both) are the families of quadrics of any given dimension; this was proved by Minkowski for quadrics over $\mathbf{Q}$ and by Hasse for quadrics over an arbitrary algebraic number field. They also both hold for Severi-Brauer varieties, which are varieties biregularly equivalent to some $\mathbf{P}^n$ over $\mathbf{C}$. But many families, even of very simple varieties, do not satisfy either the Hasse Principle or weak approximation. (For example, neither of them holds for nonsingular cubic surfaces.) It is therefore natural to ask

Question 2.3. For a given family $\mathscr{F}$, what are the obstructions to the Hasse Principle and to weak approximation?

For weak approximation there is a variant of this question which may be more interesting and and is certainly easier to answer. For another way of stating weak approximation on V is to say that if $V(\mathbf{Q})$ is not empty then it is dense in the adelic space $V(\mathbf{A}) = \prod_v V(\mathbf{Q}_v)$. This suggests the following:

Question 2.4. For a given V, or family $\mathscr{F}$, what can be said about the closure of $V(\mathbf{Q})$ in the adelic space $V(\mathbf{A})$?

For the example of cubic surfaces, see [5]. However, there are families for which Question 2.3 does not seem to be a sensible question to ask; these probably include for example all families of varieties of general type. So one should also back up Question 2.3 with

Question 2.5. For what kinds of families is either part of Question 2.3 a sensible question to ask?

The only known systematic obstruction to the Hasse Principle or to weak approximation is the Brauer-Manin obstruction, though obstructions can be found in the literature which are not Brauer-Manin. (See for example Skorobogatov [6].) It is defined as follows. Let A be a *central simple algebra* – that is, a simple algebra which is finite dimensional over a field K which is its centre. Each such algebra consists, for fixed D and n, of all $n \times n$ matrices with elements in a division algebra D with centre K. Two central simple algebras over K are *equivalent* if they have the same underlying division algebra. Formation of tensor products over K gives the set of equivalence classes the structure of a commutative group, called the *Brauer group* of K and written $\mathrm{Br}(K)$. There is a canonical isomorphism $\iota_p : \mathrm{Br}(\mathbf{Q}_p) \simeq \mathbf{Q}/\mathbf{Z}$ for each p; and there is a canonical isomorphism $\iota_\infty : \mathrm{Br}(\mathbf{R}) \simeq \{0, \frac{1}{2}\}$, the nontrivial division algebra over $\mathbf{R}$ being the classical quaternions.

Let B be an element of $\mathrm{Br}(\mathbf{Q})$. Tensoring B with any $\mathbf{Q}_v$ gives rise to an element of $\mathrm{Br}(\mathbf{Q}_v)$, and this element is trivial for almost all v. There is an exact sequence

$$0 \to \mathrm{Br}(\mathbf{Q}) \to \bigoplus \mathrm{Br}(\mathbf{Q}_v) \to \mathbf{Q}/\mathbf{Z} \to 0,$$

due to Hasse, in which the third map is the sum of the ι_v; it tells us when a set of elements, one in each $\mathrm{Br}(\mathbf{Q}_v)$ and almost all trivial, can be generated from some element of $\mathrm{Br}(\mathbf{Q})$.

Now let V be a complete nonsingular variety defined over $\mathbf{Q}$ and A an *Azumaya algebra* on V – that is, a simple algebra with centre $\mathbf{Q}(V)$ which has a good specialization at every point of V. The group of equivalence classes of Azumaya algebras on V is denoted by $\mathrm{Br}(V)$. If P is any point of V, with field of definition $\mathbf{Q}(P)$, we obtain a simple algebra $A(P)$ with centre $\mathbf{Q}(P)$ by specializing at P. For all but finitely many p, we have $\iota_p(A(P_p)) = 0$ for all p-adic points P_p on V. Thus, as was first noticed by Manin, a necessary condition for the existence of a rational point P on V is that for every v there should be a v-adic point P_v on V such that

$$\sum \iota_v(A(P_v)) = 0 \quad \text{for all } A. \tag{2}$$

Similarly, a necessary condition for V with $V(\mathbf{Q})$ not empty to admit weak approximation is that (2) should hold for all Azumaya algebras A and all adelic points $\prod_v P_v$. In each case this is the *Brauer-Manin condition*. It is clearly unaffected if we add to A a constant algebra – that is, an element of $\mathrm{Br}(\mathbf{Q})$. So what we are really interested in is $\mathrm{Br}(V)/\mathrm{Br}(\mathbf{Q})$.

All this can be put into highbrow language. For any V there is an injection of $\mathrm{Br}(V)$ into the étale cohomology group $\mathrm{H}^2(V, \mathbf{G}_m)$; and if for example V is a complete nonsingular surface, this injection is an isomorphism. If we write

$$\mathrm{Br}_1(V) = \ker(\mathrm{Br}(V) \to \mathrm{Br}(\bar{V})) = \ker(\mathrm{H}^2(V, \mathbf{G}_m) \to \mathrm{H}^2(\bar{V}, \mathbf{G}_m)),$$

there is a filtration

$$\mathrm{Br}(\mathbf{Q}) \subset \mathrm{Br}_1(V) \subset \mathrm{Br}(V).$$

However, not even the abstract structure of $\mathrm{Br}(V)/\mathrm{Br}_1(V)$ is known; and there is no known systematic way of finding Azumaya algebras which represent nontrivial elements of this quotient, though in a particular case Harari [7] has exhibited a Brauer-Manin obstruction coming from such an algebra. In contrast, provided the Picard variety of V is trivial there is an isomorphism

$$\mathrm{Br}_1(V)/\mathrm{Br}(\mathbf{Q}) \simeq \mathrm{H}^1(\mathrm{Gal}(\bar{\mathbf{Q}}/\mathbf{Q}), \mathrm{Pic}(V \otimes \bar{\mathbf{Q}})),$$

and this is computable in both directions provided the Néron-Severi group of V over $\bar{\mathbf{Q}}$ is known and is torsion-free. (For details of this, see [8].)

There is no known systematic way of determining the Néron-Severi group for arbitrary V, and there is strong reason to suppose that this is really a number-theoretic rather than a geometric problem. One may need to approach this question through the Tate conjectures, for which see Sect. 3; but this is a very long-term strategy. However, it is usually possible to determine it for any given V, even if one cannot prove that this determination is correct.

Question 2.6. Is there a general algorithm (even conjectural) for determining the Néron-Severi group of V for varieties V defined over an algebraic number field?

Lang has conjectured that if V is a variety of general type defined over an algebraic number field K then there is a finite union $\mathscr{S}$ of proper subvarieties of V such that every point of $V(K)$ lies in $\mathscr{S}$. (Faltings' theorem, for which see Sect. 4, is the special case of this for curves.) This raises another question, similar to Question 2.6 but probably somewhat easier:

Question 2.7. Is there an algorithm for determining $\text{Pic}(V)$ where V is a variety defined over an algebraic number field?

The Brauer-Manin obstruction was introduced by Manin [9] in order to bring within a single framework various sporadic counterexamples to the Hasse principle. The theory of this obstruction has been extensively developed, largely by Colliot-Thélène and Sansuc. In particular, for rational varieties they have shown how to go back and forth between the Brauer-Manin condition and the descent condition for torsors under tori. They also defined universal torsors and showed that if there is no Brauer-Manin obstruction to the Hasse principle on a variety V then there exists a universal torsor over V which has points everywhere locally. This suggests that one should pay particular attention to Diophantine problems on universal torsors. Unfortunately, it is usually not easy to exploit what is known about the geometric structure of universal torsors. Indeed there are very few families for which the Brauer-Manin obstruction can be nontrivial but for which it has been shown that it is the only obstruction to the Hasse principle. (See however [10] and, subject to Schinzel's hypothesis, [11, 12].) Colliot-Thélène and Sansuc have conjectured that the Brauer-Manin obstruction is the only obstruction to the Hasse principle for rational surfaces – that is, surfaces birationally equivalent to $\mathbf{P}^2$ over $\bar{\mathbf{Q}}$. On the other hand, Skorobogatov ([6], and see also [13]) has exhibited on a bielliptic surface an obstruction to the Hasse principle which is definitely not Brauer-Manin.

Question 2.8. Is the Brauer-Manin obstruction the only obstruction to the Hasse principle for all unirational (or all Fano) varieties?

For the method of universal torsors, the immediate question to address must be the following:

Question 2.9. Does the Hasse principle hold for universal torsors over a rational surface?

We can of course ask similar questions for weak approximation. Both for the Hasse principle and for weak approximation one can alternatively ask what is the most general class of varieties for which the Brauer-Manin obstruction is the only one. Colliot-Thélène has suggested that this class probably includes all rationally connected varieties.

There are families $\mathscr{F}$ whose universal torsors appear to be too complicated to be systematically investigated, but for which it is still possible to identify the obstruction to the Hasse principle. It is sometimes possible to start from the absence of a Brauer-Manin obstruction (the most impressive example being Chap. 3 of Wittenberg [14]); but there are also alternative strategies. Implementing these falls naturally into two parts:

1. Assuming that V in $\mathscr{F}$ satisfies the Hasse condition, one finds a necessary and sufficient condition for V to have a rational point, or to admit weak approximation.

2. One then shows that this necessary and sufficient condition is equivalent to the Brauer-Manin condition.

Both parts of this strategy have been applied to pencils of conics, where one uses Schinzel's Hypothesis to implement (1); see [12, 11]. Except for Skorobogatov's example above, I know of no families for which it has been possible to carry out (1) but not (2). But there are families for which it has been possible to find a sufficient condition for solubility (additional to the Hasse condition) which appears rather weak but which is definitely stronger than the Brauer-Manin condition. The obvious examples of such a condition are the various forms of what is called Conditions D or E in [15, 16, 17, 18]. However, in these cases it is not obvious that a condition stronger than the Brauer-Manin condition is actually necessary; and I attribute the gap to clumsiness in the proofs.

Question 2.10. When the Brauer-Manin condition is trivial, how can one make use of this fact?

In addition to the work of Wittenberg cited above, there are at least two known approaches to this question: by descent using torsors, and by the fibration method exploited in particular by Harari.

3 Zeta-Functions and L-Series

Let $W \subset \mathbf{P}^n$ be a nonsingular and absolutely irreducible projective variety of dimension d defined over the finite field $k = \mathbf{F}_q$, and denote by $\phi(q)$ the Frobenius automorphism of W given by

$$\phi(q) : (x_0, x_1, \ldots, x_n) \mapsto (x_0^q, x_1^q, \ldots, x_n^q).$$

For any $r > 0$ the fixed points of $(\phi(q))^r$ are precisely the points of W which are defined over $\mathbf{F}_{q^r}$; suppose that there are $N(q^r)$ of them. Although the context is totally different, this is almost the formalism of the Lefschetz Fixed Point theorem, since for geometric reasons each of these fixed points has multiplicity $+1$. This analogy led Weil to conjecture that there should be a cohomology theory applicable in this context. This would imply that there were finitely many complex numbers α_{ij} such that

$$N(q^r) = \sum_{i=0}^{2d} \sum_{j=1}^{B_i} (-1)^i \alpha_{ij}^r \quad \text{for all } r > 0, \tag{3}$$

where B_i is the dimension of the ith cohomology group of W and the α_{ij} are the characteristic roots of the map induced by $\phi(q)$ on the ith cohomology. For each i duality asserts that $B_i = B_{2d-i}$ and the $\alpha_{2d-i,j}$ are a permutation of the q^d/α_{ij}. If we define the local zeta-function $Z(t, W)$ by either of the equivalent relations

$$\log Z(t) = \sum_{r=1}^{\infty} N(q^r)t^r/r \quad \text{or} \quad tZ'(t)/Z(t) = \sum_{r=1}^{\infty} N(q^r)t^r,$$

then (3) is equivalent to

$$Z(t) = \frac{P_1(t,W)\cdots P_{2d-1}(t,W)}{P_0(t,W)P_2(t,W)\cdots P_{2d}(t,W)}$$

where $P_i(t,W) = \prod_j(1 - \alpha_{ij}t)$. Each $P_i(t,W)$ must have coefficients in $\mathbf{Z}$, and the analogue of the Riemann hypothesis is that $|\alpha_{ij}| = q^{i/2}$. (For a fuller account of Weil's conjectures and their motivation, see the excellent survey [19].) All this has now been proved, the main contributor being Deligne.

Now let V be a nonsingular and absolutely irreducible projective variety defined over an algebraic number field K. If V has good reduction at a prime $\mathfrak{p}$ of K we can form $\tilde{V}_{\mathfrak{p}}$, the reduction of V mod $\mathfrak{p}$, and hence form the $P_i(t, \tilde{V}_{\mathfrak{p}})$. For s in $\mathbf{C}$, we can now define the ith global L-series $L_i(s, V)$ of V as a product over all places of K, the factor at a prime $\mathfrak{p}$ of good reduction being $(P_i(q^{-s}, \tilde{V}_{\mathfrak{p}}))^{-1}$ where $q = \mathrm{Norm}_{K/\mathbf{Q}}\mathfrak{p}$. The rules for forming the factors at the primes of bad reduction and at the infinite places can be found in [20]. These L-series of course depend on K as well as on V. In particular, $L_0(s,V)$ is just the zeta-function of the algebraic number field K.

To call a function $F(s)$ a (global) zeta-function or L-series ought to carry with it certain implications, though some authors have used these terms very loosely:

- $F(s)$ should be the product of a Dirichlet series and possibly some Gamma-functions, and the half-plane of absolute convergence for the Dirichlet series should have the form $\mathfrak{R}s > \sigma_0$ with $2\sigma_0$ in $\mathbf{Z}$.
- The Dirichlet series should be expressible as an Euler product $\prod_p f_p(p^{-s})$ where the f_p are rational functions.
- $F(s)$ should have an analytic continuation to the entire s-plane as a meromorphic function all of whose poles are in $\mathbf{Z}$.
- There should be a functional equation relating $F(s)$ and $F(2\sigma_0 - 1 - s)$.
- The zeroes of $F(s)$ in the critical strip $\sigma_0 - 1 < \mathfrak{R}s < \sigma_0$ should lie on $\mathfrak{R}s = \sigma_0 - \frac{1}{2}$.

In our case, the first two implications are trivial; and fortunately one is not expected to prove the last three, but only to state them as conjectures. The last one is the Riemann Hypothesis, which appears to be out of reach even in the simplest case, which is the classical Riemann zeta-function; and the third and fourth have so far only been proved in a few favourable cases.

Question 3.1. Can one extend the list of V for which analytic continuation and the functional equation can be proved?

It seems likely that any proof of analytic continuation will carry a proof of the functional equation with it.

It has been said about the zeta-functions of algebraic number fields that "the zeta-function knows everything about the number field; we just have to prevail on it to tell us". If this is so, we have not yet unlocked the treasure-house. Apart from the classical formula which relates hR to $\zeta_K(0)$ all that has so far been proved are certain results of Borel [21] which relate the behaviour of $\zeta_K(s)$ near $s = 1 - m$ for integers $m > 1$ to the K-groups of $\mathfrak{O}_K$. One might hope that when a mysterious number turns up in the study of Diophantine problems on V, some L-series contains information about it; and this is certainly sometimes true, the most spectacular examples being the Birch/Swinnerton-Dyer conjecture and the far-reaching generalizations of it due to Bloch and Kato. But it appears to be false for the order of the Chow group of a rational surface; this is always finite, but two such surfaces can have the same L-series while having Chow groups of different orders.

Suppose for convenience that V is defined over $\mathbf{Q}$, and let its dimension be d. Even for varieties with $B_1 = 0$ we do not expect a product like

$$\prod_p N(p)/p^d \quad \text{or} \quad \prod_p N(p) \bigg/ \left(\frac{p^{d+1} - 1}{p - 1} \right) \tag{4}$$

to be necessarily absolutely convergent. But in some contexts there is a respectable expression which is formally equivalent to one of these, with appropriate modifications of the factors at the bad primes. The idea that such an expression should have number-theoretic significance goes back to Siegel (for genera of quadratic forms) and Hardy/Littlewood (for what they called the *singular series*). Using the ideas above, we are led to replace the study of the products (4) by a study of the behaviour of $L_{2d-1}(s, V)$ and $L_{2d-2}(s, V)$ near $s = d$. By duality, this is the same as studying $L_1(s, V)$ near $s = 1$ and $L_2(s, V)$ near $s = 2$. The information derived in this way appears to relate to the Picard group of V, defined as the group of divisors defined over $\mathbf{Q}$ modulo linear equivalence. By considering simultaneously both V and its Picard variety (the abelian variety which parametrises divisors algebraically equivalent to zero modulo linear equivalence), one concludes that $L_1(s, V)$ should be associated with the Picard variety and $L_2(s, V)$ with the group of divisors modulo algebraic equivalence – that is, with the Néron-Severi group of V. These ideas motivated the weak forms of the Birch/Swinnerton-Dyer conjecture (for which see Sect. 4) and the case $m = 1$ of the Tate conjecture below. For the strong forms (which give expressions for the leading coefficients of the relevent Laurent series expansions) heuristic arguments are less convincing; but one can formulate conjectures for these coefficients by asking what other mysterious numbers turn up in the same context and should therefore appear in the formulae for the leading coefficients.

The weak form of the Tate conjecture asserts that the order of the pole of $L_{2m}(s, V)$ at $s = m + 1$ is equal to the rank of the group of classes of m-cycles on V defined over K, modulo algebraic equivalence; it is a natural generalization of the case $m = 1$ for which the heuristics have just been shown. For a more detailed account of both of these, including the conjectural formulae for the leading coefficients, see [22] or [23].

Question 3.2. What information about V is contained in its L-series?

There is in the literature a beautiful edifice of conjecture, lightly supported by evidence, about the behaviour of the $L_i(s,V)$ at integral points. The principal architects of this edifice are Beilinson, Bloch and Kato. Beilinson's conjectures relate to the order and leading coefficients of the Laurent series expansions of the $L_i(s,V)$ at integer values of s; in them the leading coefficients are treated as elements of $\mathbf{C}^*/\mathbf{Q}^*$. (For a full account see [24] or [25].) Bloch and Kato [26, 27] have strengthened these conjectures by treating the leading coefficients as elements of $\mathbf{C}^*$. But I do not believe that anything like the full story has yet been revealed.

4 Curves

The most important invariant of a curve is its genus g. In the language of algebraic geometry over $\mathbf{C}$, curves of genus 0 are called *rational*, curves of genus 1 are called *elliptic* and curves of genus greater than 1 are *of general type*. But note that for a number theorist an elliptic curve is defined to be a curve of genus 1 with a distinguished point P_0 on it, both being defined over the ground field K. The effect of this is that the points on an elliptic curve form an abelian group with P_0 as its identity element, the sum of P_1 and P_2 being the other zero of the function (defined up to multiplication by a constant) with poles at P_1 and P_2 and a zero at P_0.

A canonical divisor on a curve Γ of genus 0 has degree -2; hence by the Riemann-Roch theorem Γ is birationally equivalent over the ground field to a conic. The Hasse principle holds for conics, and therefore for all curves of genus 0; this gives a complete answer to Question (A) at the beginning of these notes. But it does not give an answer to Question (B). Over $\mathbf{Q}$, a very simple answer to Question (B) is as follows:

Theorem 4.1. *Let a_0, a_1, a_2 be nonzero elements of* $\mathbf{Z}$. *If the equation*

$$a_0 X_0^2 + a_1 X_1^2 + a_2 X_2^2 = 0$$

has a nontrivial solution in $\mathbf{Z}$, *then it has a solution for which each $a_i X_i^2$ is absolutely bounded by* $|a_0 a_1 a_2|$.

Siegel [28] has given an answer to Question (B) over arbitrary algebraic number fields, and Raghavan [29] has generalized Siegel's work to quadratic forms in more variables.

The knowledge of one rational point on Γ enables us to transform Γ birationally into a line; so if Γ is soluble there is a parametric solution which gives explicitly all the points on Γ defined over the ground field. This answers Question (C).

If Γ is a curve of general type defined over an algebraic number field K, Mordell conjectured and Faltings proved that $\Gamma(K)$ is finite; and a number of other proofs have appeared since then. But it does not seem that any of them enable one to compute $\Gamma(K)$, though some of them come tantalizingly close. For a survey of several such proofs, see [30].

Question 4.2. Is there an algorithm for computing $\Gamma(\mathbf{Q})$ when Γ is a curve of general type defined over $\mathbf{Q}$?

The study of rational points on elliptic curves is now a major industry, almost entirely separate from the study of other Diophantine problems. If Γ is an elliptic curve defined over an algebraic number field K, the group $\Gamma(K)$ is called the *Mordell-Weil group*. Mordell proved that $\Gamma(K)$ is finitely generated and Weil extended this to all Abelian varieties. Thanks to Mazur [31] and Merel [32] the theory of the torsion part of the Mordell-Weil group is now reasonably complete; but for the non-torsion part all that was known before 1960 is that for any $n > 1$ $\Gamma(K)/n\Gamma(K)$ could be embedded into a certain group (the n-Selmer group, for which see Sect. 10) which is finitely generated and computable. The process involved, which is known as the method of infinite descent, goes back to Fermat; various forms of this for $n = 2$ will be described in Sect. 10. By means of this process one can always compute an upper bound for the rank of the Mordell-Weil group of any particular Γ, and the upper bound thus obtained can frequently be shown to be equal to the actual rank by exhibiting enough elements of $\Gamma(K)$. It was also conjectured that the difference between the upper bound thus computed and the actual rank was always an even integer, but apart from this the actual rank was mysterious. This not wholly satisfactory state of affairs has been radically changed by the Birch/Swinnerton-Dyer conjecture, the weak form of which is described at the end of this section. A survey of what is currently known or conjectured about the ranks of Mordell-Weil groups can be found in [33].

Suppose now that Γ is a curve of genus 1 defined over K but not necessarily containing a point defined over K. Let J be the Jacobian of Γ, defined as a curve whose points are in one-one correspondence with the divisors of degree 0 on Γ modulo linear equivalence. Then J is also a curve of genus 1 defined over K, and $J(K)$ contains the point which corresponds to the trivial divisor. So J is an elliptic curve in our sense.

Conversely, if we fix an elliptic curve J defined over K we can consider the equivalence classes (for birational equivalence over K) of curves Γ of genus 1 defined over K which have J as Jacobian. For number theory, the only ones of interest are those which contain points defined over each completion K_v. These form a commutative torsion group, called the *Tate-Shafarevich group* and usually denoted by III; the identity element of this group is the class which contains J itself, and it consists of those Γ which have J as Jacobian and which contain a point defined over K. (The simplest example of a nontrivial element of a Tate-Shafarevich group is the curve

$$3X_0^3 + 4X_1^3 + 5X_2^3 = 0 \quad \text{with Jacobian} \quad Y_0^3 + Y_1^3 + 60Y_2^3 = 0.)$$

Thus for curves of genus 1 the Tate-Shafarevich group is by definition the obstruction to the Hasse principle.

The weak form of the Birch/Swinnerton-Dyer conjecture states that the rank of the Mordell-Weil group of an elliptic curve J is equal to the order of the zero of $L_1(s,J)$ at $s = 1$; the conjecture also gives an explicit formula for the leading coefficient of the power series expansion at that point. Note that this point is at the centre of the critical strip, so that the conjecture pre-supposes the analytic continuation of $L_1(s,J)$. At present there are two well-understood cases in which analytic continuation is known: when $K = \mathbf{Q}$, so that J can be parametrised by means of modular functions, and when J admits complex multiplication. In consequence, these two cases are likely to be easier than the general case; but even here I do not expect much further progress in the next decade. In each of these two cases, if one assumes the Birch/Swinnerton-Dyer conjecture one can derive an algorithm for finding the Mordell-Weil group and the order of the Tate-Shafarevich group; and in the first of the two cases this algorithm has been implemented by Gebel [34]. Without using the Birch/Swinnerton-Dyer conjecture, Heegner long ago produced a way of generating a point on J whenever $K = \mathbf{Q}$ and J is modular; and Gross and Zagier [35, 36] have shown that this point has infinite order precisely when $L'(1,J) \neq 0$. Building on their work, Kolyvagin (see [37]) has shown the following.

Theorem 4.3. *Suppose that the Heegner point has infinite order; then the group $J(\mathbf{Q})$ has rank 1 and $\mathrm{III}(J)$ is finite.*

Kolyvagin [38] has also obtained sufficient conditions for both $J(\mathbf{Q})$ and $\mathrm{III}(J)$ to be finite. The following result is due to Nekovar and Plater.

Theorem 4.4. *If the order of $L(s,J)$ at $s = 1$ is odd then either $J(\mathbf{Q})$ is infinite or the p-part of $\mathrm{III}(J)$ is infinite for every good ordinary p.*

If J is defined over an algebraic number field K and can be parametrized by modular functions for some arithmetic subgroup of $\mathrm{SL}_2(\mathbf{R})$ then analytic continuation and the functional equation for $L_1(s,J)$ follow; but there is not even a plausible conjecture identifying the J which have this property, and there is no known analogue of Heegner's construction.

In the complex multiplication case, what is known is as follows.

Theorem 4.5. *Let K be an imaginary quadratic field and J an elliptic curve defined and admitting complex multiplication over K. If $L(1,J) \neq 0$, then*

(i) *$J(K)$ is finite;*
(ii) *For every prime $p > 7$ the p-part of $\mathrm{III}(J)$ is finite and has the order predicted by the Birch/Swinnerton-Dyer conjecture.*

Here (i) is due to Coates and Wiles, and (ii) to Rubin. For an account of the proofs, see [39]. Katz has generalized (i) and part of (ii) to behaviour over an abelian extension of $\mathbf{Q}$, but with the same J as before.

In general we do not know how to compute III. It is conjectured that it is always finite; and indeed this assertion can be regarded as part of the Birch/Swinnerton-Dyer conjecture, for the formula for the leading coefficient of the power series for $L_1(s,J)$ at $s = 1$ contains the order of $\mathrm{III}(J)$ as a factor. If indeed this order is finite, then it must be a square; for Cassels has proved the existence of a skew-symmetric bilinear form on III with values in $\mathbf{Q}/\mathbf{Z}$, which is nonsingular on the quotient of III by its maximal divisible subgroup. In particular, finiteness implies that if III contains at most $p - 1$ elements of order exactly p for some prime p then it actually contains no such elements; hence an element which is killed by p is trivial, and the curves of genus 1 in that equivalence class contain points defined over K. For use later, we state the case $p = 2$ as a lemma.

Lemma 4.6. *Suppose that* $\text{III}(J)$ *is finite and the quotient of the 2-Selmer group of* J *by its soluble elements has order at most 2; then that quotient is actually trivial.*

5 Varieties of Higher Dimension and the Hardy-Littlewood Method

A first coarse classification of varieties of dimension n is given by the *Kodaira dimension* κ, which can take the values $-\infty$ or $0, 1, \ldots, n$. Denote the genus of a curve by g; then for curves $\kappa = -\infty$ corresponds to $g = 0$, $\kappa = 0$ to $g = 1$ and $\kappa = 1$ to $g > 1$; so the major split in the Diophantine theory of curves corresponds to the possible values of κ.

Over **C** a full classification of surfaces can be found in [40]. But what is also significant for the number theory (and cuts across this classification) is whether the surface is *elliptic* – that is, whether over **C** there is a map $V \to C$ for some curve C whose general fibre is a curve of genus 1. The case when the map $V \to C$ is defined over the ground field K and C has genus 0 is discussed below; in this case the Diophantine problems for V are only of interest when $C(K)$ is nonzero, so that C can be identified with $\mathbf{P}^1$. When C has genus greater than 1, the map $V \to C$ is essentially unique and it and C are therefore both defined over K. By Faltings' theorem, $C(K)$ is then finite; thus each point of $V(K)$ lies on one of a finite set of fibres, and it is enough to study these. In contrast, we know nothing except in very special cases when C is elliptic.

The surfaces with $\kappa = -\infty$ are precisely the *ruled surfaces* – that is, those which are birationally equivalent over **C** to $\mathbf{P}^1 \times C$ for some curve C. Among these, by far the most interesting are the *rational surfaces*, which are birationally equivalent to $\mathbf{P}^2$ over **C**. From the number-theoretic point of view, there are two kinds of rational surface:

- Pencils of conics, given by an equation of the form

$$a_0(u,v)X_0^2 + a_1(u,v)X_1^2 + a_2(u,v)X_2^2 = 0 \tag{5}$$

  where the $a_i(u,v)$ are homogeneous polynomials of the same degree. Pencils of conics can be classified in more detail according to the number of bad fibres.
- Del Pezzo surfaces of degree d, where $0 < d \leq 9$. Over **C**, such a surface is obtained by blowing up $(9 - d)$ points of $\mathbf{P}^2$ in general position – except when $d = 8$, in which case the construction is more complicated. It is known that Del Pezzo surfaces of degree $d > 4$ satisfy the Hasse principle and weak approximation; indeed those of degree 5 or 7 necessarily contain rational points. Del Pezzo surfaces of degree 2 or 1 have attracted relatively little attention; it seems sen-

sible to ignore them until the problems coming from those of degrees 4 and 3 have been solved. The Del Pezzo surfaces of degree 3 are the nonsingular cubic surfaces, which have an enormous but largely irrelevant literature, and those of degree 4 are the nonsingular intersections of two quadrics in $\mathbf{P}^4$. For historical reasons, attention has been concentrated on the Del Pezzo surfaces of degree 3; but the problems presented by those of degree 4 are necessarily simpler.

Surfaces with $\kappa = 0$ fall into four families:

- Abelian surfaces. These are the analogues in two dimensions of elliptic curves, and there is no reason to doubt that their number-theoretical properties largely generalize those of elliptic curves.
- K3 surfaces, including in particular Kummer surfaces. Some but not all K3 surfaces are elliptic.
- Enriques surfaces, whose number theory has been very little studied. Enriques surfaces are necessarily elliptic.
- Bielliptic surfaces.

Surfaces with $\kappa = 1$ are necessarily elliptic.

Surfaces with $\kappa = 2$ are called *surfaces of general type* – which in mathematics is generally a derogatory phrase. About them there is currently nothing to say beyond Lang's conjecture stated in Sect. 2.

For varieties of higher dimension (other than quadrics and Severi-Brauer varieties) there seem to be at the moment only two ways of obtaining results: by deduction from special results for surfaces, and by the Hardy/Littlewood method. The latter differs from most geometric methods in that it is not concerned with an equivalence class of varieties under birational or biregular transformation, but with a particular embedding of a variety V in projective or affine space. A point P in $\mathbf{P}^n$ defined over $\mathbf{Q}$ has a representation $(x_0, \ldots, x_n)$ where the x_i are integers with no common factor; and this representation is unique up to changing the signs of all the x_i. We define the *height* of P to be $h(P) = \max |x_i|$; a linear transformation on the ambient space multiplies heights by numbers which lie between two positive constants depending on the linear transformation. Denote by $N(H,V)$ the number of points of $V(\mathbf{Q})$ whose height is less than H; then it is natural to ask how $N(H,V)$ behaves as $H \to \infty$. This is the core question for the Hardy-Littlewood method, which when it is applicable is the best (and often the only) way of proving that $V(\mathbf{Q})$ is not empty. In very general circumstances that method provides estimates of the form

$$N(H,V) = \text{leading term} + \text{error term.} \qquad (6)$$

The leading term is usually the same as one would obtain by probabilistic arguments. But such results are only valuable when it can be shown that the error term is small compared to the leading term, and to achieve this the dimension of V needs to be large compared to its degree. The extreme case of this is the following theorem, due to Birch [41].

Theorem 5.1. *Suppose that the* $F_i(X_0, \ldots, X_N)$ *are homogeneous polynomials with coefficients in* $\mathbf{Z}$ *and* $\deg F_i = r_i$ *for* $i = 1, \ldots, m$, *where* $r_1, \ldots, r_m$ *are positive odd* *integers. Then there exists* $N_0(r_1, \ldots, r_m)$ *such that if* $N \geq N_0$ *the* F_i *have a common nontrivial zero in* $\mathbf{Z}^{N+1}$.

The proof falls into two parts. First, the Hardy-Littlewood method is used to prove the result in the special case when $m = 1$ and F_1 is diagonal – that is, to show that if r is odd and $N \geq N_1(r)$ then

$$c_0 X_0^r + \ldots + c_N X_N^r = 0$$

has a nontrivial integral solution. Then the general case is reduced to this special case by purely elementary methods. The requirement that all the r_i should be odd arises from difficulties connected with the real place; over a fixed totally complex algebraic number field there is a similar theorem for which the r_i can be any positive integers.

The Hardy-Littlewood method was designed for a single equation in which the variables are separated – for example, an equation of the form

$$f_1(X_1) + \ldots + f_N(X_N) = c$$

where the f_i are polynomials, the X_i are integers, and one wishes to prove solubility in $\mathbf{Z}$ for all integers c, or all large enough c, or almost all c. But it has also been applied both to several simultaneous equations and to equations in which the variables are not separated. The following theorem of Hooley [42] is the most impressive result in this direction.

Theorem 5.2. *Homogeneous nonsingular nonary cubics over* $\mathbf{Q}$ *satisfy both the Hasse principle and weak approximation.*

6 Manin's Conjecture

Even on the most optimistic view, one can only hope to make the Hardy-Littlewood method work for families for which $N(H, V)$ is asymptotically equal to its probabilistic value; in particular it seems unlikely that it can be made to work for families for which weak approximation fails. On the other hand, one can hope that the leading term in (6) will still have the correct shape for other families, even if it is in error by a constant factor. Manin has put forward a conjecture about the asymptotic density of rational solutions for certain geometrically interesting families of varieties for which weak approximation is unlikely to hold: more precisely, for Fano varieties embedded in $\mathbf{P}^n$ by means of their anticanonical divisors. A general survey of the present state of the Manin conjecture can be found in [43]. In the full generality in which he stated the conjecture, it is known to be false; and in what follows I consider it only for Del Pezzo surfaces V of degrees 3 and 4. These are the most natural ones for the number theorist to consider, because of the simplicity of the equations which define them – one cubic and two quadratics respectively. The analogy of the Hardy-Littlewood method suggests an estimate $AH \prod (N(p)/(p+1))$ for $N(H,V)$, where the product is taken over all primes less than a certain bound which depends on H. In view of what is said in Sect. 3, this product ought to be replaced by something which depends on the behaviour of $L_2(s,V)$ near $s = 1$. The way in which the leading term in the Hardy-Littlewood method is obtained suggests that here we should take $s - 1$ to be comparable with $(\log H)^{-1}$. Remembering the Tate conjecture, this gives the right hand side of (7) as a conjectural estimate for $N(H,V)$. But to ask about $N(H,V)$ is the wrong question, for V may contain lines L defined over $\mathbf{Q}$, and for any line $N(H,L) \sim AH^2$ for some nonzero constant A. This is much greater than the order-of-magnitude estimate for $N(H,V)$ given by a probabilistic argument. Manin's way to resolve this absurdity is to study not $N(H,V)$ but $N(H,U)$, where U is the open subset of V obtained by deleting the finitely many lines on V. He therefore conjectured that

$$N(H,U) \sim AH(\log H)^{r-1} \text{ where } r \text{ is the rank of Pic}(V). \qquad (7)$$

Peyre [44] has given a conjectural formula for A. Unfortunately there are no nonsingular Del Pezzo surfaces of degrees 3 or 4, and very few singular ones, for which (7) has been proved.

Question 6.1. Are there nonsingular Del Pezzo surfaces V of degree 3 or 4 for which the Manin conjecture can be proved by present methods?

In the first instance, it would be wise to address this problem under rather restrictive hypotheses about V, not least because the Brauer-Manin obstruction to weak approximation occurs in the conjectural formula for A and therefore the problem is likely to be easier for families of V for which weak approximation holds. The simplest cases of all are likely to be among those for which V is birationally equivalent to $\mathbf{P}^2$ over $\mathbf{Q}$. For nonsingular cubic surfaces, for example, it has long been known that this happens if and only if V is everywhere locally soluble and contains a divisor defined over $\mathbf{Q}$ which is the union of 2, 3 or 6 skew lines. In the case when V contains two skew lines each defined over $\mathbf{Q}$, a lower bound for $N(H,U)$ of the correct order of magnitude was proved in [45].

An alternative method of describing the statistics of rational points on U is by means of the *height zeta function*

$$Z(h,U,s) = \sum_{P \in U(\mathbf{Q})} (h(P))^{-s}$$

where h is some height function – for example, the classical one defined in Sect. 5. (Note that, despite the name, we do not expect this function to have the properties listed in Sect. 3.) Now (7) is more or less equivalent to

$$Z(h,U,s) \sim A'(s-1)^{-r} \text{ as } s \text{ tends to 1 from above.}$$

It is now natural to hope that $Z(h,U,z)$ can be analytically continued to some halfplane $\Re s > c$ for some $c < 1$, subject to a pole of order r at $s = 1$. If this is so, we can derive $N(H,U)$ from $Z(U,s)$ by means of Perron's formula

$$N(H,U) = \frac{1}{2\pi i} \int_{c-i\infty}^{c+i\infty} B^s \frac{Z(U,s)}{s} ds$$

where $H - 1 < B < H$ and $c > 1$. Now (7) can be strengthened to

$$N(H,U) = Hf(\log H) + O(H^{c+\varepsilon}) \tag{8}$$

where f is a polynomial of degree $r - 1$.

De la Bretèche and Browning [46, 47, 48] have proved results of the form (8) for several singular Del Pezzo surfaces of degrees 3 and 4. Their methods are intricate, and it would be interesting to know what features of the geometry of their particular surfaces underlie them. The simplest surface of this kind, and the one about which most is known, is the toric surface

$$X_0^3 = X_1 X_2 X_3. \tag{9}$$

Let U be the open subset of (9) given by $X_0 \neq 0$. Building on earlier work of de la Bretèche [49] and assuming the Riemann Hypothesis, he and I have proved [50] that

$$N(U,H) = Hf(\log H) + CH^{9/11} + \Re \sum \gamma_n H^{3/4 + \rho_n/8} + O(H^{4/5}) \tag{10}$$

where C and the γ_n are constants, ρ_n runs through the zeros of the Riemann zeta function, and f is a certain polynomial of degree 6. Some bracketing of terms for which the ρ_n are nearly equal may be needed to ensure convergence. The associated height zeta function can be meromorphically continued to $\Re s > \frac{3}{4}$ but no further.

The key idea in the proofs of (10) and of analytic continuation is to introduce the multiple Dirichlet series

$$\phi(s_1, s_2, s_3) = \sum_{P \in U(\mathbf{Q})} |x_1|^{-s_1} |x_2|^{-s_2} |x_3|^{-s_3}$$

where (x_0, x_1, x_2, x_3) is a primitive integral representation of P. At the cost of a factor 4, we can confine ourselves in the definition of ϕ to points with all coordinates positive. We have

$$|x_1|^{-s_1} |x_2|^{-s_2} |x_3|^{-s_3} = \prod_p p^{-\{s_1 v_p(x_1) + s_2 v_p(x_2) + s_3 v_p(x_3)\}}$$

from which it follows that $\phi(s_1, s_2, s_3) = 4 \prod_p \phi^*(p^{-s_1}, p^{-s_2}, p^{-s_3})$ where the factor associated with p is the sum over the points all of whose coordinates are powers of p. A straightforward calculation shows that

$$\phi^*(z_1, z_2, z_3) = \frac{1 + \sum z_i^2 z_j (1 - z_k^3) - z_1^3 z_2^3 z_3^3}{(1 - z_1^3)(1 - z_2^3)(1 - z_3^3)},$$

the sum being taken over the six permutations i, j, k of 1, 2, 3. This expresses $\phi(s_1, s_2, s_3)$ as an Euler product and enables its meromorphic continuation to the open set in which $\Re s_i > 0$ for $i = 1, 2, 3$. (A simpler example of the same process will be found in the next paragraph.) Moreover

$$Z(h, U, s) = \frac{1}{(2\pi i)^2} \int_{c_2 - i\infty}^{c_2 + i\infty} \int_{c_3 - i\infty}^{c_3 + i\infty} \frac{s\phi(s - s_2 - s_3, s_2, s_3)}{s_2 s_3 (s - s_2 - s_3)} ds_2 ds_3$$

provided $\Re s, c_2$ and c_3 are chosen so that the series for ϕ is absolutely convergent. One can now move the contours of integration to the left, though the reader is warned that this imvolves technical problems as well as some tedious calculation. In the end (10) and the meromorphic continuation of $Z(h, U, s)$ follow.

The estimate (10) is reminiscent of the explicit formula of prime number theory. But the second term on the right is unexpected, and one would have hoped that the exponent in the third term would have been ρ_n rather than $\frac{3}{4} + \frac{1}{8}\rho_n$. Both these blemishes are caused by the fact that $h(P)$, though classical, is not the most natural height function. For comparison, we now consider what happens if we use as our height function $h_1(P) = |x_0|$ where (x_0, x_1, x_2, x_3) is a primitive integral representation of P. Now we obtain

$$Z(h_1, U, s) = 4 \prod_p \frac{1 + 7p^{-s} + p^{-2s}}{(1 - p^{-s})^2}.$$

The factor corresponding to p on the right is

$$(1 - p^{-s})^{-9}\{(1 - p^{-s})^7 (1 + 7p^{-s} + p^{-2s})\}$$

where the expression in curly brackets is $1 + O(p^{-2s})$; so using the known analytic continuation of the Riemann zeta function, this gives the continuation of $Z(h_1, U, s)$ to $\Re s > \frac{1}{2}$. The expression in curly brackets is actually

$$1 - 27p^{-2s} + O(p^{-3s});$$

so we can take out a factor $(1 - p^{-2s})^{27}$ and obtain the continuation of $Z(h_1, U, s)$ to $\Re s > \frac{1}{3}$ – and so on. The eventual conclusion is that $Z(h_1, U, s)$ can be meromorphically continued to $\Re s > 0$ but that $\Re s = 0$ is a natural boundary. Using Perron's formula we can obtain a complicated formula for the corresponding counting function $N_1(H, U)$ of which the leading terms are

$$N_1(H, U) = H f_1(\log H) + H^{1/3} g_1(\log H) + \sum \gamma_{1n}(\log H) H^{\rho_n/2} + O(H^{1/5 + \varepsilon}).$$

Here f_1 and g_1 are polynomials of degrees 8 and 104 respectively, ρ_n runs through the complex zeros of the Riemann zeta function and the γ_{1n} are polynomials of degree 26.

The second term here is much smaller than that in (10). This raises the question of what is the best height function to choose, and indeed whether there is a canonical height function on at least some Del Pezzo surfaces. (Recall that on abelian varieties there certainly is a canonical height function.) A very partial answer to this can be found in [51], where reasons are given for using for (9) the height function $h^*(P) = \prod_p p^{-s\alpha_p}$ where

$$\alpha_p = \frac{1}{2}((v_p(x_1))^2 + (v_p(x_2))^2 + (v_p(x_3))^2 - 3(v_p(x_0))^2).$$

With this choice, again assuming the Riemann Hypothesis, the natural boundary for the height zeta function $Z(h^*, U, s)$ is $\Re s = 0$ and we can exhibit a (very complicated) formula for the corresponding counting function $N^*(H, U)$ with error $O(H^\varepsilon)$. This time the polynomial f^* in the leading term $Hf^*(\log H)$ has degree 5 and the remaining terms contribute $O(H^{1/2+\varepsilon})$.

In the light of these results it is natural to wonder what is the shape of the error term in (8) when V is nonsingular. At present, the only way to approach this question is by computation. It is advantageous to study varieties whose equations have the form $g_1(X_0, X_1) = g_2(X_2, X_3)$ because counting rational points is then much faster than for general cubic surfaces. Some computations have been made for V of the form

$$a_0 X_0^3 + a_1 X_1^3 + a_2 X_2^3 + a_3 X_3^3 = 0.$$

Now $r = 1$ and the evidence strongly suggests that in (8) we can take $c = \frac{1}{2}$. The results are indeed compatible with a conjecture of the form

$$N(H, V) = AH + \sum \gamma_n H^{1/2+it_n} + O(H^{1/2-\varepsilon}) \tag{11}$$

for a discrete sequence of real t_n. But the evidence available so far, which is for $H \leq 10^5$, is too scanty for one to be able to estimate the first few t_n with any great accuracy. However, the way in which they appear in (11) suggests that they should be the zeros of some L-series – and of course there is one L-series naturally associated with V.

There is no reason why one should not also ask about the density of rational points on surfaces which are not Fano. For Del Pezzo surfaces, the conjectural value of c for which $N(H, U) \sim AH(\log H)^c$ is defined by the geometry rather than by the number theory, though that is not true of A. For other varieties, the corresponding statement need no longer be true. We start with curves. For a curve of genus 0 and degree d, we have $N(H, V) \sim AH^{2/d}$; and for a curve of genus greater than 1 Faltings' theorem is equivalent to the statement that $N(H, V) = O(1)$. But if V is an elliptic curve then $N(H, V) \sim A(\log H)^{r/2}$ where r is the rank of the Mordell-Weil group. (For elliptic curves there is a more canonical definition of height, which is invariant under bilinear transformation; this is used to prove the result above.)

For pencils of conics, Manin's question is probably not the best one to ask, and it would be better to proceed as follows. A pencil of conics is a surface V together with a map $V \to \mathbf{P}^1$ whose fibres are conics. Let $N^*(H, V)$ be the number of points on $\mathbf{P}^1$ of height less than H for which the corresponding fibre contains rational points.

Question 6.2. What is the conjectural estimate for $N^*(H,V)$ and under what conditions can one prove it?

It may be worth asking the same questions for pencils of curves of genus 1.

For surfaces of general type, Lang's conjecture implies that questions about $N(H,V)$ are really questions about certain curves on V; and for Abelian surfaces (and indeed Abelian varieties in any dimension) the obvious generalisation of the theorem for elliptic curves holds. The new case of greatest interest is that of K3 surfaces, and in particular that of nonsingular quartic surfaces. The same heuristics which led to (7) for Del Pezzo surfaces now lead one to

$$N(H,V) \sim A(\log H)^r \tag{12}$$

where r is as before the rank of $\mathrm{Pic}(V)$. Unfortunately, if V contains at least one soluble curve of genus 0 it contains infinitely many; and on each one of them the rational points will outnumber the estimate given by (12). To delete all these curves and count the rational points on what is left appears neither sensible nor feasible; so we have to assume that V contains no such curves. If V contains a pencil of curves of genus 1 it again seems unlikely that (12) can hold. Van Luijk has tabulated $N(H,V)$ for certain quartic surfaces which have neither of these properties and which have $r = 1$ or 2, and his results fit the conjecture (12) very well.

7 Schinzel's Hypothesis and Salberger's Device

Schinzel's Hypothesis gives a conjectural answer to the following question: given finitely many polynomials $F_1(X),\ldots,F_n(X)$ in $\mathbf{Z}[X]$ with positive leading coefficients, is there an arbitrarily large integer x at which they all take prime values? There are two obvious obstructions to this:

- One or more of the $F_i(X)$ may factorize in $\mathbf{Z}[X]$.
- There may be a prime p such that for any value of x mod p at least one of the $F_i(x)$ is divisible by p.

If the congruence $F_i(x) \equiv 0 \bmod p$ is non-trivial, it has at most $\deg(F_i)$ solutions; so the second obstruction can only happen for $p \le \sum \deg(F_i)$ or if p divides every coefficient of some F_i. Schinzel's Hypothesis is that these are the only obstructions: in other words, if neither of them happens then we can choose an arbitrarily large x so that every $F_i(x)$ is a prime.

If one assumes Schinzel's Hypothesis the corresponding result over any algebraic number field follows easily. But in most applications there is a predetermined set $\mathfrak{B}$ of bad places, and we need to impose local conditions on x at some or all of them. These conditions constrain the values of the $F_i(x)$ at those places, and therefore we cannot necessarily require these values to be units at the bad primes; nor in the applications do we need to. I have stated Lemma 7.1 in a form which applies to homogeneous polynomials G_i in two variables; but the reader who wishes to do so will have no difficulty in stating and proving the corresponding (stronger) result for polynomials in one variable. Just as with the original version of Schinzel's Hypothesis, provided that the coefficients of G_i for each i have no common factor we need only verify the existence of the $y_\mathfrak{p}, z_\mathfrak{p}$ in the statement of the lemma when the absolute norm of $\mathfrak{p}$ does not exceed $\sum \deg(G_i)$.

Lemma 7.1. *Let k be an algebraic number field and $\mathfrak{o}$ the ring of integers of k. Let $G_1(Y,Z), \ldots, G_n(Y,Z)$ be homogeneous irreducible elements of $\mathfrak{o}[Y,Z]$ and $\mathfrak{B}$ a finite set of primes of k. Suppose that for each $\mathfrak{p}$ not in $\mathfrak{B}$ there exist $y_\mathfrak{p}, z_\mathfrak{p}$ in $\mathfrak{o}$ such that none of the $G_i(y_\mathfrak{p}, z_\mathfrak{p})$ is in $\mathfrak{p}$. For each $\mathfrak{p}$ in $\mathfrak{B}$, let $V_\mathfrak{p}$ be a non-empty open subset of $k_\mathfrak{p} \times k_\mathfrak{p}$; and for each infinite place v of k let V_v be a non-empty open subset of k_v^*. Assume Schinzel's Hypothesis; then there is a point $\eta \times \zeta$ in $k^* \times k^*$, with η, ζ integral outside $\mathfrak{B}$, such that*

- *$\eta \times \zeta$ lies in $V_\mathfrak{p}$ for each $\mathfrak{p}$ in $\mathfrak{B}$;*
- *η / ζ lies in V_v for each infinite place v;*
- *Each ideal $(G_i(\eta, \zeta))$ is the product of a prime ideal not in $\mathfrak{B}$ and possibly powers of primes in $\mathfrak{B}$.*

Proof. Choose α, β in $\mathfrak{o}$ so that α/β lies in V_v for each infinite place v and no $G_i(\alpha, \beta)$ vanishes. We can repeatedly adjoin a further prime $\mathfrak{p}$ to $\mathfrak{B}$ provided we define the corresponding $V_\mathfrak{p}$ to be the set of all $y \times z$ in $\mathfrak{o}_\mathfrak{p} \times \mathfrak{o}_\mathfrak{p}$ such that each $G_i(y,z)$ is a unit at $\mathfrak{p}$. We can therefore assume that $\mathfrak{B}$ contains all ramified primes $\mathfrak{p}$ and all primes $\mathfrak{p}$ such that

- The absolute norm of $\mathfrak{p}$ is not greater than $[k : \mathbf{Q}] \sum \deg(G_i)$; or
- $\mathfrak{p}$ divides any $G_i(\alpha, \beta)$.

Let $\mathscr{B}$ be the set of primes in $\mathbf{Q}$ which lie below some prime of $\mathfrak{B}$, and further adjoin to $\mathfrak{B}$ all the primes of k not already in $\mathfrak{B}$ which lie above some prime of $\mathscr{B}$. By the Chinese Remainder Theorem we can choose η_0, ζ_0 in k, integral outside $\mathfrak{B}$ and such that each $G_i(\eta_0, \zeta_0)$ is nonzero and $\eta_0 \times \zeta_0$ lies in $V_\mathfrak{p}$ for each $\mathfrak{p}$ in $\mathfrak{B}$. For reasons which will become clear after (13), we also need to ensure that $\beta \eta_0 \neq \alpha \zeta_0$; this can be done by varying η_0 or ζ_0 by a suitable element of $\mathfrak{o}$ divisible by large powers of each $\mathfrak{p}$ in $\mathfrak{B}$. As an ideal, write

$$(G_i(\eta_0, \zeta_0)) = \mathfrak{a}_i \mathfrak{b}_i$$

where the prime factors of each $\mathfrak{a}_i$ are outside $\mathfrak{B}$ and those of each $\mathfrak{b}_i$ are in $\mathfrak{B}$; thus $\mathfrak{a}_i$ is integral. Let N_i be the absolute norm of $\mathfrak{b}_i$. Now choose $\gamma \neq 0$ in $\mathfrak{o}$ to be a unit at all the primes outside $\mathfrak{B}$ which divide any $G_i(\eta_0, \zeta_0)$ and to be divisible by such large powers of each $\mathfrak{p}$ in $\mathfrak{B}$ that

$$\eta \times \zeta = (\alpha \gamma \xi + \eta_0) \times (\beta \gamma \xi + \zeta_0)$$

is in $V_\mathfrak{p}$ for all $\xi \in \mathfrak{o}$ and all $\mathfrak{p} \in \mathfrak{B}$, and that if we write

$$g_i(X) = G_i(\alpha \gamma X + \eta_0, \beta \gamma X + \zeta_0), \tag{13}$$

then every coefficient of $g_i(X)$ is divisible by at least as great a power of $\mathfrak{p}$ as is $\mathfrak{b}_i$. We have arranged that the two arguments of G_i in (13), considered as linear forms in X, are not proportional; thus if $g_i(X)$ factorizes in $k[X]$ then $G_i(\alpha\gamma U + \eta_0 V, \beta\gamma U + \zeta_0 V)$ would factorize in $k[U,V]$, contrary to the irreducibility of $G_i(Y,Z)$. We shall also require for each i that $g_i(X)$ is prime to all its conjugates as elements of $\bar{k}[X]$; since the zeros of $g_i(X)$ have the form $\gamma^{-1}\xi_{ij}$ for some ξ_{ij} independent of γ, this merely requires the ratios of γ to its conjugates to avoid finitely many values. Write

$$R_i(X) = \mathrm{Norm}_{k(X)/\mathbf{Q}(X)}(g_i(X))/N_i;$$

then $R_i(X)$ has all its coefficients integral, for at each prime it is the norm of a polynomial with locally integral coefficients. An irreducible factor of $R_i(X)$ in $\mathbf{Q}[X]$ cannot be prime to $g_i(X)$, because then it would also be prime to all the conjugates of $g_i(X)$ and therefore to their product – which is absurd. If $R_i(X)$ had two coprime factors in $\mathbf{Q}[X]$, their highest common factors with $g_i(X)$ would be non-trivial coprime factors of $g_i(X)$ in $k[X]$, whence $g_i(X)$ would not be irreducible in $k[X]$. Finally, $R_i(X)$ cannot have a repeated factor because the conjugates of $g_i(X)$ are pairwise coprime. So $R_i(X) = A_i H_i(X)$ in $\mathbf{Z}[X]$, with $H_i(X)$ irreducible. Clearly we can require the leading coefficient of each $H_i(X)$ to be positive. But the only primes which divide the constant term in $R_i(X)$ are the primes outside $\mathscr{B}$ which divide $G_i(\eta_0,\zeta_0)$, and none of them divide the leading coefficient of $R_i(X)$; hence $A_i = \pm 1$. Now apply Schinzel's Hypothesis to the $H_i(X)$, which we can do because no $H_i(0)$ is divisible by any prime in $\mathscr{B}$. But if $H_i(x)$ is equal to a prime not in $\mathscr{B}$ then the ideal $(g_i(x))$ must be equal to the product of $\mathfrak{b}_i$ and a prime ideal not in $\mathfrak{B}$. $\qquad\square$

If we are content to obtain results about 0-cycles of degree 1 instead of results about points, it would be enough to prove solubility in some field extension of each large enough degree. Arguments of this type were pioneered by Salberger. Unfortunately neither of the recipes below enables us to control either the units or the ideal class group of the field involved, so at present the usefulness of this idea is rather limited.

Lemma 7.2. *Let k be an algebraic number field and $P_1(X),\ldots,P_n(X)$ monic irreducible non-constant polynomials in $k[X]$; and let $N \geq \sum \deg(P_i)$ be a given integer. Let $\mathfrak{B}$ be a finite set of places of k which contains the infinite places, the primes at which some coefficient of some P_i is not integral and any other primes $\mathfrak{p}$ at which $\prod P_i(X)$ does not remain separable when reduced mod $\mathfrak{p}$. Let $b > 1$ be in $\mathbf{Z}$ and such that no prime of k which divides b is in $\mathfrak{B}$. For each v in $\mathfrak{B}$ let U_v be a non-empty open set of separable monic polynomials of degree N in $k_v[X]$. Let $M > 0$ be a fixed rational integer. Then we can find an irreducible monic polynomial $G(X)$ in $k[X]$ of degree N which lies in each U_v and for which λ, the image of X in $K = k[X]/G(X)$, satisfies*

$$(P_i(\lambda)) = \mathfrak{P}_i \mathfrak{A}_i \mathfrak{C}_i^M \tag{14}$$

for each i, where the $\mathfrak{P}_i$ are distinct first degree primes in K not lying above any prime in $\mathfrak{B}$, the $\mathfrak{A}_i$ are products of bad primes in K and the $\mathfrak{C}_i$ are integral ideals

*in K. (Here we call a prime in K bad if it divides b or any prime in $\mathfrak{B}$.) Moreover
we can arrange that $\lambda = \alpha/\beta$ where α is integral and β is an integer all of whose
prime factors are bad.*

Lemma 7.3. *Let k be an algebraic number field and $P_1(X),\ldots,P_n(X)$ monic irreducible non-constant polynomials in $k[X]$; and let $N \geq \sum \deg(P_i)$ be a given integer.
Let $\mathfrak{B}$ be a finite set of places of k which contains the infinite places, the primes at
which some coefficient of some P_i is not integral and any other primes $\mathfrak{p}$ at which
$\prod P_i(X)$ does not remain separable when reduced mod $\mathfrak{p}$.*

*Let L be a finite extension of k in which all the polynomials P_i split completely,
and which is Galois over $\mathbf{Q}$. Let V be an infinite set of finite primes of k lying over
primes in $\mathbf{Q}$ which are totally split in L. Suppose that we are given for each $v \in \mathfrak{B}$ a
non-empty open set U_v of separable monic polynomials in $k_v[X]$ of degree N. Then
we can find an irreducible monic polynomial $G(X)$ in $k[X]$ of degree N such that if
θ is the image of X in $k[X]/G(X)$ then*

(i) θ is an integer except perhaps at primes in $k(\theta)$ above those in $\mathfrak{B} \cup V$;
(ii) $G(X)$ is in U_v for each v in $\mathfrak{B}$;
*(iii) For each i there is a finite prime w_i in $k(\theta)$, of absolute degree one, such that
$P_i(\theta)$ is a uniformizing parameter for w_i and a unit at all primes except w_i and
possibly some of those above some prime in $\mathfrak{B} \cup V$.*

The existence of V follows from Tchebotarov's density theorem. The proof of
Lemma 7.3 can be found in [52]. The proof of Lemma 7.2 is currently unpub-
lished. The idea underlying the proofs of both these Lemmas is as follows. Write
$R(X) = \prod P_i(X)$ and $R_i(X) = R(X)/P_i(X)$. Any polynomial $G(X)$ in $k[X]$ can be
written in just one way in the form

$$G(X) = R(X)Q(X) + \sum R_i(X)\psi_i(X) \tag{15}$$

with $\deg \psi_i < \deg P_i$; for if λ_i is a zero of $P_i(X)$ this is just the classical partial
fractions formula

$$\frac{G(X)}{\prod P_i(X)} = Q(X) + \sum \frac{\psi_i(X)}{P_i(X)}$$

with $\psi_i(\lambda_i) = G(\lambda_i)/R_i(\lambda_i)$. This property determines for each i a unique $\psi_i(X)$ in
$k[X]$ of degree less than $\deg P_i$. The same result holds over any k_v. If the coefficients
of G are integral at v, for some v not in $\mathfrak{B}$, then so are those of Q and each ψ_i because
R and the R_i are monic and $R_i(\lambda_i)$ is a unit outside $\mathfrak{B}$. For each v in $\mathfrak{B}$ let $G_v(X)$ be
a polynomial of degree N lying in U_v, and write

$$G_v(X) = R(X)Q_v(X) + \sum R_i(X)\psi_{iv}(X)$$

with $\deg \psi_{iv} < \deg P_i$. We adjoin to $\mathfrak{B}$ a further finite place w at which b is a unit,
and associate with it a monic irreducible polynomial $G_w(X)$ in $k_w[X]$ of degree N;
the only purpose of G_w is to ensure that the $G(X)$ which we construct is irreducible
over k. We then build $G(X)$, close to $G_v(X)$ for every $v \in \mathfrak{B}$ including w.

Let $\mathfrak{p}_i$ be the prime in k below $\mathfrak{P}_i$. By computing the resultant of $P_i(X)$ and $G(X)$ in two different ways, we obtain

$$\text{Norm}_{K/k}P_i(\lambda) = \pm\text{Norm}_{k_i/k}G(\lambda_i) = \pm\text{Norm}_{k_i/k}(\phi_i R_i(\lambda_i)) \tag{16}$$

where λ_i is a zero of $P_i(X)$. By hypothesis $R_i(\lambda_i)$ is a unit at every place of $k(\lambda_i)$ which does not lie above a place in $\mathfrak{B}$; and we can arrange that the denominator of $\text{Norm}_{k_i/k}\phi_i$ is only divisible by bad primes, and its numerator is the product of the first degree prime $\mathfrak{p}_i$, powers of primes in $\mathfrak{B}$ and other factors which we can largely control by the way in which we build $G(X)$. That depends on which Lemma we are trying to prove, and it is the presence of these factors that lead to the complications in the statements of the two Lemmas.

8 The Legendre-Jacobi Function

If α, β are elements of k^* and v is a place of k, the multiplicative *Hilbert symbol* $(\alpha, \beta)_v$ is defined by

$$(\alpha, \beta)_v = \begin{cases} 1 & \text{if } \alpha X^2 + \beta Y^2 = Z^2 \text{ is soluble in } k_v, \\ -1 & \text{otherwise.} \end{cases}$$

The additive Hilbert symbol is defined in the same way except that it takes the values 0 and 1 in $\mathbf{F}_2$ instead of 1 and -1. The Hilbert symbol is effectively a replacement for the quadratic residue symbol, with the advantage that it treats the even primes and the infinite places in just the same way as any other prime. It is symmetric in α, β and its principal properties are

- $(\alpha_1\alpha_2, \beta)_v = (\alpha_1, \beta)_v(\alpha_2, \beta)_v$ and $(\alpha, \beta_1\beta_2)_v = (\alpha, \beta_1)_v(\alpha, \beta_2)_v$;
- For fixed α, β, $(\alpha, \beta)_v = 1$ for almost all v, and $\prod(\alpha, \beta)_v = 1$ where the product is taken over all places v of k.

The second of these is one of the main results of class field theory.

The Legendre-Jacobi function L is crucial to much of what follows. Its theory is described in some detail here, because there is no adequate source for it in print. Let $F(U,V), G(U,V)$ be homogeneous coprime square-free polynomials in $k[U,V]$. Let $\mathscr{B}$ be a finite set of places of k containing the infinite places, the primes dividing 2, those at which any coefficient of F or G is not integral, and any primes $\mathfrak{p}$ at which FG does not remain separable when reduced mod $\mathfrak{p}$.

Let $\mathscr{N}^2 = \mathscr{N}^2(k)$ be the set of $\alpha \times \beta$ with α, β integral and coprime outside $\mathscr{B}$, and let $\mathscr{N}^1 = \mathscr{N}^1(k)$ be $k \cup \{\infty\}$. For $\alpha \times \beta$ in $k \times k$ with α, β not both zero, we shall consistently write $\lambda = \alpha/\beta$ with λ in $\mathscr{N}^1(k)$. Provided $F(\alpha, \beta)$ and $G(\alpha, \beta)$ are nonzero, we define the function

$$L(\mathscr{B}; F, G; \alpha, \beta) : \alpha \times \beta \mapsto \prod_{\mathfrak{p}} (F(\alpha, \beta), G(\alpha, \beta))_{\mathfrak{p}} \tag{17}$$

on $\mathcal{N}^2$, where the outer bracket on the right is the multiplicative Hilbert symbol and the product is taken over all primes $\mathfrak{p}$ of k outside $\mathscr{B}$ which divide $G(\alpha,\beta)$. By the definition of $\mathscr{B}$, $F(\alpha,\beta)$ is a unit at any such prime. We can restrict the product in (17) to those $\mathfrak{p}$ which divide $G(\alpha,\beta)$ to an odd power; thus we can also write it as $\prod \chi_{\mathfrak{p}}(F(\alpha,\beta))$ where $\chi_{\mathfrak{p}}$ is the quadratic character mod $\mathfrak{p}$ and the product is taken over all $\mathfrak{p}$ outside $\mathscr{B}$ which divide $G(\alpha,\beta)$ to an odd power. This relationship with the quadratic residue symbol underlies the proof of Lemma 8.1. The function L does depend on $\mathscr{B}$, but the effect on the right hand side of (17) of increasing $\mathscr{B}$ is obvious. Some of the more interesting properties of L depend on $\deg F$ being even, but this usually holds in applications. In the course of the proofs, however, we need to consider functions (17) with $\deg F$ odd; and for this reason it is expedient to introduce

$$M(\mathscr{B};F,G;\alpha,\beta) = L(\mathscr{B};F,G;\alpha,\beta)(\prod(\alpha,\beta)_v)^{(\deg F)(\deg G)},$$

where the product is taken over all $\mathfrak{p}$ outside $\mathscr{B}$ which divide β and therefore do not divide α.

Lemma 8.1. *The value of M is continuous in the topology induced on $\mathcal{N}^2$ by $\mathscr{B}$. For each v in $\mathscr{B}$ there is a function $m(v;F,G;\alpha,\beta)$ with values in $\{\pm 1\}$ which is continuous on $\mathcal{N}^2$ in the v-adic topology, such that*

$$M(\mathscr{B};F,G;\alpha,\beta) = \prod_{v\in\mathscr{B}} m(v;F,G;\alpha,\beta). \tag{18}$$

Proof. If $\deg F$ is even, so that $M = L$, the neatest proof of the lemma is by means of the evaluation formula in [11], Lemma 7.2.4. When $\deg G$ is even but $\deg F$ may not be, the result follows from (20), and (19) then gives the general case. (The proof in [11] is for $k = \mathbf{Q}$, but there is not much difficulty in modifying it to cover all k.) However, the proof which we shall give, using the ideas of [53], provides a more convenient method of evaluation.

For this proof we have to impose on $\mathscr{B}$ the additional condition that it contains all primes whose absolute norm does not exceed $\deg(FG)$. As the proof in [11] shows, this condition is not needed for the truth of Lemma 8.1 itself; but we use it in the proof of (25) below, and the latter is crucial to the subsequent argument. In any case, to classify all small enough primes as bad is quite usual. We repeatedly use the fact that $L(\mathscr{B};F,G)$ and $M(\mathscr{B};F,G)$ are multiplicative in both F and G; the effect of this is that we can reduce to the case when both F and G are irreducible in $\mathfrak{o}_{\mathscr{B}}[U,V]$, where $\mathfrak{o}_{\mathscr{B}}$ is the ring of elements of k integral outside $\mathscr{B}$. Introducing M and dropping the parity condition on $\deg F$ are not real generalizations since if we increase $\mathscr{B}$ so that the leading coefficient of F is a unit outside $\mathscr{B}$ then

$$M(\mathscr{B};F,G) = L(\mathscr{B};F,GV^{\deg G}) \tag{19}$$

by (21), and we can apply (20) to the right hand side.

It follows from the product formula for the Hilbert symbol that

$$L(\mathscr{B}; f, g; \alpha, \beta) L(\mathscr{B}; g, f; \alpha, \beta) = \prod_{v \in \mathscr{B}} (f(\alpha, \beta), g(\alpha, \beta))_v, \qquad (20)$$

provided that $f(\alpha, \beta)$, $g(\alpha, \beta)$ are nonzero. The right hand side of (20) is the product of continuous terms each of which only depends on a single v in $\mathscr{B}$. This formula enables us to interchange F and G when we want to, and in particular to require that $\deg F \geq \deg G$ in the reduction process which follows. We also have

$$L(\mathscr{B}; f, g; \alpha, \beta) = L(\mathscr{B}; f - gh, g; \alpha, \beta) \qquad (21)$$

for any homogeneous h in $k[U, V]$ with $\deg h = \deg f - \deg g$ provided the coefficients of h are integral outside $\mathscr{B}$, because corresponding terms in the two products are equal. Both (20) and (21) also hold for M.

We deal first with two special cases:

- G is a constant. Now $M(\mathscr{B}; F, G) = 1$ because all the prime factors of G must be in $\mathscr{B}$, so that $M(\mathscr{B}; F, G) = L(\mathscr{B}; F, G)$ and the product in the definition of $L(\mathscr{B}; F, G)$ is empty.
- $G = V$. Choose H so that $F - GH = \gamma U^{\deg F}$ for some nonzero γ. Now $M(\mathscr{B}; F, G) = 1$ follows from the previous case and (21), since all the prime factors of γ must be in $\mathscr{B}$.

We now argue by induction on $\deg(FG)$. Since we can assume that F and G are irreducible, we need only consider the case when

$$\deg F \geq \deg G > 0, \quad G = \gamma U^{\deg G} + \dots, \quad F = \delta U^{\deg F} + \dots$$

for some nonzero γ, δ. Let $\mathscr{B}_1$ be obtained by adjoining to $\mathscr{B}$ those primes of k not in $\mathscr{B}$ at which γ is not a unit. By (21) we have

$$M(\mathscr{B}_1; F, G) = M(\mathscr{B}_1; F - \gamma^{-1} \delta G U^{\deg F - \deg G}, G). \qquad (22)$$

By taking a factor V out of the middle argument on the right, and using (20), the second special case above and the induction hypothesis, we see that $M(\mathscr{B}_1; F, G)$ is continuous in the topology induced by $\mathscr{B}_1$ and is a product taken over all v in $\mathscr{B}_1$ of continuous terms each one of which depends on only one of the v. Hence the same is true of $M(\mathscr{B}; F, G)$, because this differs from $M(\mathscr{B}_1; F, G)$ by finitely many continuous factors, each of which depends only on one prime in $\mathscr{B}_1 \setminus \mathscr{B}$.

But $\mathscr{B}_1 \setminus \mathscr{B}$ only contains primes whose absolute norm is greater than $\deg(FG)$. Thus by an integral unimodular transformation from U, V to U, V_1 we can arrange that $G = \gamma_1 U^{\deg G} + \dots$ and $F = \delta_1 U^{\deg F} + \dots$ where γ_1 is a unit at each prime in $\mathscr{B}_1 \setminus \mathscr{B}$. Let $\mathscr{B}_2$ be obtained from $\mathscr{B}$ by adjoining all the primes at which γ_1 is not a unit; then $M(\mathscr{B}; F, G)$ has the same properties with respect to $\mathscr{B}_2$ that we have already shown that it has with respect to $\mathscr{B}_1$. Since $\mathscr{B}_1 \cap \mathscr{B}_2 = \mathscr{B}$, this implies that $M(\mathscr{B}; F, G)$ already has these properties with respect to $\mathscr{B}$.   □

Of course there will be finitely many values of α/β for which at some stage of the argument the right hand side of (20) appears to be indeterminate; but by means of a preliminary linear transformation on U, V one can avoid this and ensure that the formula (18) is meaningful except when $F(\alpha, \beta)$ or $G(\alpha, \beta)$ vanishes.

When deg F is even, the value of $L(\mathscr{B}; F, G; \alpha, \beta)$ is already determined by $\lambda = \alpha/\beta$ regardless of the values of α and β separately; here λ lies in $k \cup \{\infty\}$ with the roots of $F(\lambda, 1)$ and $G(\lambda, 1)$ deleted. We shall therefore also write this function as $L(\mathscr{B}; F, G; \lambda)$. But note that it is not necessarily a continuous function of λ; see the discussions in [12] and Sect. 9 of [11], or Lemma 8.4 below. Moreover if $\mathscr{B}$ does not contain a base for the ideal class group of k then not all elements of $k \cup \{\infty\}$ can be written in the form α/β with α, β integers coprime outside $\mathscr{B}$; so we have not yet defined $L(\mathscr{B}; F, G; \lambda)$ for all λ. To go further in the case when deg F is even, we modify the definition (17) so that it extends to all $\alpha \times \beta$ in $k \times k$ such that $F(\alpha, \beta)$ and $G(\alpha, \beta)$ are nonzero. For any such α, β and any $\mathfrak{p}$ not in $\mathscr{B}$, choose $\alpha_{\mathfrak{p}}, \beta_{\mathfrak{p}}$ integral at $\mathfrak{p}$, not both divisible by $\mathfrak{p}$ and such that $\alpha/\beta = \alpha_{\mathfrak{p}}/\beta_{\mathfrak{p}}$. Write

$$L(\mathscr{B}; F, G; \alpha, \beta) = \prod (F(\alpha_{\mathfrak{p}}, \beta_{\mathfrak{p}}), G(\alpha_{\mathfrak{p}}, \beta_{\mathfrak{p}}))_{\mathfrak{p}} \qquad (23)$$

where the product is taken over all $\mathfrak{p}$ not in $\mathscr{B}$ such that $\mathfrak{p} | G(\alpha_{\mathfrak{p}}, \beta_{\mathfrak{p}})$. This is a finite product whose value does not depend on the choice of the $\alpha_{\mathfrak{p}}$ and $\beta_{\mathfrak{p}}$; indeed it only depends on $\lambda = \alpha/\beta$ and when α, β are integers coprime outside $\mathscr{B}$ it is the same as the function given by (17). Thus we can again write it as $L(\mathscr{B}; F, G; \lambda)$. This generalization is not really needed until we come to (27); but at that stage we cannot take account of the ideal class group of K because we need $\mathscr{B}$ to be independent of K. Its disadvantage is that L is no longer necessarily a continuous function of $\alpha \times \beta$.

If deg F or deg G is 0 or 1, it is easy to obtain an evaluation formula; so the first case of interest is when deg $F = \deg G = 2$. Suppose that

$$F = a_1 U^2 + b_1 UV + c_1 V^2, \quad G = a_2 U^2 + b_2 UV + c_2 V^2 \qquad (24)$$

and that $\mathscr{B}$ contains the infinite places and the primes which divide 2 or

$$R = (a_1 c_2 - a_2 c_1)^2 - b_1 b_2 (a_1 c_2 + a_2 c_1) + a_1 c_1 b_2^2 + a_2 c_2 b_1^2,$$

the resultant of F and G. Suppose also that $\eta \times \zeta$ and $\rho \times \sigma$ are in $\mathscr{N}^2$. Then

$$\begin{aligned} L(\mathscr{B}; F, G; \eta, \zeta) L(\mathscr{B}; F, G; \rho, \sigma) \\ = \prod_{v \in \mathscr{B}} \{ (f/(\sigma \eta - \rho \zeta), R)_v (fG(\rho, \sigma), -fG(\eta, \zeta))_v \} \end{aligned}$$

where

$$f = F(\eta, \zeta) G(\rho, \sigma) - F(\rho, \sigma) G(\eta, \zeta).$$

If we set ρ, σ to convenient values, this gives the value of $L(\mathscr{B}; F, G; \eta, \zeta)$.

The proof of Lemma 8.1 constructs an evaluation formula all of whose terms come from the right hand side of (20) for various pairs f, g. For $\alpha \times \beta$ in $\mathcal{N}^2$, the formula can therefore be described by an equation of the form

$$m(v; F, G; \alpha, \beta) = \prod_j (\phi_j(\alpha, \beta), \psi_j(\alpha, \beta))_v. \tag{25}$$

Here the ϕ_j, ψ_j are homogeneous elements of $k[U, V]$ which depend only on F and G and not on v or $\mathcal{B}$. The decomposition (25) is not unique, but we can display an invariant aspect of it.

Let $\theta = \gamma_1 U + \gamma_2 V$ be a linear form with γ_1, γ_2 coprime integers in k. By using $(\phi, \psi)_v = (\phi, \theta\psi)_v (\phi, \theta)_v$ and $(-\theta, \theta)_v = 1$, we can ensure that all the ϕ_j, ψ_j in (25) have even degree except perhaps that $\psi_0 = \theta$. Denote by Θ the group of elements of k^* which are not divisible to an odd power by any prime of k outside $\mathcal{B}$, and by $\Theta_0 \subset \Theta$ the subgroup consisting of those ξ which are quadratic residues mod $\mathfrak{p}$ for all $\mathfrak{p}$ in $\mathcal{B}$; thus we are free to multiply ϕ_0 by any element of Θ_0.

Lemma 8.2. *Provided* $\deg F$ *is even, if* $\psi_0 = \theta$ *we can take* ϕ_0 *to be in* Θ.

The evaluation formula for (24) shows that $(\phi_0, \psi_0)_v$ may not be trivial even when F and G both have even degree.

Proof. Let γ in k^* be a unit outside $\mathcal{B}$, and apply (25) to the identity

$$L(\mathcal{B}; F, G; \gamma\alpha, \gamma\beta) = L(\mathcal{B}; F, G; \alpha, \beta),$$

where $\alpha \times \beta$ is in $\mathcal{N}^2$. On cancelling common factors, we obtain

$$\prod_{v \in \mathcal{B}} (\phi_0(\alpha, \beta), \gamma)_v = 1. \tag{26}$$

If we can choose $\alpha \times \beta$ in $\mathcal{N}^2$ so that $\phi_0(\alpha, \beta)$ is not in Θ, this gives a contradiction. For let δ prime to $\phi_0(\alpha, \beta)$ be such that $\prod(\phi_0(\alpha, \beta), \delta)_{\mathfrak{p}} = -1$ where the product is taken over all primes $\mathfrak{p}$ outside $\mathcal{B}$ at which $\phi_0(\alpha, \beta)$ is not a unit. Let $\mathcal{B}_1$ be obtained by adjoining to $\mathcal{B}$ all the primes at which δ is not a unit; then $\prod(\phi_0(\alpha, \beta), \delta)_v = -1$ by the Hilbert product formula, where the product is taken over all places v in $\mathcal{B}_1$. Recalling that ϕ_0 does not depend on $\mathcal{B}$ and writing $\mathcal{B}_1, \delta$ for $\mathcal{B}, \gamma$ in (26), we obtain a contradiction. It follows that $\phi_0(\alpha, \beta)$ lies in Θ for all α, β; this can only happen if $\phi_0(U, V)$ is itself in Θ modulo squares of elements of $k[U, V]$. □

In practice, what we usually need to study is the subspace of $\mathcal{N}^2$ given by n conditions $L(\mathcal{B}; F_v, G_v; \alpha, \beta) = 1$, or the subspace of $\mathcal{N}^1$ given by the $L(\mathcal{B}; F_v, G_v; \lambda) = 1$, where the $\deg F_v$ are all even. Let Λ be the abelian group of order 2^n whose elements are the n-tuples each component of which is ± 1; then there is a natural identification, which we shall write τ, of each element of Λ with a partial product of the $L(\mathcal{B}; F_v, G_v)$. Thus each element of Λ can be interpreted as a condition, which we shall write as $\mathcal{L} = 1$. If ϕ_0 is as in Lemma 8.2, there is a homomorphism

$$\phi_0 \circ \tau : \Lambda \to \Theta/\Theta_0;$$

let Λ_0 denote its kernel. It turns out that the conditions which are continuous in λ are just those which come from Λ_0. The following lemma corresponds to Harari's Formal Lemma (Theorem 3.2.1 of [11]); it shows that for most purposes we need only consider the conditions coming from the elements of Λ_0. For obvious reasons, we call these the *continuous* conditions.

Lemma 8.3. *Suppose that every* $\deg F_v$ *is even and all the conditions corresponding to* Λ_0 *hold at some given* λ_0. *Then there exists* λ *arbitrarily close to* λ_0 *such that all the conditions* $L(\mathscr{B}; F_v, G_v) = 1$ *hold at* λ.

Proof. Let $\lambda_0 = \alpha_0/\beta_0$. For a suitably chosen γ we show that we can take $\lambda = \alpha/\beta$, where $\alpha \times \beta$ is close to $\gamma\alpha_0 \times \gamma\beta_0$ at every finite prime in $\mathscr{B}$ and α/β is close to α_0/β_0 at the infinite places. For any c in Λ, write $\phi_{0c} = \phi_0 \circ \tau(c)$ for the corresponding element of Θ/Θ_0. If θ is as defined just before Lemma 8.2, the corresponding partial product $\mathscr{L}$ of the $L(\mathscr{B}; F_v, G_v; \lambda)$ is equal to

$$f_c(\lambda) \prod_{v \in \mathscr{B}} (\phi_{0c}, \theta(\alpha_0, \beta_0))_v \prod_{v \in \mathscr{B}} (\phi_{0c}, \gamma)_v$$

where f_c comes from the ϕ_j, ψ_j with $j > 0$ and is therefore continuous. The map $c \mapsto f_c(\lambda)$ is a homomorphism $\Lambda \to \{\pm 1\}$ for any fixed λ; moreover if two distinct c give rise to the same ϕ_{0c} their quotient comes from an element of Λ_0; so the quotient of the corresponding f_c takes the value 1 at λ_0. In other words, if λ is close enough to λ_0 then $f_c(\lambda)$ only depends on the class of c in Λ/Λ_0. The map $c \mapsto \phi_{0c}$ induces an embedding $\Lambda/\Lambda_0 \to \Theta/\Theta_0$. The homomorphism $\mathrm{Image}(\Lambda/\Lambda_0) \to \{\pm 1\}$ induced by $c \mapsto f_c(\lambda)$ can be extended to a homomorphism $\Theta/\Theta_0 \to \{\pm 1\}$ because Θ/Θ_0 is killed by 2; and any such homomorphism can be written in the form

$$\theta \to \prod_{v \in \mathscr{B}} (\theta, \gamma)_v$$

for a suitably chosen γ, because the Hilbert symbol induces a nonsingular form on Θ/Θ_0. But given any such γ we can construct $\lambda = \alpha/\beta$ having the properties listed above. $\square$

In circumstances in which we wish to use Salberger's device, we need analogues of these last statements for positive 0-cycles. To state these, we introduce more notation. We continue to assume that $\deg F$ is even. Let K be the direct product of finitely many fields k_i each of finite degree over k, and let $\mathfrak{B}$ be the set of places of K lying over some place v in $\mathscr{B}$, and $\mathfrak{B}_i$ the corresponding set of places of k_i. (The place $\prod v_i$, where v_i is a place of k_i, lies over v if each v_i does so.) For λ in $\mathbf{P}^1(K)$ write $\lambda = \prod \lambda_i$ with λ_i in $\mathbf{P}^1(k_i)$; for each place w in k_i write $\lambda_i = \alpha_{iw}/\beta_{iw}$ where α_{iw}, β_{iw} are in k_i and integral at w and at least one of them is a unit at w. For any λ in K such that each $F(\lambda_i, 1)$ and $G(\lambda_i, 1)$ is nonzero, we define the function

$$L^*(\mathscr{B}; K; F, G; \lambda) : \lambda \mapsto \prod_{\mathfrak{P}_i} (F(\alpha_{iw}, \beta_{iw}), G(\alpha_{iw}, \beta_{iw}))_{\mathfrak{P}_i} \qquad (27)$$

where w is the place associated with the prime $\mathfrak{P}_i$ in k_i and the product is taken over all i and all primes $\mathfrak{P}_i$ of k_i not lying in $\mathfrak{B}_i$ and such that $G(\alpha_{iw}, \beta_{iw})$ is divisible by $\mathfrak{P}_i$. As with (17), we can restrict the product to those $\mathfrak{P}_i$ which divide $G(\alpha_{iw}, \beta_{iw})$ to an odd power. Note that the functions ϕ_j, ψ_j in the evaluation formula (25) are the same for $k_i \supset k$ as they are for k. Now let $\mathfrak{a}$ be a positive 0-cycle on $\mathbf{P}^1$ defined over k and let $\mathfrak{a} = \cup \mathfrak{a}_i$ be its decomposition into irreducible components. Let λ_i be a point of $\mathfrak{a}_i$ and write $k_i = k(\lambda_i)$. If $K = \prod k_i$ and $\lambda = \prod \lambda_i$, write

$$L^*(\mathscr{B}; F, G; \mathfrak{a}) = L^*(\mathscr{B}; K; F, G; \lambda) = \prod_i L(\mathfrak{B}_i; F, G; \lambda_i). \tag{28}$$

This is legitimate, because the right hand side does not depend on the choice of the λ_i. If $K = k$ this L^* is the same as the previous function L. Moreover $L^*(\mathfrak{a} \cup \mathfrak{b}) = L^*(\mathfrak{a})L^*(\mathfrak{b})$. We can define a topology on the set of positive 0-cycles $\mathfrak{a}$ of given degree N by means of the isomorphism between that set and the points on the N-fold symmetric power of $\mathbf{P}^1$. With this topology, it is straightforward to extend to L^* the results already obtained for L.

The product in (27) is finite; so there is a finite set $\mathscr{S}$ of primes of k, disjoint from $\mathscr{B}$ and such that every $\mathfrak{P}_i$ which appears in this product lies above a prime in $\mathscr{S}$. For each i we can write $\lambda_i = \alpha_i / \beta_i$ with α_i, β_i integers in k_i. Let $(\alpha_i, \beta_i) = \mathfrak{a}_i$ and choose an integral ideal $\mathfrak{b}_i$ in k_i which is prime to $\mathfrak{a}_i$, in the same ideal class as $\mathfrak{a}_i$ and such that no prime of k_i which divides $\mathfrak{b}_i$ also divides $G(\alpha_i, \beta_i)$ or any $\phi_j(\alpha_i, \beta_i)$ or $\psi_j(\alpha_i, \beta_i)$ or lies above any prime in $\mathscr{S}$. Let γ_i be such that $(\gamma_i) = \mathfrak{b}_i / \mathfrak{a}_i$ and let $\mathscr{B}_1$ be obtained from $\mathscr{B}$ by adjoining all the primes of k which lie below any prime of k_i which divides $\mathfrak{b}_i$. For most purposes it costs us nothing to replace $\mathscr{B}$ by $\mathscr{B}_1$, and we then have

$$\lambda = \prod \lambda_i = \prod (\alpha_i \gamma_i / \beta_i \gamma_i) \text{ where } \alpha_i \gamma_i \times \beta_i \gamma_i \text{ is in } \mathscr{N}^2(k_i).$$

The following lemma is a trivial consequence of earlier results.

Lemma 8.4. *Suppose that* $\deg F$ *is even, and let* $\mathscr{L} = 1$ *be a continuous condition derived from the L and $\mathscr{L}^* = 1$ the corresponding condition derived from the L^*. For each v in $\mathscr{B}$ there is a function* $\ell^*(v; F, G; \mathfrak{a})$ *with values in* $\{\pm 1\}$ *which is a continuous function of $\mathfrak{a}$ in the v-adic topology and is such that*

$$\mathscr{L}^*(\mathscr{B}; F, G; \mathfrak{a}) = \prod_{v \in \mathscr{B}} \ell^*(v; F, G; \mathfrak{a}). \tag{29}$$

9 Pencils of Conics

Let W be the surface fibred by the pencil of conics

$$a_0(U, V)Y_0^2 + a_1(U, V)Y_1^2 + a_2(U, V)Y_2^2 = 0. \tag{30}$$

We normally expect this pencil to be presented in a form in which a_0, a_1, a_2 are homogeneous of the same degree. But this is not the most convenient form for the arguments which follow. Instead we shall call the pencil *reduced* if a_0, a_1, a_2 are homogeneous elements of $k[U,V]$ square-free and coprime in pairs and such that

$$\deg a_0 \equiv \deg a_1 \equiv \deg a_2 \bmod 2.$$

After a linear transformation on U, V if necessary, we can also assume that $a_0 a_1 a_2$ is not divisible by V. Clearly any pencil of conics can be put into reduced form; for if a_i has a squared factor f^2 we write $f^{-1} Y_i$ for Y_i, and if for example a_0 and a_1 have a common factor g we write $g Y_2$ for Y_2 and divide (30) by g. Suppose that (30) is reduced and everywhere locally soluble. Let $\lambda = \alpha/\beta$ be a point of $\mathbf{P}^1(k)$; whether (30) is soluble at $\alpha \times \beta$ depends only on λ and not on the choice of α, β. Similar statements hold for local solubility at a place v and for solubility in the adeles. Let $\mathscr{B}$ be a finite set of places of k containing the infinite places, the primes dividing 2, those whose absolute norm does not exceed $\deg(a_0 a_1 a_2)$, those at which any coefficient of any a_i is not integral, and any other primes $\mathfrak{p}$ at which $a_0 a_1 a_2$ does not remain separable when reduced mod $\mathfrak{p}$. We also assume that $\mathscr{B}$ contains a base for the ideal class group of k. Denote by $c(U,V)$ an irreducible factor of $a_0 a_1 a_2$ in $k[U,V]$; we can assume that $c(U,V)$ has integer coefficients whose highest common factor is not divisible by any prime outside $\mathscr{B}$. To prove local solubility, we need only check it at the places of $\mathscr{B}$, because it is trivial at any other prime. Local solubility of (30) at the place v is equivalent to $(-a_0 a_1, -a_0 a_2)_v = 1$, which can be written in the more symmetric form

$$(a_0, -a_1)_v (a_1, -a_2)_v (a_2, -a_0)_v = (-1, -1)_v. \tag{31}$$

The singular fibres of the pencil are given by the values of λ at which $a_0 a_1 a_2$ vanishes. If there is a singular fibre defined over k, then (30) is certainly soluble on it; but little if any of the argument which follows makes sense there. We therefore work not on $\mathbf{P}^1$ but on the subset $\mathbf{L}^1$ obtained by deleting the zeros of $a_0 a_1 a_2$, and not on W but on W_0, the inverse image of $\mathbf{L}^1$ in W. Let $\lambda \in k \cup \{\infty\}$ be a point of $\mathbf{L}^1(k)$, and write $\lambda = \alpha/\beta$ where α, β are integers of k coprime outside $\mathscr{B}$; it will not matter which pair α, β we choose.

There is a non-empty set $\mathscr{N} \subset \mathbf{L}^1(k)$, open in the topology induced by $\mathscr{B}$, such that the conic (30) is soluble at every place of $\mathscr{B}$ if and only if λ lies in $\mathscr{N}$. Let $\mathfrak{p}$ be a prime of k not in $\mathscr{B}$ and consider the solubility of (30) in $k_{\mathfrak{p}}$ at the point λ. If none of the $a_i(\alpha, \beta)$ is divisible by $\mathfrak{p}$, then local solubility of (30) is trivial. Otherwise there is just one c such that $c(\alpha, \beta)$ is divisible by $\mathfrak{p}$; to fix ideas, suppose that this c divides a_2. The condition for local solubility at $\mathfrak{p}$ is then

$$(-a_0(\alpha, \beta) a_1(\alpha, \beta), c(\alpha, \beta))_{\mathfrak{p}} = 1 \tag{32}$$

where the outer bracket is the multiplicative Hilbert symbol. Hence necessary conditions for the local solubility of (30) at λ for all $\mathfrak{p}$ outside $\mathscr{B}$ are the conditions like

$$L(\mathscr{B}; -a_0 a_1, c; \lambda) = \prod (-a_0(\alpha,\beta)a_1(\alpha,\beta), c(\alpha,\beta))_{\mathfrak{p}} = 1 \qquad (33)$$

where the product is taken over all $\mathfrak{p}$ outside $\mathscr{B}$ which divide $c(\alpha,\beta)$, and the function L is well defined since $-a_0 a_1$ has even degree. There is one of these conditions for each irreducible c which divides $a_0 a_1 a_2$.

What makes the set of conditions (33) interesting is that they give not merely a necessary but also a sufficient condition for solubility – at least if one assumes Schinzel's Hypothesis. In view of Lemma 8.3, it is enough to require the continuous conditions derived from the conditions (33) to hold. The following theorem provides the exact obstruction both to the Hasse principle and to weak approximation.

Theorem 9.1. *Assume Schinzel's Hypothesis. Let $\mathscr{A} \subset \mathscr{N}$ be the subset of $\mathbf{L}^1(k)$ at which all the continuous conditions derived from (33) hold and (30) is locally soluble at each place in $\mathscr{B}$. Then the λ in $\mathbf{L}^1(k)$ at which (30) is soluble form a dense subset of $\mathscr{A}$ in the topology induced by $\mathscr{B}$.*

Proof. Let $\alpha_0 \times \beta_0$ correspond to a point λ_0 in $\mathscr{A}$, and let $\mathscr{N}_0 \subset \mathscr{A}$ be an open neighbourhood of λ_0. We have to show that we can find λ_2 in $\mathscr{N}_0$ such that (30) is soluble at λ_2; for this it is enough to show that (30) is everywhere locally soluble there. Let c_i run through the factors c. By Lemma 8.3 we can find α_1, β_1 in k^*, integral and coprime outside $\mathscr{B}$ and such that $\lambda_1 = \alpha_1/\beta_1$ is in $\mathscr{N}_0$ and all the conditions (33) hold at $\alpha_1 \times \beta_1$. By Lemma 7.1 we can now find $\alpha_2 \times \beta_2$ close to $\alpha_1 \times \beta_1$ and such that each ideal $(c_i(\alpha_2,\beta_2))$ is the product of a prime ideal $\mathfrak{p}_i$ not in $\mathscr{B}$ and prime ideals in $\mathscr{B}$. We claim that (30) is everywhere locally soluble at $\alpha_2 \times \beta_2$. Since $\mathscr{N}_0 \subset \mathscr{A}$, local solubility at each place of $\mathscr{B}$ is automatic. If $\mathfrak{p}$ is a prime outside $\mathscr{B}$ which does not divide any of the $a_j(\alpha_2,\beta_2)$ then (30) at $\alpha_2 \times \beta_2$ is certainly soluble at $\mathfrak{p}$; so it only remains to consider the $\mathfrak{p}_i$. To fix ideas, suppose that $c_i(U,V)$ is a factor of $a_2(U,V)$. Taking $\alpha = \alpha_2, \beta = \beta_2$ and $c = c_i$, the product in (33) reduces to the single term with $\mathfrak{p} = \mathfrak{p}_i$. In other words, (32) holds in this case, and this proves local solubility at $\mathfrak{p}_i$. □

An apparently weaker result, but one for which it is easier to check the hypotheses, is the following. Here the hypotheses give us the existence of the $\alpha_1 \times \beta_1$ generated in the proof of Theorem 9.1, and the rest of the proof is as there. The advantage of this is that we do not need the arguments which follow (25).

Corollary 9.2. *Assume Schinzel's Hypothesis. Let $\mathscr{A}_1 \subset k \times k$ be the open set in which none of the a_i vanish, the conditions (33) hold and (30) is locally soluble at each place in $\mathscr{B}$. Then the $\alpha \times \beta$ for which (30) is soluble form a dense subset of $\mathscr{A}_1$ in the topology induced by $\mathscr{B}$.*

The corresponding theorem for positive 0-cycles, or equivalently for 0-cycles of degree 1, does not require Schinzel's Hypothesis; instead we use Lemma 7.2 and the notation introduced at (27). We apply Lemma 7.2 to the surface W_0 fibred by the pencil (30), again assuming that $\mathscr{B}$ satisfies the conditions listed after (30) and that $\mathbf{L}^1$ has the same meaning as there.

Lemma 9.3. *With the notation above, let $N \geq \deg(a_0 a_1 a_2)$ be a fixed integer, and for each v in $\mathscr{B}$ let $\mathfrak{b}'_v$ be a positive 0-cycle on W_0 of degree N and defined over k_v.*

Then we can find a positive 0-cycle $\mathfrak{a}$ of degree N on $\mathbf{L}^1$ defined over k and for each v in $\mathscr{B}$ a positive 0-cycle $\mathfrak{b}_v$ on W_0 of degree N and defined over k_v, close to $\mathfrak{b}'_v$ and such that the projection of each $\mathfrak{b}_v$ on $\mathbf{P}^1$ is $\mathfrak{a}$.

The proof of this Lemma is a straightforward application of the Chinese Remainder Theorem. Its purpose is to show that the hypotheses of the following Theorem are less restrictive than might appear.

Theorem 9.4. *With the notation above, let $N \geq \deg(a_0 a_1 a_2)$ be a fixed integer. Let $\mathfrak{a}$ be a positive 0-cycle of degree N on $\mathbf{L}^1$ defined over k, and for each place v of k suppose that $\mathfrak{b}_v$ is a positive 0-cycle on W_0 of degree N and defined over k_v; for v in $\mathscr{B}$ suppose further that the projection of $\mathfrak{b}_v$ on $\mathbf{L}^1$ is $\mathfrak{a}$. If all the continuous conditions derived from the conditions*

$$L^*(\mathscr{B}; -a_0 a_1, c; \mathfrak{a}) = 1 \tag{34}$$

hold, then there is a positive 0-cycle of degree N on W_0 defined over k whose projection is arbitrarily close to $\mathfrak{a}$ in the topology induced by $\mathscr{B}$.

Proof. We must first show that for the purpose of proving this theorem we are allowed to increase $\mathscr{B}$. Suppose that $\mathscr{B}_0$ satisfies the conditions which were imposed on $\mathscr{B}$ after (30), and let $\mathfrak{p}$ be a prime of k not in $\mathscr{B}_0$. Suppose also that the hypotheses of the theorem hold for $\mathscr{B} = \mathscr{B}_0$ and $\mathfrak{a} = \mathfrak{a}_0$. Having chosen $\mathfrak{b}_\mathfrak{p}$ we can find a positive 0-cycle $\mathfrak{a}'$ on $\mathbf{L}^1$ of degree N and defined over k which is close at every v in $\mathscr{B}_0$ to $\mathfrak{a}$ and close at $\mathfrak{p}$ to the projection of $\mathfrak{b}_\mathfrak{p}$. Now

$$L^*(\mathscr{B}_0 \cup \{\mathfrak{p}\}; -a_0 a_1, c; \mathfrak{a}') = L^*(\mathscr{B}_0; -a_0 a_1, c; \mathfrak{a}');$$

for writing both sides as products by means of (27), if there is a factor on the right hand side which is not present on the left, that factor must come from $\mathfrak{p}$ and is therefore equal to 1. But a continuous condition for $\mathscr{B}_0$ holds at $\mathfrak{a}'$ if and only if it holds at $\mathfrak{a}$, which it does by hypothesis. Hence the continuous conditions for $\mathscr{B}_0 \cup \{\mathfrak{p}\}$ hold at $\mathfrak{a}'$. Now suppose that the theorem holds for $\mathscr{B}_0 \cup \{\mathfrak{p}\}$; then there is a positive 0-cycle $\mathfrak{b}$ of degree N on W_0 defined over k whose projection on $\mathbf{L}^1$ is close to $\mathfrak{a}'$ in the topology induced by $\mathscr{B}_0 \cup \{\mathfrak{p}\}$. The same projection is close to $\mathfrak{a}$ in the topology induced by $\mathscr{B}_0$. So the theorem also holds for $\mathscr{B}_0$.

Note that if $\mathfrak{a}$ is actually the projection of a positive 0-cycle of degree N in W_0, then the continuous conditions certainly hold in view of (28); thus imposing the hypothesis that they all hold costs us nothing. To simplify the notation, we assume henceforth that K is an algebraic number field; this will be true for the application in this article because K will be constructed by means of Lemma 7.2. In view of the previous paragraph, we can assume that $\mathscr{B}$ is so large that it satisfies the conditions imposed on $\mathfrak{B}$ in the statement of Lemma 7.2 and contains the additional place w which was adjoined to $\mathfrak{B}$ in the first paragraph of the proof of Lemma 7.2; and if b is as in Lemma 7.2 we also adjoin to $\mathscr{B}$ all the primes in k which divide b. By the analogue of Lemma 8.3, we can now choose $\mathfrak{a}''$ close to $\mathfrak{a}$ so that all the conditions like $L^*(\mathscr{B}; -a_0 a_1, c; \mathfrak{a}'') = 1$ hold. As was remarked in the previous paragraph, we can now increase $\mathscr{B}$ so that if $\lambda_0 = \alpha_0/\beta_0$ is a point of $\mathbf{L}^1(K)$ in $\mathfrak{a}''$ then α_0, β_0 are coprime and integral except perhaps at primes of K above a prime in $\mathscr{B}$. Now apply Lemma 7.2 with $M = 2$, where we take the $c(X,1)$, normalized to be monic, to be the $P_i(X)$ and each U_v to be a small neighbourhood of the monic polynomial whose roots determine $\mathfrak{a}''$. Let $G(X)$ be given by Lemma 7.2, and let $\mathfrak{a}'$ be the associated 0-cycle on $\mathbf{L}^1(k)$ and λ a point of $\mathbf{L}^1(K)$ in $\mathfrak{a}'$. For each v in $\mathscr{B}$, the cycle $\mathfrak{a}'$ is close to $\mathfrak{a}''$ in the v-adic topology; so (30) at λ is soluble in K_w for each w above v, by continuity. But $\lambda = \alpha/\beta$ with α, β coprime except at primes of K above a prime of $\mathscr{B}$. So

$$\prod_{\mathfrak{P}}(-a_0(\alpha,\beta)a_1(\alpha,\beta), c(\alpha,\beta))_{\mathfrak{P}} = L^*(\mathscr{B}; -a_0 a_1, c; \alpha, \beta) = 1,$$

where the product is taken over all primes $\mathfrak{P}$ not above a prime in $\mathscr{B}$ and such that $c(\alpha,\beta)$ is divisible to an odd power by $\mathfrak{P}$. Here the first equality holds by definition and the second one follows from the evaluation formula (25) by continuity. But if $c(X,1) = P_i(X)$ then the product on the left reduces to the single term for which $\mathfrak{P}$ is the prime of K above $\mathfrak{p}_i$ whose existence was proved by means of (16). Hence (30) at λ is locally soluble at this prime; and because these are the only primes not lying above a prime of $\mathscr{B}$ which divide any $c(\alpha,\beta)$ or any $a_i(\alpha,\beta)$ to an odd power, they are the only primes not lying above a prime of $\mathscr{B}$ at which local solubility might present any difficulty. Thus λ can be lifted to a point of the fibre above λ, which is a conic, and the theorem now follows because weak approximation holds on conics.

$\square$

Since (30) contains positive 0-cycles of degree 2 defined over k, it is trivial to deduce from Theorem 9.4 the corresponding result for 0-cycles of degree 1; conversely, if we know the analogue of Theorem 9.4 for 0-cycles of degree 1 we can deduce that (30) contains positive 0-cycles of some odd degree defined over k. It is tempting to hope that if a pencil of conics contains 0-cycles of degree 1 then it contains points; indeed, the corresponding result is true for Del Pezzo surfaces of degree 4, as is proved in Theorem 14.3. But this hope is false. A simple counterexample is given by the pencil

$$Y_0^2 + Y_1^2 - 7(U^2 - UV - V^2)(U^2 + UV - V^2)(U^2 - 2V^2)Y_2^2 = 0. \qquad (35)$$

This is insoluble in $\mathbf{Q}$. For we can take $\mathscr{B} = \{\infty, 2, 3, 5, 7\}$, and the three possible $c(U,V)$ are $U^2 - UV - V^2$, $U^2 + UV - V^2$ and $U^2 - 2V^2$. By (20) we have

$$L(\mathscr{B}; -1, c) = (-1, c)_\infty (-1, c)_2 (-1, c)_7,$$

the factors at 3 and 5 being trivial. Local solubility of (35) holds at each place; at $\alpha \times \beta$ local solubility at 2 and at 7 requires respectively that $4 | \alpha$ and $\alpha^2 - 2\beta^2$ is divisible by an odd power of 7. Hence

$$(-1, \alpha^2 \pm \alpha\beta - \beta^2)_2 = -1, \quad (-1, \alpha^2 - 2\beta^2)_2 = -1$$

and

$$(-1, \alpha^2 \pm \alpha\beta - \beta^2)_7 = 1, \quad (-1, \alpha^2 - 2\beta^2)_7 = -1.$$

To satisfy the conditions (33) we therefore need

$$(-1, \alpha^2 \pm \alpha\beta - \beta^2)_\infty = -1, \quad (-1, \alpha^2 - 2\beta^2)_\infty = 1;$$

but this is equivalent to $\alpha^2 \pm \alpha\beta - \beta^2 < 0 < \alpha^2 - 2\beta^2$, which is impossible. Now let $K = \mathbf{Q}(\rho)$ where $\rho = 2\cos(2\pi/7)$, so that $\rho^3 + \rho^2 - 2\rho - 1 = 0$. If $U = \rho^2 + 2\rho - 3$ and $V = \rho^2 + \rho - 2$ then

$$Y_0 = (\rho - 2)^2 (\rho^2 - \rho + 1), \ Y_1 = (\rho - 2)^2 (\rho^2 - 1), \ Y_2 = 1$$

gives a solution in K.

On pencils of conics the appropriate Brauer-Manin condition is a necessary and sufficient condition for the Hasse principle and for weak approximation (in each case subject to Schinzel's Hypothesis) and for the existence of positive 0-cycles of degree N for all large enough N. This is the same as saying that the appropriate Brauer-Manin condition is equivalent to the necessary and sufficient conditions stated in Theorems 9.1 and 9.4. That is the content of the following lemma.

Lemma 9.5. *Let W_0 be everywhere locally soluble. Then the continuous conditions derived from (30) are collectively equivalent to the Brauer-Manin conditions for the existence of points of W_0 defined over k. The continuous conditions similarly derived from the $L^*(\mathfrak{a})$ are collectively equivalent to the Brauer-Manin conditions for the existence of positive 0-cycles of degree N on W_0 defined over k.*

Proof. The first assertion is proved for $k = \mathbf{Q}$ in [11], Sect. 8; as with Lemma 8.1, the proof there can easily be extended to our more general case. The second sentence follows trivially from the first in the light of (28). $\square$

10   2-Descent on Elliptic Curves

In this section we describe the process of 2-descent on elliptic curves defined over an algebraic number field k which have the form

$$\Gamma : y^2 = (x - c_1)(x - c_2)(x - c_3)$$

– that is, elliptic curves all of whose 2-division points are rational. We can clearly take the c_i to lie in $\mathfrak{o}$, the ring of integers of k. Let $\mathscr{B}$, the set of bad places, be any finite set of places containing the even primes, the infinite places, all the odd primes dividing $(c_1 - c_2)(c_1 - c_3)(c_2 - c_3)$ and a set of generators for the ideal class group of k; thus $\mathscr{B}$ contains the primes of bad reduction for Γ.

The basic version of 2-descent, which over $\mathbf{Q}$ goes back to Fermat, is as follows. To any point (x, y) on $\Gamma(k)$ there correspond m_1, m_2, m_3 in k^* with $m_1 m_2 m_3 = m^2 \neq 0$

such that the three equations

$$m_i y_i^2 = x - c_i \quad \text{for} \quad i = 1, 2, 3 \tag{36}$$

are simultaneously soluble. We can multiply the m_i by non-zero squares; indeed we should really think of them as elements of k^*/k^{*2}, with a suitable interpretation of the equations which involve them. Denote by $\mathscr{C}(\mathbf{m})$ the curve given by the three equations (36), where $\mathbf{m} = (m_1, m_2, m_3)$. Looking for points of $\Gamma(k)$ is the same as looking for quadruples x, y_1, y_2, y_3 which satisfy (36) for some $\mathbf{m}$. If for example $\mathfrak{p}$ divides m_1 and m_2 to an odd power and therefore m_3 to an even power, then x must be an integer at $\mathfrak{p}$ and therefore $\mathfrak{p}|(c_1 - c_2)$. Hence in looking for soluble $\mathscr{C}(\mathbf{m})$ we need only consider the finitely many $\mathbf{m}$ for which the m_i are units at all primes outside $\mathscr{B}$.

One question of interest is the effect of *twisting* on the arithmetic properties of the curve Γ. If b is in k^*, the *quadratic twist* of Γ by b is defined to be the curve

$$\Gamma_b : y^2 = (x - bc_1)(x - bc_2)(x - bc_3),$$

where we can regard b as an element of k^*/k^{*2}. The curve Γ_b is often written in the alternative form

$$v^2 = b(u - c_1)(u - c_2)(u - c_3).$$

The analogue of (36) for Γ_b is

$$m_i y_i^2 = x - bc_i \quad \text{for} \quad i = 1, 2, 3;$$

we shall call the curve given by these three equations $\mathscr{C}_b(\mathbf{m})$. It is often useful to compare $\mathscr{C}(\mathbf{m})$ and $\mathscr{C}_b(\mathbf{m})$ for the same $\mathbf{m}$.

Provided one treats the m_i as elements of k^*/k^{*2}, the triples $\mathbf{m}$ form an abelian group under componentwise multiplication:

$$\mathbf{m}' \times \mathbf{m}'' \mapsto \mathbf{m}'\mathbf{m}'' = (m_1' m_1'', m_2' m_2'', m_3' m_3'').$$

The $\mathbf{m}$ for which $\mathscr{C}(\mathbf{m})$ is everywhere locally soluble form a finite subgroup, called the 2-*Selmer group*. This is computable, and it contains the group of those $\mathbf{m}$ for which $\mathscr{C}(\mathbf{m})$ is actually soluble in k. This smaller group is $\Gamma(k)/2\Gamma(k)$, where $\Gamma(k)$ is the *Mordell-Weil* group of Γ. The quotient of the 2-Selmer group by this smaller group is $_2\text{III}$, the group of those elements of the *Tate-Safarevic group* which are killed by 2. One of the key conjectures in the subject is that the order of III is finite and hence a square.

The process of going from the curve Γ to the set of curves $\mathscr{C}(\mathbf{m})$, or the finite subset which is the 2-Selmer group, is called a 2-*descent*, or sometimes a *first descent*, and the curves $\mathscr{C}(\mathbf{m})$ themselves are called 2-*coverings*. The reason for this terminology is that there is a commutative diagram

$$\begin{array}{ccc} \Gamma & \longrightarrow & \Gamma \\ \| & \nearrow & \\ \mathscr{C}(\mathbf{m}) & & \end{array} \qquad (37)$$

in which the left hand map is biregular (but defined over $\bar{k}$ rather than k), the top map is multiplication by 2 and the diagonal map is given by $y = m y_1 y_2 y_3$. A 2-covering which is everywhere locally soluble, and therefore in the 2-Selmer group, can also be written in the form

$$\eta^2 = f(\xi) \quad \text{where} \quad f(\xi) = a\xi^4 + b\xi^3 + c\xi^2 + d\xi + e,$$

and many 2-coverings do arise in this way; but a 2-covering which is not in the 2-Selmer group cannot always be put into this form.

We now put this process into more modern language. In what follows, italic capitals will denote vector spaces over $\mathbf{F}_2$, the finite field of two elements, and each of $\mathfrak{p}$ and $\mathfrak{q}$ will be either a finite prime or an infinite place. Write

$$Y_\mathfrak{p} = k_\mathfrak{p}^* / k_\mathfrak{p}^{*2}, \quad Y_\mathscr{B} = \oplus_{\mathfrak{p} \in \mathscr{B}} Y_\mathfrak{p}.$$

Let $V_\mathfrak{p}$ denote the vector space of all triples (μ_1, μ_2, μ_3) with each μ_i in $Y_\mathfrak{p}$ and $\mu_1 \mu_2 \mu_3 = 1$; and write $V_\mathscr{B} = \oplus_{\mathfrak{p} \in \mathscr{B}} V_\mathfrak{p}$. This is the best way to introduce these spaces, because it preserves symmetry; but the reader should note that the prevailing custom in the literature is to define $V_\mathfrak{p}$ as $Y_\mathfrak{p} \times Y_\mathfrak{p}$, which is isomorphic to the $V_\mathfrak{p}$ defined above but not in a canonical way. Next, write $X_\mathscr{B} = \mathfrak{o}_\mathscr{B}^* / \mathfrak{o}_\mathscr{B}^{*2}$ where $\mathfrak{o}_\mathscr{B}^*$ is the group of elements of k^* which are units outside $\mathscr{B}$; and let $U_\mathscr{B}$ be the image in $V_\mathscr{B}$ of the group of triples (m_1, m_2, m_3) such that the m_i are in $X_\mathscr{B}$ and $m_1 m_2 m_3 = 1$. It is known that the map $X_\mathscr{B} \to Y_\mathscr{B}$ is an embedding and $\dim U_\mathscr{B} = \frac{1}{2} \dim V_\mathscr{B}$; both these depend on the requirement that $\mathscr{B}$ contains the even primes and the infinite places, and the first of them depends also on the fact that $\mathscr{B}$ contains a base for the ideal class group. Finally, if (x, y) is a point of $\Gamma(k_\mathfrak{p})$ other than a 2-division point then the product of the three components in the triple $(x - c_1, x - c_2, x - c_3)$ is y^2 which is in $k_\mathfrak{p}^{*2}$; so this triple has a natural image in $V_\mathfrak{p}$. We can supply the images of the 2-division points by continuity; for example the image of $(c_1, 0)$ is

$$((c_1 - c_2)(c_1 - c_3), c_1 - c_2, c_1 - c_3), \qquad (38)$$

and the image of the point at infinity is the trivial triple $(1, 1, 1)$, which is also the product of the three triples like (38). Thus we obtain a map $\Gamma(k_\mathfrak{p}) \to V_\mathfrak{p}$. This map, which is called the *Kummer map*, is a homomorphism. We denote its image by $W_\mathfrak{p}$; clearly $W_\mathfrak{p}$ is the set of those triples $\mathbf{m}$ for which (36) is soluble in $k_\mathfrak{p}$. The 2-Selmer group of Γ can now be identified with $U_\mathscr{B} \cap W_\mathscr{B}$ where $W_\mathscr{B} = \oplus_{\mathfrak{p} \in \mathscr{B}} W_\mathfrak{p}$; for as was noted above, (36) is soluble at every prime outside $\mathscr{B}$ if and only if the elements of $\mathbf{m}$ are in $X_\mathscr{B}$.

Over the years, many people must have noticed that

$$\dim W_{\mathscr{B}} = \dim U_{\mathscr{B}} = \tfrac{1}{2}\dim V_{\mathscr{B}}. \tag{39}$$

The next major step, which explains and may well have been inspired by this relation, was taken by Tate. He introduced the bilinear form e_{p} on $V_{\mathrm{p}} \times V_{\mathrm{p}}$, defined by

$$e_{\mathrm{p}}(\mathbf{m}', \mathbf{m}'') = (m_1', m_1'')_{\mathrm{p}} (m_2', m_2'')_{\mathrm{p}} (m_3', m_3'')_{\mathrm{p}}.$$

Here $(u, v)_{\mathrm{p}}$ is the multiplicative Hilbert symbol already defined in Sect. 8.

The bilinear form e_{p} is non-degenerate and alternating on $V_{\mathrm{p}} \times V_{\mathrm{p}}$, so that $e_{\mathscr{B}} = \prod_{\mathrm{p} \in \mathscr{B}} e_{\mathrm{p}}$ is a non-degenerate alternating bilinear form on $V_{\mathscr{B}} \times V_{\mathscr{B}}$. (For a bilinear form with values in $\{\pm 1\}$, "symmetric" and "skew-symmetric" are the same and they each mean that $e(\mathbf{m}', \mathbf{m}'') = e(\mathbf{m}'', \mathbf{m}')$; "alternating" means that also $e(\mathbf{m}, \mathbf{m}) = 1$.) It is known from class field theory that $U_{\mathscr{B}}$ is a maximal isotropic subspace of $V_{\mathscr{B}}$. Tate showed that W_{p} is a maximal isotropic subspace of V_{p}, and therefore $W_{\mathscr{B}}$ is a maximal isotropic subspace of $V_{\mathscr{B}}$. (The proof of this, which is difficult, can be found in Milne [54].) This explains (39); and it also shows that the 2-Selmer group of Γ can be identified with both the left and the right kernel of the restriction of $e_{\mathscr{B}}$ to $U_{\mathscr{B}} \times W_{\mathscr{B}}$.

For both aesthetic and practical reasons, one would like to show that this restriction is symmetric or skew-symmetric – these two properties being the same. But to make such a statement meaningful we need an isomorphism between $U_{\mathscr{B}}$ and $W_{\mathscr{B}}$; and though they have the same structure as vector spaces it is not obvious that there is a natural isomorphism between them. The way round this obstacle was first shown in [16]. It requires the construction inside each V_{p} of a maximal isotropic subspace K_{p} such that $V_{\mathscr{B}} = U_{\mathscr{B}} \oplus K_{\mathscr{B}}$ where $K_{\mathscr{B}} = \oplus_{\mathrm{p} \in \mathscr{B}} K_{\mathrm{p}}$. Assuming that such spaces K_{p} can be constructed, let $t_{\mathscr{B}} : V_{\mathscr{B}} \to U_{\mathscr{B}}$ be the projection along $K_{\mathscr{B}}$ and write

$$U_{\mathscr{B}}' = U_{\mathscr{B}} \cap (W_{\mathscr{B}} + K_{\mathscr{B}}), \quad W_{\mathscr{B}}' = W_{\mathscr{B}}/(W_{\mathscr{B}} \cap K_{\mathscr{B}}) = \bigoplus_{\mathrm{p} \in \mathscr{B}} W_{\mathrm{p}}'$$

where $W_{\mathrm{p}}' = W_{\mathrm{p}}/(W_{\mathrm{p}} \cap K_{\mathrm{p}})$. The map $t_{\mathscr{B}}$ induces an isomorphism

$$\tau_{\mathscr{B}} : W_{\mathscr{B}}' \to U_{\mathscr{B}}',$$

and the bilinear function $e_{\mathscr{B}}$ induces a bilinear function

$$e_{\mathscr{B}}' : U_{\mathscr{B}}' \times W_{\mathscr{B}}' \to \{\pm 1\}.$$

The bilinear functions $U_{\mathscr{B}}' \times U_{\mathscr{B}}' \to \{\pm 1\}$ and $W_{\mathscr{B}}' \times W_{\mathscr{B}}' \to \{\pm 1\}$ defined respectively by

$$\theta_{\mathscr{B}}^{\flat} : u_1' \times u_2' \mapsto e_{\mathscr{B}}'(u_1', \tau_{\mathscr{B}}^{-1}(u_2')) \quad \text{and} \quad \theta_{\mathscr{B}}^{\sharp} : w_1' \times w_2' \mapsto e_{\mathscr{B}}'(\tau_{\mathscr{B}} w_1', w_2') \tag{40}$$

are symmetric. (For the proof, see [16].) Here the images of $w'_1 \times w'_2$ under the second map and of $\tau_{\mathscr{B}} w'_1 \times \tau_{\mathscr{B}} w'_2$ under the first map are the same. The 2-Selmer group of Γ is isomorphic to both the left and the right kernel of $e'_{\mathscr{B}}$, and hence also to the kernels of the two maps (40).

There is considerable freedom in choosing the $K_{\mathfrak{p}}$, and this raises three obvious questions:

- Is there a canonical choice of the $K_{\mathfrak{p}}$?
- How small can we make U' and W'?
- Can we ensure that the functions (40) are not merely symmetric but alternating?

These questions were first raised and also to a large extent answered in [55]; proofs of the assertions which follow can be found there. The motive for ensuring that the functions (40) are alternating is that it implies that the ranks of these functions are even; this means that their coranks, which are equal to the dimension of the 2-Selmer group, are congruent mod 2 to $\dim U'_{\mathscr{B}}$ and $\dim W'_{\mathscr{B}}$.

The answer to the first question appears to be negative, though there is little freedom in the optimum choice of the $K_{\mathfrak{p}}$ – particularly if one wishes to obtain not merely Lemma 10.1 but Theorem 10.2. Since $U'_{\mathscr{B}} \supset U_{\mathscr{B}} \cap W_{\mathscr{B}}$, the best possible answer to the second question would be that we can achieve $U'_{\mathscr{B}} = U_{\mathscr{B}} \cap W_{\mathscr{B}}$; we do this by satisfying the stronger requirement

$$W_{\mathscr{B}} = (U_{\mathscr{B}} \cap W_{\mathscr{B}}) \oplus (K_{\mathscr{B}} \cap W_{\mathscr{B}}). \tag{41}$$

For suppose that (41) holds; then $W_{\mathscr{B}} + K_{\mathscr{B}} = (U_{\mathscr{B}} \cap W_{\mathscr{B}}) + K_{\mathscr{B}}$ and it follows immediately that

$$U'_{\mathscr{B}} = U_{\mathscr{B}} \cap (W_{\mathscr{B}} + K_{\mathscr{B}}) = U_{\mathscr{B}} \cap W_{\mathscr{B}}. \tag{42}$$

The motivation for (41) is that we want to make $W_{\mathscr{B}} \cap K_{\mathscr{B}}$ as large as possible – that is, to choose $K_{\mathscr{B}}$ so that as much of it as possible is contained in $W_{\mathscr{B}}$. But because $K_{\mathscr{B}}$ must be complementary to $U_{\mathscr{B}}$, only the part of $W_{\mathscr{B}}$ which is complementary to $W_{\mathscr{B}} \cap U_{\mathscr{B}}$ is available for this purpose.

Since the 2-Selmer group $U_{\mathscr{B}} \cap W_{\mathscr{B}}$ is identified with the left and right kernels of each of the functions (40), if (42) holds then these functions are trivial and therefore alternating. The formal statement of all this is as follows.

Lemma 10.1. *We can choose maximal isotropic subspaces $K_{\mathfrak{p}} \subset V_{\mathfrak{p}}$ for each $\mathfrak{p}$ in $\mathscr{B}$ so that $V_{\mathscr{B}} = U_{\mathscr{B}} \oplus K_{\mathscr{B}}$. We can further ensure that*

$$W_{\mathscr{B}} = (U_{\mathscr{B}} \cap W_{\mathscr{B}}) \oplus (K_{\mathscr{B}} \cap W_{\mathscr{B}}),$$

which implies $U'_{\mathscr{B}} = U_{\mathscr{B}} \cap W_{\mathscr{B}}$. If so, the functions $\theta^{\flat}_{\mathscr{B}}$ and $\theta^{\sharp}_{\mathscr{B}}$ defined in (40) are trivial.

But the other properties of the $K_{\mathfrak{p}}$ chosen in this way are not at all obvious. Hence it is advantageous to consider other recipes for choosing the $K_{\mathfrak{p}}$, for which (41) does not hold but we can still prove that the functions (40) are alternating.

For this purpose we write $\mathscr{B}$ as the disjoint union of $\mathscr{B}'$ and $\mathscr{B}''$, where we shall always suppose that the even primes and the infinite places are all in $\mathscr{B}'$. For any odd prime $\mathfrak{p}$ we denote by $T_{\mathfrak{p}}$ the subset of $V_{\mathfrak{p}}$ consisting of those triples (μ_1, μ_2, μ_3) with $\mu_1\mu_2\mu_3 = 1$ for which each μ_i is in $\mathfrak{o}_{\mathfrak{p}}^*/\mathfrak{o}_{\mathfrak{p}}^{*2}$ – that is, each μ_i is the image of a $\mathfrak{p}$-adic unit. The main point of the following theorem is that for $\mathfrak{p}$ in $\mathscr{B}''$ it enables us to replace the complicated inductive definition of $K_{\mathfrak{p}}$ used in the proof of Lemma 10.1 by the much simpler choice $K_{\mathfrak{p}} = T_{\mathfrak{p}}$. How one chooses $\mathscr{B}''$ depends on the particular application which one has in mind.

Theorem 10.2. *Let $\mathscr{B}$ be the disjoint union of $\mathscr{B}'$ and $\mathscr{B}''$, and suppose that $\mathscr{B}'$ contains the even primes and the infinite places. We can construct maximal isotropic subspaces $K_{\mathfrak{p}} \subset V_{\mathfrak{p}}$ such that $V_{\mathscr{B}} = U_{\mathscr{B}} \oplus K_{\mathscr{B}}$,*

$$W_{\mathscr{B}'} = (U_{\mathscr{B}'} \cap W_{\mathscr{B}'}) \oplus (K_{\mathscr{B}'} \cap W_{\mathscr{B}'}) \tag{43}$$

and $K_{\mathfrak{p}} = T_{\mathfrak{p}}$ for all $\mathfrak{p}$ in $\mathscr{B}''$; and (43) implies that $U'_{\mathscr{B}'} = U_{\mathscr{B}'} \cap W_{\mathscr{B}'}$. Moreover

$$U'_{\mathscr{B}} = j_* U'_{\mathscr{B}'} \oplus \tau_{\mathscr{B}} W'_{\mathscr{B}''} = j_* U'_{\mathscr{B}'} \oplus \left(\oplus_{\mathfrak{p} \in \mathscr{B}''} \tau_B W'_{\mathfrak{p}} \right), \tag{44}$$

and the restriction of $\theta^{\flat}_{\mathscr{B}}$ to $j_* U'_{\mathscr{B}'} \times j_* U'_{\mathscr{B}'}$ is trivial.

If $\mathscr{B}'$ also contains all the odd primes $\mathfrak{p}$ such that the $v_{\mathfrak{p}}(c_i - c_j)$ are not all congruent mod 2, then we can choose the $K_{\mathfrak{p}}$ for $\mathfrak{p}$ in $\mathscr{B}'$ so that also $\theta^{\flat}_{\mathscr{B}}$ is alternating on $U'_{\mathscr{B}}$.

The appearance of $j_* U'_{\mathscr{B}'}$ in and just after (44) calls for some explanation. Let u be any element of $U_{\mathscr{B}'}$; then u is in $U_{\mathscr{B}}$. Moreover, for $\mathfrak{p}$ in $\mathscr{B}''$ the image of u in $V_{\mathfrak{p}}$ is in $T_{\mathfrak{p}} = K_{\mathfrak{p}}$ and therefore in $K_{\mathfrak{p}} + W_{\mathfrak{p}}$; hence u is in $U'_{\mathscr{B}}$. In this way we define a map $U'_{\mathscr{B}'} \to U'_{\mathscr{B}}$ which is clearly an injection and which we denote by j_*.

Lemma 10.1 is the special case of Theorem 10.2 in which $\mathscr{B}' = \mathscr{B}$ and $\mathscr{B}''$ is empty. But the proof of Lemma 10.1 is a necessary step (and indeed the most substantial step) in the proof of Theorem 10.2.

The main application of Theorem 10.2 is to twisted curves Γ_b, where we can clearly take b to be an integer. Let $\mathscr{S}$ denote the set of bad primes for Γ itself and let $\mathscr{B} \supset \mathscr{S}$ be the set of bad primes for Γ_b. If we are to apply any part of Theorem 10.2, $\mathscr{B}''$ must in practice consist entirely of primes which divide b and are not in $\mathscr{S}$. To describe the effect of twisting, we shall denote by d_b the dimension of the 2-Selmer group of Γ_b regarded as a vector space over $\mathbf{F}_2$; we write $d = d_1$ for the dimension of the 2-Selmer group of Γ itself. It is now possible to prove results about $d_b - d$, the change in the dimension of the 2-Selmer group as one goes from Γ to Γ_b. There is reason to expect that statements about the parities of d and d_b will be simpler and much easier to prove than statements about their actual values. The two major statements known about d_b are Lemmas 10.3 and 10.4; both of these are easy consequences of Theorem 10.2.

Lemma 10.3. *If b is in $\mathfrak{o}_{\mathfrak{p}}^*$ for every $\mathfrak{p} \in \mathscr{S}$, then $d_b \equiv \dim(U_{\mathscr{S}} \cap W_{\mathscr{S}})$ mod 2 where $W_{\mathscr{S}} = \oplus_{\mathfrak{p} \in \mathscr{S}} W_{\mathfrak{p}}$ and the $W_{\mathfrak{p}}$ must be defined with respect to Γ_b and not with respect to Γ. Thus d_b mod 2 only depends on the classes of b in the $k_{\mathfrak{p}}^*/k_{\mathfrak{p}}^{*2}$ for $\mathfrak{p}$ in $\mathscr{S}$.*

Lemma 10.4. *Let $\mathfrak{p}$ be an odd prime in $\mathscr{S}$ such that*

$$v_{\mathfrak{p}}(c_1 - c_2) > 0, \quad v_{\mathfrak{p}}(c_1 - c_3) = v_{\mathfrak{p}}(c_2 - c_3) = 0.$$

Let b in k^* be such that b is in $k_{\mathfrak{q}}^{*2}$ for all $\mathfrak{q}$ in $\mathscr{S}$ other than $\mathfrak{p}$ and b is a quadratic non-residue at $\mathfrak{p}$. Then d and d_b have opposite parities.

11 Pencils of Curves of Genus 1

In this section we shall be concerned with pencils of 2-coverings of elliptic curves defined over an algebraic number field k, where the underlying pencil of elliptic curves has the form

$$E: Y^2 = (X - c_1(U,V))(X - c_2(U,V))(X - c_3(U,V)). \tag{45}$$

Here the $c_i(U,V)$ are homogeneous polynomials in $\mathfrak{o}[U,V]$ all having the same even degree. By means of a linear transformation on U,V we can ensure that the leading coefficients of the $c_i(U,V)$ are nonzero. Write

$$R(U,V) = p_{12}(U,V)p_{23}(U,V)p_{31}(U,V)$$

where $p_{ij} = c_i - c_j$.

The 2-coverings of (45) are given by

$$m_i(U,V)Y_i^2 = X - c_i(U,V) \text{ for } i = 1,2,3 \tag{46}$$

where the $m_i(U,V)$ are square-free homogeneous polynomials in $\mathfrak{o}[U,V]$ of even degree such that $m_1 m_2 m_3$ is a square. We should really regard the m_i as homogeneous polynomials modulo squares, but this complicates the notation. Equation (46) are equivalent to the three equations

$$m_i Y_i^2 - m_j Y_j^2 = (c_j - c_i)Y_0^2 \tag{47}$$

of which only two are independent. The sum of two 2-coverings is obtained by multiplying the corresponding triples $\mathbf{m} = (m_1, m_2, m_3)$ componentwise and then removing squared factors. Denote by $\Gamma = \Gamma(\mathbf{m}; U, V)$ the curve given by the three equations (46) or the three equations (47) for particular values of $\mathbf{m}, U, V$, and by $C_{ij} = C_{ij}(\mathbf{m}; U, V)$ the conic given by a single equation (47). There are natural maps $\Gamma \to C_{ij}$. Equation (47) also imply

$$m_1(c_2 - c_3)Y_1^2 + m_2(c_3 - c_1)Y_2^2 + m_3(c_1 - c_2)Y_3^2 = 0, \tag{48}$$

and for Γ to be soluble so too must be this conic. These are Brauer-Manin conditions; they do not appear explicitly in the statement of Theorem 11.2 but they are implied by the condition that $\mathcal{N}$ is not empty.

Our objective is to provide sufficient conditions for the solubility of a particular pencil of curves Γ, where the pencil is assumed to be everywhere locally soluble. We shall use a superscript 0 to denote a curve of this pencil or other objects connected with it. We shall need to distinguish between $\mathcal{S}$, the set of bad places for the pencil of curves Γ, and the larger set $\mathcal{B}$ of bad places for the particular curve $\Gamma^0(\alpha, \beta)$ on which we want to prove that there are rational points. Thus $\mathcal{S}$ is a finite set containing the infinite places, the primes above 2, those which divide the resolvent of any two coprime factors of $R(U, V)$ in $\mathfrak{o}[U, V]$ or have norm not greater than $\deg(R(U, V))$, and those which are bad in the sense of Sect. 9 for any of the pencils of conics C_{ij}. (In particular, this ensures that $\mathcal{S}$ contains a base for the ideal class group of k.) In terms of the definitions below, $\mathcal{B}$ must contain $\mathcal{S}$ and all the $\mathfrak{p}_{k\tau}$. The additional prime $\mathfrak{p}$ which we introduce at each step of the algorithm should be thought of as being thereby adjoined to $\mathcal{S}$.

We denote the irreducible factors of $p_{ij}(U, V)$ in $k[U, V]$ by $f_{k\tau}(U, V)$, and we assume that the coefficients of any $f_{k\tau}$ are integers and that there is no prime outside $\mathcal{S}$ which divides all of them. When we apply the results of Sect. 10 it will be with $U = \alpha, V = \beta$ where $\alpha \times \beta$ is so chosen that each ideal $(f_{k\tau}(\alpha, \beta))$ is the product of primes in $\mathcal{S}$ and one prime $\mathfrak{p}_{k\tau}$ outside $\mathcal{S}$; to do this we appeal to Lemma 7.1. In what follows, we shall call the $\mathfrak{p}_{k\tau}$ the *Schinzel primes*. The arguments of Sect. 10 show that we can confine ourselves to those triples $\mathbf{m}$ whose components take values in $\mathfrak{o}^*_{\mathcal{B}}$ when $U = \alpha, V = \beta$. Because of the constraint just stated on the choice of α, β, this means that we can restrict the components of $\mathbf{m}$ to be products of some of the $f_{k\tau}(U, V)$ by elements of $\mathfrak{o}^*_{\mathcal{S}}$. In view of the description of 2-descents in Sect. 10, we can further restrict ourselves to the triples $\mathbf{m}$ such that $m_1 m_2 m_3$ divides R^2 and m_i is prime to p_{jk} in $k[U, V]$ up to factors in $\mathcal{S}$, where here and throughout this section i, j, k is any permutation of $1, 2, 3$.

We shall also assume that the $p_{ij}(U, V)$ are coprime in $k[U, V]$. The case when this condition fails is also of interest, but the methods used and the conclusions are quite different; for a more detailed account see [56]. This assumption is weaker than that in [16], which was that $R(U, V)$ is square-free in $k[U, V]$, and it enables us to bring the example of diagonal quartics within the scope of the general theory.

The parity conditions on the degrees of the c_i and m_i are needed to ensure that the curves (45) and Γ with $U = \alpha, V = \beta$ only depend on $\lambda = \alpha/\beta$ and not on α, β separately; otherwise we would not be dealing with pencils. But even if two of the m_i have odd degree, which can happen if R has factors of odd degree, the curve Γ given by (46) or (47) is a 2-covering of E; and such 2-coverings do play a part in our arguments. For given E, let G be the group of all triples (m_1, m_2, m_3) satisfying the conditions above, including that the degrees of the m_i are even, and define $G^* \supset G$ by dropping the condition that the m_i have even degree. Provided we take the m_i modulo squares, both G and G^* are finite; and either G or G^* can be regarded as defining those pencils of 2-coverings of the pencil E which are of number-theoretic interest.

Now suppose that we are given a triple $\mathbf{m}^0 = (m_1^0, m_2^0, m_3^0)$ in G. Denote by $\Gamma^0 = \Gamma(\mathbf{m}^0, U, V)$ the curve of genus 1 given by the three equations (47) with $\mathbf{m} = \mathbf{m}^0$, and similarly for the C_{ij}^0. For simplicity we assume that the elliptic curve (45) has no primitive 4-division points defined over $k(U,V)$, and to avoid trivialities we also assume that the 2-covering Γ^0 does not correspond to a 2-division point.

The only values of U/V for which Γ^0 can be soluble are ones for which Γ^0 is everywhere locally soluble; so for any such value of U/V the 2-Selmer group of E must contain the subgroup of order 8 generated by Γ^0 and the 2-coverings coming from the 2-division points. We shall call this the *inescapable* part of the 2-Selmer group. The essential tool in proving solubility will be the special case $p = 2$ of Lemma 4.6, which we restate for ease of reference.

Lemma 11.1. *Suppose that the Tate-Shafarevich group of E/k is finite and the 2-Selmer group of E has order 8. Then every curve representing an element of the 2-Selmer group contains rational points.*

As this shows, everything in this section will depend on the finiteness of III; and everything will also depend on Schinzel's Hypothesis.

As in Sect. 9, we need to work not in $\mathbf{P}^1$ but in the subset $\mathbf{L}^1$ obtained by deleting the points $\lambda = \alpha/\beta$ at which $R(\alpha, \beta)$ vanishes. The topology on $\mathbf{L}^1(k)$ will be that induced by $\mathscr{S}$. There is an open set $\mathscr{N} \subset \mathbf{L}^1(k)$ such that $\Gamma^0(\alpha, \beta)$ is soluble at every place of $\mathscr{S}$ if and only if λ lies in $\mathscr{N}$. Let us assume temporarily that we are going to apply Lemma 7.1 to choose α, β so that each ideal $(f_{k\tau}(\alpha, \beta))$ is a prime $\mathfrak{p}_{k\tau}$ up to possible (and well determined) factors in $\mathscr{S}$. Until we have chosen α, β we do not know the $\mathfrak{p}_{k\tau}$; but we do already know a set of generators of $U_{\mathscr{B}}$ as polynomials in U, V, and in the notation of Sect. 10 we also know the bilinear form $e_{\mathscr{B}}$ because of the results of Sect. 8. It is therefore possible to implement all the apparatus of Sect. 10. Solubility of $\Gamma^0(\alpha, \beta)$ at a particular Schinzel prime $\mathfrak{p}_{k\tau}$ is equivalent to the bilinear form $e_{\mathscr{B}}$ defined in Sect. 10 taking the value 1 at each $\mathbf{m}^0 \times w$, where w is either of the two generators of W associated with $\mathfrak{p}_{k\tau}$. This is a Legendre-Jacobi condition, so it determines a certain open set $\mathscr{N}_{\mathfrak{p}} \subset \mathbf{L}^1(k)$ where $\mathfrak{p} = \mathfrak{p}_{k\tau}$. If we take any α/β not in $\mathscr{N}_{\mathfrak{p}}$ and make no assumption about $f_{k\tau}(\alpha, \beta)$, then $\prod e_{\mathscr{B}}(\mathbf{m}^0, w)$ taken over the w coming from the prime factors of $f_{k\tau}(\alpha, \beta)$ not in $\mathscr{S}$ will be the same Legendre-Jacobi function which we have just studied and will therefore have the same value -1. In other words, $\Gamma^0(\alpha, \beta)$ will be locally insoluble at some prime dividing $f_{k\tau}(\alpha, \beta)$. Thus for studying the solubility of (46) we can replace $\mathscr{N}$ by the intersection of $\mathscr{N}$ with all the $\mathscr{N}_{\mathfrak{p}}$. In what follows we assume that this new $\mathscr{N}$ is not empty.

In contrast to what happened in Sect. 9, nothing is gained by simply applying Lemma 7.1 to choose α, β so that all the $f_{k\tau}(\alpha, \beta)$ are prime up to possible factors in $\mathscr{S}$, because this might give rise to a 2-Selmer group too big for us to be able to apply Lemma 11.1. What we do instead is most conveniently described as an algorithm, which consists of repeatedly introducing a further well-chosen prime $\mathfrak{p}$ into $\mathscr{S}$, with a corresponding extra condition on the set $\mathscr{N}$ of possible values of $U \times V$, in such a way that if we then apply Lemma 7.1 the dimension of the 2-Selmer group is one less than it would have been before. If we can go on doing this as long as the 2-Selmer group remains too big, we shall eventually reach a situation to which we can apply Lemma 11.1. However, this process cannot be always possible; for otherwise we would be able to prove that the Hasse principle held for the pencil (46), and this is known to be false. Hence there must be a potential obstruction to the argument. This is provided by Condition D, which will be introduced below. What we thereby obtain is Theorem 11.2 below.

The process of introducing a new prime $\mathfrak{p}$ is as follows. We choose an $f_{k\tau}$ and integers $\theta_{\mathfrak{p}}, \phi_{\mathfrak{p}}$ not both divisible by $\mathfrak{p}$ and such that $\mathfrak{p} \| f_{k\tau}(\theta_{\mathfrak{p}}, \phi_{\mathfrak{p}})$. Without loss of generality we can assume that $\theta_{\mathfrak{p}}, \phi_{\mathfrak{p}}$ are coprime. Choose integers $\gamma_{\mathfrak{p}}, \delta_{\mathfrak{p}}$ such that $\theta_{\mathfrak{p}} \delta_{\mathfrak{p}} - \phi_{\mathfrak{p}} \gamma_{\mathfrak{p}} = 1$, write

$$U = \theta_{\mathfrak{p}} U_1 + \gamma_{\mathfrak{p}} V_1, \quad V = \phi_{\mathfrak{p}} U_1 + \delta_{\mathfrak{p}} V_1$$

and impose on $\mathcal{N}$ the additional condition $\mathfrak{p}^2 | V_1$. Thus at any point of $\mathcal{N}$ the value of $f_{k\tau}$ is exactly divisible by $\mathfrak{p}$, and the values of all the other functions $f_{..}$ are prime to $\mathfrak{p}$.

For given $f_{k\tau}$ which $\mathfrak{p}$ satisfy the condition that there exist $\alpha_{\mathfrak{p}}, \beta_{\mathfrak{p}}$ as above? Let $K_{k\tau} = k[X]/f_{k\tau}(X,1)$ be the field obtained by adjoining to k a root of $f_{k\tau}$, and let $\xi_{k\tau}$ be the class of X in $K_{k\tau}$; thus $f_{k\tau}(\xi_{k\tau}, 1) = 0$. The singular fibres of the pencil of elliptic curves (45), as also those of the pencil of 2-coverings (46), correspond to the roots of the $f_{k\tau}$. The reason for being interested in the singular fibres is as follows. Let $\mathfrak{p}$ be a prime of k not in $\mathcal{S}$, and let $\alpha_{\mathfrak{p}}, \beta_{\mathfrak{p}}$ in $\mathfrak{o}$ be such that $\mathfrak{p} \| f_{k\tau}(\alpha_{\mathfrak{p}}, \beta_{\mathfrak{p}})$; such $\alpha_{\mathfrak{p}}, \beta_{\mathfrak{p}}$ exist if and only if there is a prime $\mathfrak{P}$ in $K_{k\tau}$ whose relative norm over k is $\mathfrak{p}$. This last condition may appear tiresome. But what one really does is to choose a first-degree prime $\mathfrak{P}$ in $K_{k\tau}$ and define $\mathfrak{p}$ to be the prime below it in k. Now norm $\mathfrak{P} = \mathfrak{p}$ is automatic.

The arguments needed to validate each step of the algorithm are lengthy, and we list them as (i)–(v) below. We impose further conditions on the additional prime $\mathfrak{p}$ which ensure (i); we then deduce (ii), (iii) and (iv). Finally we use Condition D to show that unless the process is complete, we can choose $\mathfrak{p}$ so that (v) holds. After all this we choose $\alpha \times \beta$ according to the recipe in Lemma 7.1 for the $f_{k\tau}$, and with the additional property that $L(\mathcal{S}; U, V; \alpha, \beta) = 1$ if there is any $f_{k\tau}$ of odd degree. One can satisfy this additional requirement by a slight modification of the construction used to prove Lemma 7.1. Alternatively, one can render it unnecessary by replacing U, V by homogeneous quadratic forms in U_1, V_1; this does not alter the values of the functions L.

(i) We determine necessary and sufficient conditions for $\Gamma(\alpha, \beta)$ to be locally soluble at $\mathfrak{p}$. We use these immediately to choose $\mathfrak{p}$-adic conditions on $\mathcal{N}$ such that $\Gamma^0(\alpha, \beta)$ is locally soluble at $\mathfrak{p}$; but in (v) we shall also need them to ensure for a particular $\mathbf{m}$ that the corresponding $\Gamma(\alpha, \beta)$ is not locally soluble at $\mathfrak{p}$.

For (ii)–(iv) we assume that $\alpha \times \beta$ satisfies the conditions of Lemma 7.1.

(ii) The bilinear form $\theta_{\mathcal{B}}^{\sharp} : W_{\mathcal{B}}' \times W_{\mathcal{B}}' \to \{\pm 1\}$ defined in (40) does not depend on the choice of $\alpha \times \beta$ and hence of the $\mathfrak{p}_{k\tau}$.

By this we mean that if we change α, β, thereby replacing the old W' by a new W' canonically isomorphic to it and replacing the old $\mathfrak{p}_{k\tau}$ in $\mathscr{B}$ by the new ones, then this isomorphism preserves $\theta^{\sharp}_{\mathscr{B}}$. The next result which we need, which is only meaningful once we have proved (ii), is as follows:

(iii) We determine the effect on the function $\theta^{\sharp}_{\mathscr{B}}$ of introducing a new prime $\mathfrak{p}$ in the way described above.

(iv) The curve $\Gamma^0(\alpha, \beta)$ is locally soluble at $\mathfrak{p}_{k\tau}$.

By requiring that $\lambda = \alpha/\beta$ is in $\mathcal{N}$ we ensure that $\Gamma^0(\alpha, \beta)$ is soluble in k_v for every v in $\mathcal{S}$ including $\mathfrak{p}$; and it is also soluble at all the Schinzel primes other than possibly $\mathfrak{p}_{k\tau}$. Thus (i) and (iv) prove that the class of $\Gamma^0(\alpha, \beta)$ is in the 2-Selmer group of the curve $E(\alpha, \beta)$ given by (45) provided that α, β are chosen according to the recipe in Lemma 7.1. The $\mathfrak{p}_{k\tau}$ are not determined until we know α and β; but this is unimportant because of (ii). Finally, the condition which we need for our algorithm to achieve what we want is as follows:

(v) If $\mathbf{m}$ is in the kernel of the old $\theta^{\sharp}_{\mathscr{B}}$ but not in the inescapable part of it, then we can introduce a new prime $\mathfrak{p}$ which removes $\mathbf{m}$ from the kernel and does not put anything new into it.

It is in the proof of (v) that we need Condition D. Once we have (v), we can after a sufficient number of steps satisfy the conditions of Lemma 11.1, and this implies that $\Gamma^0(\alpha, \beta)$ has rational solutions. The result of this process is Theorem 11.2. A more sophisticated treatment of the solubility of pencils (46) can be found in Chap. I of [14].

Theorem 11.2. *Assume Schinzel's Hypothesis and the finiteness of* III, *and suppose that the three* $p_{ij}(U,V)$ *are coprime in* $k[U,V]$. *Suppose that the* $\mathcal{N}$ *constructed above is not empty and that Condition D holds. Then we can construct a non-empty set* $\mathscr{A}$ *which lies in the closure of the set of* λ *in* $\mathbf{L}^1(k)$ *at which* $\Gamma^0(\alpha, \beta)$ *is soluble in* k.

Theorem 11.2 gives a sufficient condition for the Hasse principle to hold, though the condition is not always necessary. Indeed, we shall see at the end of this section that we can replace Condition D by a potentially weaker Condition E; but probably even the latter is not always necessary for solubility. The relation between Condition D and the Brauer-Manin obstructions is addressed in [16].

Achieving (i). The condition that any particular Γ is soluble in $k_{\mathfrak{p}}$ throughout some neighbourhood of $\alpha_{\mathfrak{p}} \times \beta_{\mathfrak{p}}$ is that the reduction of $\Gamma(\alpha_{\mathfrak{p}}, \beta_{\mathfrak{p}})$ mod $\mathfrak{p}$ should contain a point defined over $\mathfrak{o}/\mathfrak{p}$ which is liftable to a point on Γ defined over $k_{\mathfrak{p}}$. Denote by $L_{k\tau}$ the least extension of $K_{k\tau}$ over which some absolutely irreducible component of the singular fibre at $\xi_{k\tau} \times 1$ is defined; conveniently, all these components are defined over the same least extension, which is normal over $K_{k\tau}$. The decomposition of $\Gamma(\alpha_{\mathfrak{p}}, \beta_{\mathfrak{p}})$ mod $\mathfrak{p}$ corresponds to the decomposition of the fibre $\Gamma(\xi_{k\tau}, 1)$; so we can solve Γ in $k_{\mathfrak{p}}$ in a suitable neighbourhood of $\alpha_{\mathfrak{p}} \times \beta_{\mathfrak{p}}$ if and only if $\mathfrak{P}$ splits completely in $L_{k\tau}$.

If $f_{k\tau} \| p_{ij}$, each singular fibre given by $f_{k\tau} = 0$ of the pencil of curves Γ splits as a pair of irreducible conics which meet in two points and are each defined over the field $L_{k\tau} = K_{k\tau}(\sqrt{g_{k\tau}(\xi_{k\tau}, 1)})$; here $g_{k\tau} = m_k$ if $f_{k\tau}$ divides neither of m_i and m_j or $g_{k\tau} = m_k p_{jk}$ if $f_{k\tau}$ divides both of them. The same holds if $f_{k\tau}^2 | p_{ij}$ and $f_{k\tau}$ divides neither m_i nor m_j, and again we have $g_{k\tau} = m_k$. If $f_{k\tau}^2 | p_{ij}$ and $f_{k\tau}$ divides both m_i and m_j, then each singular fibre given by $f_{k\tau} = 0$ splits as a set of four lines which form a skew quadrilateral, and each of these lines is defined over

$$L_{k\tau} = K_{k\tau}\left(\sqrt{m_k(\xi_{k\tau}, 1)}, \sqrt{p_{jk}(\xi_{k\tau}, 1)} \right). \tag{49}$$

Write $L_{k\tau}^0$ for the field corresponding to Γ^0 under this construction. To test for Condition D, we need to list those $\mathbf{m}$ for which $L_{k\tau}$ is contained in $L_{k\tau}^0$. It is easy to verify that they form a group, which contains m^0 and the triples coming from the 2-division points.

Proof of (ii). We are allowed to choose $\alpha \times \beta$ only within a set which is small in the topology induced by $\mathscr{S}$. In particular, this means that the power of any prime in $\mathscr{S}$ which divides any $f_{k\tau}(\alpha, \beta)$ is independent of α and β. Since the only other prime which divides any particular $f_{k\tau}(\alpha, \beta)$ is $\mathfrak{p}_{k\tau}$, which does so to the first power, the ideal class of $\mathfrak{p}_{k\tau}$ is fixed. If the place v is given by some $\mathfrak{p}_{k\tau}$ then a generator of W_v' can be lifted back to $\sigma \times \tau$ where each of σ and τ is either 1 or $f_{k\tau}(\alpha, \beta)$; and if v is in $\mathscr{S}$ the elements of a base for W_v' can be lifted back to elements $\sigma \times \tau$ independent of α, β with σ, τ in $\mathfrak{o}_{\mathscr{S}}^*$. We choose a base for $W_{\mathscr{B}}'$ composed of these two kinds of elements; then the value of $\theta_{\mathscr{B}}^{\sharp}$ at any pair of elements of this base is a product of expressions of the form $(\sigma'(\alpha, \beta), \tau'(\alpha, \beta))_v$ where v is in $\mathscr{B}$ and each of σ' and τ' is the product of an element of $\mathfrak{o}_{\mathscr{S}}^*$ and possibly an $f_{k\tau}$. If v is in $\mathscr{S}$ the value of this expression is independent of α, β. If v is given by $\mathfrak{p}_{k\tau}$ then using symmetry and $(\xi, -\xi)_v = 1$ if necessary we can reduce to the case when σ' is not divisible by $f_{k\tau}$. If also τ' is not divisible by $f_{k\tau}$ then $(\sigma'(\alpha, \beta), \tau'(\alpha, \beta))_v = 1$; otherwise $(\sigma'(\alpha, \beta), \tau'(\alpha, \beta))_v = L(\mathscr{S}; \sigma', \tau'; \alpha, \beta)$ is continuous.

Achieving (iii). When we introduce $\mathfrak{p}$ we adjoin two more generators to W, and the description in terms of U, V of the product of $\mathfrak{p}$ and the new $\mathfrak{p}_{k\tau}$ is the same as the description of the old $\mathfrak{p}_{k\tau}$. We use Theorem 10.2 with $\mathscr{B}'' = \{\mathfrak{p}, \mathfrak{p}_{k\tau}\}$ to describe the change in W'. In the notation of Sect. 10 all the triples in $W_{\mathfrak{p}}$ have $v_{\mathfrak{p}}(m_k)$ even. Since $K_{\mathfrak{p}} = T_{\mathfrak{p}}$, the set of triples all whose components are units at $\mathfrak{p}$, it follows that $W_{\mathfrak{p}} \cap K_{\mathfrak{p}}$ has dimension 1 and so has $W_{\mathfrak{p}}'$. A similar argument holds for the primes $\mathfrak{p}_{k\tau}$ provided by Lemma 7.1, and shows that to each such prime there corresponds one generator of W'. Hence introducing $\mathfrak{p}$ increases the dimension of W' by 1. If we regard the $\theta^{\sharp}$ defined at (40) as being given by a matrix whose last two columns are the only ones which depend on $\mathfrak{p}_{k\tau}, \mathfrak{p}$ respectively, the old $\theta^{\sharp}$ can be obtained from the new one by adding together the last two rows and the last two columns.

Proof of (iv). As we have just noted,

$$e^*(\mathbf{m}^0, w^{\text{old}}) = e^*(\mathbf{m}^0, w_{\mathfrak{p}}) e^*(\mathbf{m}^0, w^{\text{new}})$$

where w^{old} and w^{new} correspond to the old and the new $\mathfrak{p}_{k\tau}$. The first two factors here are 1, so the third must be so.

Choice of $\mathfrak{p}$. Let $w_{\mathfrak{p}}$ be a lift to $W_{\mathfrak{p}}$ of the non-trivial element of $W'_{\mathfrak{p}}$, and let $\mathbf{m}$ be an element of $U_{\mathscr{B}} \cap W_{\mathscr{B}}$ which is not in the inescapable part of the 2-Selmer group. Thus $\tau_{\mathscr{B}}^{-1}\mathbf{m}$ is in the kernel of $e_{\mathscr{B}}^*$. Suppose that we can choose $\mathfrak{p}$ so that the 2-covering corresponding to $\mathbf{m}$ is locally insoluble at $\mathfrak{p}$. On the one hand this is equivalent to $e^*(\tau_{\mathscr{B}}^{-1}\mathbf{m}, w_{\mathfrak{p}}) = -1$. On the other hand it requires $\mathfrak{P}$ to split completely in $L_{k\tau}^0$ but not in $L_{k\tau}$. The condition below, which in the literature is called Condition D, ensures that such a choice is possible. We shall see later that Condition D can be replaced by a weaker condition, but one which is less natural and sometimes less computationally convenient.

Condition D: *If* $\mathbf{m}$ *is not in the inescapable subgroup of the 2-Selmer group, then there is a pair* k, τ *such that the field* $L_{k\tau}$ *is not contained in* $L_{k\tau}^0$.

By incorporating the definitions of $L_{k\tau}$ and $L_{k\tau}^0$ into this condition, we can restate it as follows:

The kernel of the composite map

$$\mathbf{m} \mapsto \oplus_{k,\tau} g_{k\tau}(\mathbf{m}) \mapsto \oplus_{k,\tau} K_{k\tau}^* / \langle K_{k\tau}^{*2}, H_{k\tau} \rangle$$

is generated by the inescapable subgroup of the 2-Selmer group, where

$$g_{k\tau} = \begin{cases} m_k & \text{if } f_k \text{ divides neither of } m_i \text{ and } m_j, \\ m_k p_{jk} & \text{if } f_k \text{ divides both of } m_i \text{ and } m_j, \end{cases}$$

and

$$H_{k\tau} = \begin{cases} m_k(\xi_{k\tau}, 1) & \text{if } f_{k\tau} \text{ divides neither of } m_i \text{ and } m_j, \\ m_k(\xi_{k\tau}, 1)p_{jk}(\xi_{k\tau}, 1) & \text{if } f_{k\tau} \| p_{ij} \text{ and } f_{k\tau} \text{ divides } m_i \text{ and } m_j, \\ \{m_k(\xi_{k\tau}, 1), p_{jk}(\xi_{k\tau}, 1)\} & \text{if } f_{k\tau}^2 \| p_{ij} \text{ and } f_{k\tau} \text{ divides } m_i \text{ and } m_j. \end{cases}$$

The $\mathbf{m}$ for which $L_{k\tau}$ is contained in $L_{k\tau}^0$ for each subscript $k\tau$ are those which do not satisfy Condition D. If $\mathbf{m}$ satisfies Condition D we can choose k, τ and a $\mathfrak{P}$ which splits in $L_{k\tau}^0$ but not in $L_{k\tau}$. The underlying $\mathfrak{p}$ has the properties we want. This process remove $\mathbf{m}$ from the 2-Selmer group without creating any new elements of that group. So we have certainly decreased the dimension of the 2-Selmer group, which is what we needed to show to justify the algorithm. In fact it is easy to show that we have decreased it by exactly 1.

It will be seen that we have not used the full force of Condition D; indeed it is stated for all elements of G^*, but we have only used it for those elements which lie in the initial 2-Selmer group. These are the ones for which the corresponding 2-covering is locally soluble at each place in $\mathscr{B}$. The proof of (ii) above shows that local solubility in $\mathscr{S}$ implies local solubility at each $\mathfrak{p}_{k\tau}$; and the proof of (iii) shows that this 2-Selmer group, considered as a subgroup of G^*, does not vary as $\alpha \times \beta$ varies within a small enough open set. We actually use Condition D only for the $\mathbf{m}$ which lie in this 2-Selmer group; and to require merely that such $\mathbf{m}$ satisfy Condition D is weaker than the full Condition D. We call this weaker condition Condition

E. Its disadvantage is that Condition D is independent of α and β, whereas Condition E is not; however Condition E becomes independent of $\alpha \times \beta$ when $\alpha \times \beta$ is restricted to a small enough open set. A particularly favourable case is when the 2-Selmer group has order 8, because then Condition E is trivial. I do not know whether Condition E, together with the conditions imposed in Theorem 11.2, is necessary as well as sufficient for global solubility, nor whether these conditions are together equivalent to the Brauer-Manin conditions, though I doubt whether either of these is true. However, the arguments in [11] do enable one to link Conditions D and E to the Brauer-Manin obstructions.

12   Some Examples

In this section we consider three particular families of surfaces to which the ideas of the previous section (suitably modified in the last two examples) can be applied. The first family consists of diagonal quartic surfaces (51), subject to the additional condition (52) which ensures that (51) contains a pencil of curves of genus 1 whose Jacobian has rational 2-division points. The second family is a particular family of Kummer surfaces, and the third consists of diagonal cubic surfaces. What these last two examples have in common is that the argument does not use Schinzel's Hypothesis; more precisely, we only need to force one linear polynomial to take a prime value, and this can be done by means of Dirichlet's Theorem on primes in arithmetic progression. But the price of this is that we have to apply Lemma 4.6 to two pencils of elliptic curves rather than to one, and to make the process work the constraints on the choice of additional primes associated with the two pencils must not interfere with each other. Thus the proof requires some additional (but not very restrictive) conditions which are unlikely to be actually needed for solubility. For the third example we also need to require that the field k over which we work does not contain $\sqrt{-3}$.

12.1   Diagonal Quartic Surfaces

Let V be a smooth quartic surface whose equation can be put into the form $AD = BC$, where $A, \ldots, D$ are linearly independent homogeneous quadratics in $X_0, \ldots, X_3$. Such a V is fibred by the pencil of curves of genus 1

$$yA = zB, \quad yC = zD, \tag{50}$$

which are 2-coverings of elliptic curves. Recall that if M_1, M_2 are the matrices associated with the quadratic forms $yA - zB$ and $yC - zD$ respectively, then the Jacobian of (50) can be written in the form $Y^2 = f(Z)$ where f is the resolving cubic of the quartic polynomial $\det(M_1 - XM_2)$. Hence if $A, \ldots, D$ are linear combinations of the X_i^2 then the 2-division points of the Jacobian of (50) are all rational. Over an algebraic number field k the K3 surfaces whose equations have the form

$$a_0 X_0^4 + a_1 X_1^4 + a_2 X_2^4 + a_3 X_3^4 = 0 \tag{51}$$

satisfy the condition above if and only if

$$a_0 a_1 a_2 a_3 \text{ is a square.} \tag{52}$$

Full details of the argument which follows can be found in [17]. We shall always assume that (51) is everywhere locally soluble and the a_i are integral. The surfaces (51) are very special within the family of nonsingular quartic surfaces for at least two reasons: they are Kummer surfaces, and their Néron-Severi groups over $\mathbf{C}$ have maximal rank, which is 20. But this is probably the simplest family of K3 surfaces that can be written down explicitly.

It is known that the Néron-Severi group of (51) over $\mathbf{C}$ is generated by the 48 lines on the surface. However, what is equally important for our purposes is the Néron-Severi group over k. When $k = \mathbf{Q}$ there are now 282 possibilities for the Galois group over $\mathbf{Q}$ of the least field of definition of the 48 lines; these have been tabulated by Martin Bright in his Cambridge Ph.D. thesis [57], which can be found at

http://www.boojum.org.uk/maths/quartic-surfaces/

together with a good deal of other relevent material. The large number of cases means that the calculation needed to be automated, and one interesting feature of the thesis is that it shows that this is possible.

Write $A = \alpha_0 X_0^2 + \alpha_1 X_1^2 + \alpha_2 X_2^2 + \alpha_3 X_3^2$ and so on. Eliminating each of the four variables X_ν from (50) in turn, we obtain four equations of the form

$$d_{i\ell} X_i^2 + d_{j\ell} X_j^2 + d_{k\ell} X_k^2 = 0, \tag{53}$$

only two of which are linearly independent. Here i, j, k, ℓ is any permutation of $1, 2, 3, 4$ and $d_{\mu\nu}$ is the value of the determinant formed by columns μ and ν of the matrix

$$\begin{pmatrix} \alpha_0 y - \beta_0 z & \alpha_1 y - \beta_1 z & \alpha_2 y - \beta_2 z & \alpha_3 y - \beta_3 z \\ \gamma_0 y - \delta_0 z & \gamma_1 y - \delta_1 z & \gamma_2 y - \delta_2 z & \gamma_3 y - \delta_3 z \end{pmatrix}.$$

We have the unexpected result that each $d_{k\ell}$ is a constant multiple of d_{ij}, where i, j, k, ℓ is any permutation of $0, 1, 2, 3$. We note the identity

$$d_{01} d_{23} + d_{02} d_{31} + d_{03} d_{12} = 0,$$

which is frequently useful. The Jacobian of the curve (50) has the form

$$E : Y^2 = (X - c_1)(X - c_2)(X - c_3)$$

where
$$c_1 - c_2 = d_{03}d_{21}, \quad c_2 - c_3 = d_{01}d_{32}, \quad c_3 - c_1 = d_{02}d_{13},$$

and the map from the curve (50) to its Jacobian is given by

$$Y = d_{12}d_{23}d_{31}X_1X_2X_3/X_0^3, \quad X - c_i = d_{ij}d_{ki}X_i^2/X_0^2$$

where i, j, k is any permutation of $1, 2, 3$. Although everything so far is homogeneous in y, z, we have to work in $k(y, z)$ rather than $k(y/z)$, for reasons which are already implicit in Sect. 8.

There is an obvious map from (51) to the quadric surface

$$a_0Y_0^2 + a_1Y_1^2 + a_2Y_2^2 + a_3Y_3^2 = 0. \tag{54}$$

We have assumed that (51), and therefore (54), is everywhere locally soluble; so (54) is soluble in k. From this and the fact that $a_0a_1a_2a_3$ is a square it follows that $-a_1$ is represented by $a_2Y_2^2 + a_3Y_3^2$ over k. In other words, there exist integers r_1, r_2, r_3 and h such that
$$a_1r_1^2 + a_2r_2^2 + a_3r_3^2 = 0, \quad h^2 = a_0a_1a_2a_3.$$

After rescaling (51) if necessary, we can take

$$\begin{aligned}
A(X^2) &= hr_2X_0^2 + a_1a_3(r_3X_1^2 - r_1X_3^2), \\
B(X^2) &= hr_3X_0^2 - a_1a_2(r_2X_1^2 + r_1X_2^2), \\
C(X^2) &= a_3hr_3X_0^2 - a_1a_2a_3(r_2X_1^2 - r_1X_2^2), \\
D(X^2) &= -a_2hr_2X_0^2 - a_1a_2a_3(r_3X_1^2 + r_1X_3^2);
\end{aligned}$$

and the d_{ij} are given by

$$\begin{aligned}
d_{23} &= a_1^2a_2a_3r_1^2(a_3y^2 + a_2z^2), \quad d_{01} = (h/a_2a_3)d_{23}, \\
d_{31} &= a_1^2a_2a_3r_1(a_3r_2y^2 - 2a_3r_3yz - a_2r_2z^2), \quad d_{02} = (h/a_3a_1)d_{31}, \\
d_{12} &= a_1^2a_2a_3r_1(a_3r_3y^2 + 2a_2r_2yz - a_2r_3z^2), \quad d_{03} = (h/a_1a_2)d_{12}.
\end{aligned}$$

These choices do not preserve the symmetry, but that loss appears to be unavoidable. Changing the r_i corresponds to a linear transformation on y, z; changing the sign of h gives the pencil $yA = zC, yB = zD$ instead of (50).

The 2-covering of E given by the triple (m_1, m_2, m_3) with $m_1m_2m_3 = 1$ is

$$m_iZ_i^2 = X - c_i \text{ for } i = 1, 2, 3 \quad \text{and} \quad Y^2 = Z_1Z_2Z_3.$$

As in Sect. 11, values associated with the particular 2-covering given by (50) will be denoted by a superfix 0; the 2-covering itself is given by

$$m_1^0 = -d_{21}d_{31}, \quad m_2^0 = -d_{12}d_{32}, \quad m_3^0 = -d_{13}d_{23}.$$

We shall also need to know the 2-coverings corresponding to the 2-division points. That corresponding to $(c_1, 0)$, for example, is given by

$$m_1 = -a_0 a_1, \quad m_2 = d_{03} d_{21}, \quad m_3 = d_{02} d_{31}, \tag{55}$$

which can alternatively be written

$$m_1 = -a_0 a_1, \quad m_2 = -h/a_1 a_2, \quad m_3 = h/a_3 a_1.$$

It follows from the expressions for the d_{ij} that, up to a squared factor, the discriminant of d_{ij} is equal to $-a_i a_j$; thus in particular d_{ij} has no repeated linear factor and it is a product of two linear factors over k if and only if $-a_i a_j$ is in k^{*2}. If i, j, k is a cyclic permutation of $1, 2, 3$ then

$$d_{0i}/d_{jk} = a_0 a_i/h = h/a_j a_k.$$

Moreover the resultant of d_{ij} and d_{ik} is $-4a_i^2 a_j a_k$, so that d_{ij} and d_{ik} cannot have a common root. The pencil (50) has six singular fibres, given by the roots of $d_{01} d_{02} d_{03} = 0$, and each singular fibre consists of four lines which form a skew quadrilateral. Thus each of the 48 lines on (51) is part of a singular fibre of one of the two pencils on V.

Martin Bright's thesis contains a dictionary which gives the Néron-Severi group of (51) over any field k. This group has rank at least 2 whenever (52) holds; subject to (52), it has rank greater than 2 if and only if up to fourth powers there is a relation of the form $a_j = 4a_i$ or $a_j = -a_i$ or $a_i a_j = a_k a_\ell$.

In order to apply Theorem 11.2, we must know when Condition D holds, and we must evaluate the relevent Legendre-Jacobi functions. This is where a splitting of cases becomes necessary. In what follows, we confine ourselves to the case when none of the $-a_i a_j$ is in k^{*2}, which is equivalent to requiring that all the d_{ij} are irreducible over k.

Lemma 12.1. *Suppose that no $-a_i a_j$ is in k^{*2}. Then for any $\mathbf{m}$ which does not satisfy Condition D, one of $\mathbf{m}$ and $\mathbf{mm}^0$ can be chosen to be independent of y and z. Moreover the group of such $\mathbf{m}$ has order exactly 8 (and consists of the inescapable part of the 2-Selmer group) if and only if $a_0 a_1 a_2 a_3$ is not a fourth power and no $a_i a_j$ is a square.*

What happens in the exceptional cases is as follows. If for example $a_2 a_3$ is a square then $(1, -a_1 a_2, -a_1 a_2)$ does not satisfy Condition D. Again, if h is in $-k^{*2}$ then $(a_1 a_3, a_1 a_2, a_2 a_3)$ does not satisfy Condition D, whereas if h is in k^{*2} then $(a_1 a_2, a_2 a_3, a_3 a_1)$ does not satisfy Condition D. In each of these cases, the group of inescapable elements of the 2-Selmer group acquires one extra generator, which is the $\mathbf{m}$ just listed; and this provides a straightforward description of Condition E. If some $a_i a_j$ and one of $\pm h$ are both squares, then we acquire two extra generators in this way.

We can now state the main result of this subsection, which is simply the specialization of Theorem 11.2 to our case, and which therefore requires no further proof. The set $\mathscr{S}$ of bad places consists of the infinite places, the primes which divide $2a_0a_1a_2a_3$ and a basis for the ideal class group of k. Denote by $\mathscr{A}$ the closure of the set of points $\alpha \times \beta$ in $\mathscr{N}^2$ at which (50) is locally soluble for $y = \alpha, z = \beta$ at each place of $\mathscr{S}$ and all the Legendre-Jacobi conditions associated with any pencil of conics (53) hold.

Theorem 12.2. *Suppose that (51) is everywhere locally soluble and such that $a_0a_1a_2a_3$ is a square, and that no $-a_ia_j$ is in k^{*2}. Assume Schinzel's Hypothesis and the finiteness of* Ⅲ. *If $\mathscr{A}$ is not empty and Condition D holds, then (51) contains rational points.*

As was remarked at the end of Sect. 11, we can here replace Condition D by the weaker Condition E.

The solubility of the pencil of conics (53) is equivalent to three Legendre-Jacobi conditions, of which a typical one is

$$L(\mathscr{B}; -d_{i\ell}d_{j\ell}, d_{k\ell}) = 1. \tag{56}$$

There are 12 conditions of this kind, but they are not all independent. Indeed in the notation of Lemma 8.3 the continuous conditions, which form a subgroup there called Λ_0, are all Brauer-Manin; and Bright's table shows that in the most general case satisfying (52) there is only one algebraic Brauer-Manin condition. In general the twelve conditions of the form (56) all reduce to $F_{12}F_{23}F_{31} = 1$ where

$$F_{ij} = L(\mathscr{B}; -d_{i\ell}d_{j\ell}, d_{k\ell}; \alpha, \beta) = L(\mathscr{B}; -d_{ik}d_{jk}, d_{\ell k}; \alpha, \beta).$$

If however one of the a_ia_j is a square then the corresponding condition $F_{ij} = 1$ is also in Λ_0. The remarks which follow Lemma 12.1 show that Condition D cannot then hold, but Condition E may still hold in some part of $\mathscr{A}$. One can evaluate the F_{ij} by using the formulae which follow (24). Of the surfaces (51) defined over $\mathbf{Q}$ which satisfy (52) and have the a_i integral with each $|a_i| < 16$, there are just two which are everywhere locally soluble but do not have a solution in $\mathbf{Q}$. They are

$$2X_0^4 + 9X_1^4 = 6X_2^4 + 12X_3^4 \quad \text{and} \quad 4X_0^4 + 9X_1^4 = 8X_2^4 + 8X_3^4.$$

The first of these fails the condition $F_{12}F_{23}F_{31} = 1$ and the second has a_0a_1 square and fails the condition $F_{01} = 1$.

Using the methods of Cassels [58] we can carry out a second descent on some of the surfaces considered in this subsection, and thereby prove that certain equations (51) are insoluble or do not admit weak approximation. The prettiest result that has been obtained in this way is as follows.

Lemma 12.3. *Suppose that $X_0^4 + 4X_1^4 = W_0^2 - 2W_1^2$ for X_0, X_1, W_0, W_1 in $\mathbf{Z}$ such that no prime $p \equiv 7 \mod 8$ divides both W_0 and W_1. Then $|W_0| \not\equiv 5$ or $7 \mod 8$.*

This is a weak approximation property, but of a rather unusual sort; and it appears unlikely that it corresponds to a Brauer-Manin condition.

12.2   Some Kummer Surfaces

In this subsection we consider Kummer surfaces of the form

$$Z^2 = f^{(1)}(X)f^{(2)}(Y) \tag{57}$$

defined over an algebraic number field k, where the $f^{(i)}$ are separable quartic polynomials. For (57) to be everywhere locally soluble, for each place v of k there must exist c_v in k_v^* such that both the equations

$$U^2 = c_v f^{(1)}(X) \text{ and } V^2 = c_v f^{(2)}(Y)$$

are soluble in k_v. For (57) to be soluble in k requires the stronger condition that there exists c in k^* such that both the equations

$$U^2 = c f^{(1)}(X) \text{ and } V^2 = c f^{(2)}(Y) \tag{58}$$

are everywhere locally soluble. For the existence of the c_v to imply the existence of c is a local-to-global theorem, and the obstruction to this turns out to be a Brauer-Manin obstruction. In my view, this is the most interesting feature of the whole argument.

To be able to use the methods of Sect. 11 on the pair of equations (58), we must require that their Jacobians each have all their 2-division points rational. In this case it turns out that the Brauer-Manin obstruction introduced in the previous paragraph is trivial; in other words, we can always find the c that we need. Call one such value c^0; then we can replace c^0 by any c which is close to c^0 at all the bad places of (57) and is such that the good primes $\mathfrak{p}$ which divide it to an odd power are such that both equations (58) are soluble in $k_{\mathfrak{p}}$. To prove solubility of (57) we introduce well-chosen primes into c in such a way as to reduce the orders of the 2-Selmer groups of both the underlying Jacobians to 8. This requires some intricate but elementary arguments; and for these we need to assume that for each Jacobian there are two primes which are bad in a specified way for that curve but which are good for the other. Full details can be found in [23], but this is definitely not recommended reading.

12.3   Diagonal Cubic Surfaces

In this subsection we consider diagonal cubic surfaces

$$a_0 X_0^3 + a_1 X_1^3 = a_2 X_2^3 + a_3 X_3^3 \tag{59}$$

over certain algebraic number fields k. Without loss of generality we can assume that the a_i in (59) are integers. To show that (59) has a rational solution it is enough to show that there exists c in k^* such that each of the two curves

$$a_0 X_0^3 + a_1 X_1^3 = c X^3, \quad \text{and} \quad a_2 X_2^3 + a_3 X_3^3 = c X^3 \tag{60}$$

is soluble. The hypothesis that (59) is everywhere locally soluble implies that for each place v in k there exists c_v in k_v^* such that each of

$$a_0 X_0^3 + a_1 X_1^3 = c_v X^3, \quad \text{and} \quad a_2 X_2^3 + a_3 X_3^3 = c_v X^3$$

is soluble in k_v. The first step in the argument is to deduce from the existence of the c_v the existence of c in k^* such that each of the two equations (60) is everywhere locally soluble. In contrast with what happened in the previous subsection, such a c always exists; and indeed if $\mathscr{S}$ is any given finite set of places of k, we can choose c integral and such that c/c_v is in k_v^{*3} for each v in $\mathscr{S}$. Following the methods of Sect. 11, we denote by $\mathbf{L}^1$ the affine line with the origin deleted. Let $\mathscr{S}$ be a set of bad places for the surface (59), which means that $\mathscr{S}$ must contain all the primes of k dividing $3a_0 a_1 a_2 a_3$ and a basis for the ideal class group of k; and let $\mathscr{B} \supset \mathscr{S}$ be a set of bad places for the pair of curves (60), so that $\mathscr{B}$ must also contain all the primes dividing c. Under the topology induced by $\mathscr{S}$, let $\mathscr{A}$ be the open subset of $\mathbf{L}^1(k)$ on which each of the two curves (60) is locally soluble at each place of $\mathscr{S}$, let c_0 be a given point of $\mathscr{A}$ and let $\mathscr{N}_0 \subset \mathscr{A}$ be an open neighbourhood of c_0. Because of the possible presence of Brauer-Manin obstructions, it is not necessarily true that there exists c in $\mathscr{N}_0$ such that the two equations (60) are both soluble. But one may still ask what additional assumptions are needed in order to prove solubility by the methods of Sect. 11 – always of course on the basis that III is finite.

The Jacobians of the two curves (60) are

$$Y_0^3 + Y_1^3 = a_0 a_1 c Y^3 \quad \text{and} \quad Y_2^3 + Y_3^3 = a_2 a_3 c Y^3 \tag{61}$$

respectively. The obvious descent to apply to each of them is the ρ-descent, where $\rho = \sqrt{-3}$. Applying this to the elliptic curve

$$X^3 + Y^3 = A Z^3 \tag{62}$$

replaces it by the equations

$$\rho X + \rho^2 Y = m_1 Z_1^3, \quad \rho^2 X + \rho Y = m_2 Z_2^3, \quad X + Y = A Z_3^3 / m_1 m_2$$

where $Z = Z_1 Z_2 Z_3$. Here m_1, m_2, Z_1, Z_2 are in $K = k(\rho)$ and if ρ is not in k then m_1, m_2 are conjugate over k, as are Z_1, Z_2; but Z_3 is in k. It would appear natural to work in K rather than k, since if (59) is soluble in K it is soluble in k. But actually our methods could not then be applied, for complex multiplication by ρ induces an isomorphism on (62), so that the Mordell-Weil group of (62) over K has an even number of generators of infinite order and there is no possibility of applying Lemma 4.6. Thus a prerequisite for applying the methods of Sect. 11 is the unexpected constraint:

$$\sqrt{-3} \text{ is not in } k. \tag{63}$$

This does however allow us to take $k = \mathbf{Q}$, for example. But even if (63) holds, there is considerable interplay between the descent theory over K and that over k; and it seems necessary to make use of this interplay in the argument.

The basic idea is to write c as a product of primes in $\mathscr{S}$ (which are forced on us by the choice of $\mathscr{N}_0$) and some other well-chosen primes; the latter make up the set $\mathscr{B} \setminus \mathscr{S}$. We need to choose the latter so that the ρ-Selmer group of each of the curves (61) has order 9; and following the precedent of Sect. 11 we expect to do this by adjoining additional primes one by one to $\mathscr{B}$, always preserving the local solubility of the curves (60) and keeping c within $\mathscr{N}_0$. The latter condition simply means that each new prime $\mathfrak{p}$ should be close to 1 in our topology and should be such that a_0/a_1 and a_2/a_3 are in $k_{\mathfrak{p}}^{*3}$. But here we encounter the final pair of complications. To adjoin one more prime divides or multiplies the order of each ρ-Selmer group by 3. If one of these orders has already been reduced to 9 we cannot reduce it further; so adjoining one more prime can no longer improve the situation. Instead we eventually reach the stage when we have to adjoin two more primes simultaneously, in such a way that the order of one of the ρ-Selmer groups remains unchanged, while the order of the other is divided by 9. To be able to reduce the orders of both ρ-Selmer groups to 9, we therefore need the initial choice of c to satisfy the following additional condition:

The product of the orders of the ρ-Selmer groups of the two curves (61) is a power of 9.

As should be clear from the preceding discussion, the truth or falsehood of this statement depends only on $\mathscr{N}_0$ (provided it is small enough) and not on the value of c within $\mathscr{N}_0$. In other words, it depends only on the choice of c_0; and we need to show that we can choose c_0 so that (in addition to the previous requirements) this condition holds at c_0. Having done all this, we still need the equivalent of Condition D or Condition E.

However, at the end of all these complications we do obtain Theorem 12.4 below; the full details of the proof can be found in [18]. The sufficient conditions stated in Theorem 12.4 are clumsy and could certainly be improved; but I do not believe that this method is powerful enough to replace them by the Brauer-Manin conditions.

Theorem 12.4. *Let k be an algebraic number field not containing the primitive cube roots of unity. Assume that III is always finite. If (59) is everywhere locally soluble and the a_i are all cubefree, then each of the following criteria is sufficient for the solubility of (59) in k.*

(i) *There exist primes $\mathfrak{p}_1, \mathfrak{p}_3$ of k not dividing 3 such that a_1 is a non-unit at $\mathfrak{p}_1$ and a_3 is a non-unit at $\mathfrak{p}_3$, but for $j = 1$ or 3 the three a_i with $i \neq j$ are units at $\mathfrak{p}_j$.*

(ii) *There is a prime $\mathfrak{p}$ of k not dividing 3 such that a_1 is a non-unit at $\mathfrak{p}$ but the other a_i are units there; and a_2, a_3, a_4 are not all in the same coset of $k_\mathfrak{p}^{*3}$.*

(iii) *There is a prime $\mathfrak{p}$ of k not dividing 3 such that exactly two of the a_i are units at $\mathfrak{p}$, and (59) is not birationally equivalent to a plane over $k_\mathfrak{p}$.*

13   The Case of One Rational 2-Division Point

It is possible to carry out a 2-descent without using information about a field extension provided that the elliptic curve involved has one rational 2-division point – though it is then necessary to implement the process in two stages. The details of this process have been worked out, with increasing degrees of sophistication, in [15, 59] and Chap. II of [14]. I sketch it in this section.

We are concerned with pencils of 2-coverings whose pencil of Jacobians has the form

$$Y^2 = (X - c(U,V))(X^2 - d(U,V))$$

where c, d are homogeneous polynomials in $k[U,V]$ with $\deg d = 2 \deg c$. We start by recalling the standard machinery for 2-descent on

$$E' : Y^2 = (X - c)(X^2 - d)$$

for c, d in k and d not in k^2.

If O' is the point at infinity on E' and P' the 2-division point $(c, 0)$ then there is an isogeny $\phi' : E' \to E'' = E'/\{O', P'\}$ where E'' is

$$E'' : Y_1^2 = (X_1 + 2c)(X_1^2 + 4(d - c^2));$$

the places of bad reduction for E'' are the same as those for E'. Explicitly, ϕ' is given by

$$X_1 = \frac{d - X^2}{c - X} - 2c, \quad Y_1 = \frac{Y(X^2 - 2cX + d)}{(X - c)^2}.$$

There is also a dual isogeny $\phi'' : E'' \to E'$, and $\phi'' \circ \phi'$ and $\phi' \circ \phi''$ are the doubling maps on E' and E'' respectively. We are primarily interested in the case when neither d nor $c^2 - d$ is a square in k, so that E' and E'' each contain only one primitive 2-division point defined over k.

The elements of $\mathrm{H}^1(k, \{O', P'\}) \sim k^*/k^{*2}$ classify the ϕ'-coverings of E''; the covering corresponding to the class of m' is

$$V_1^2 = m'(X_1 + 2c), \quad V_2^2 = m'(X_1^2 + 4(d - c^2)) \tag{64}$$

with the obvious two-to-one map to E''. The ϕ'-covering corresponding to P'' is given by $m' = d$. Similarly the ϕ''-coverings of E' are classified by the elements of $\mathrm{H}^1(k, \{O'', P''\}) \sim k^*/k^{*2}$, the covering corresponding to the class of m'' being

$$W_1^2 = m''(X - c), \quad W_2^2 = m''(X^2 - d). \tag{65}$$

The ϕ''-covering corresponding to P' is given by $m'' = c^2 - d$. We denote by S_2' the 2-Selmer group of E', and by S_ϕ', S_ϕ'' the ϕ'-Selmer group of E'' and the ϕ''-Selmer group of E' respectively.

Write $K = k(d^{1/2})$; then the group of 2-coverings of E' is naturally isomorphic to K^*/K^{*2}, where the 2-covering corresponding to the class of $a + bd^{1/2}$ is given by

$$Z_1^2 = (a^2 - db^2)(X - c), \quad (Z_2 \pm d^{1/2}Z_3)^2 = (a \pm bd^{1/2})(X \pm d^{1/2}).$$

In homogeneous form, this can be written

$$\left.\begin{aligned}
Z_2^2 + dZ_3^2 &= aZ_1^2/(a^2 - db^2) + (ac + bd)Z_0^2, \\
2Z_2Z_3 &= bZ_1^2/(a^2 - db^2) + (a + bc)Z_0^2.
\end{aligned}\right\} \tag{66}$$

Call this curve Γ'; then the map $\Gamma' \to E'$ has degree 4 and is given by

$$X = \frac{Z_1^2}{(a^2 - db^2)Z_0^2} + c, \quad Y = \frac{Z_1(Z_2^2 - dZ_3^2)}{(a^2 - db^2)Z_0^3}.$$

The map $\Gamma' \to E'$ can be factorized as $\Gamma' \to C'' \to E'$, where C'' is the ϕ''-covering of E' given by (65) with $m'' = a^2 - db^2$ and the map $\Gamma' \to C''$ is

$$W_1 = Z_1/Z_0, \quad W_2 = (Z_2^2 - dZ_3^2)/Z_0^2.$$

Conversely, suppose that we have a curve of genus 1 defined over k and given by the equations

$$\left.\begin{aligned}
\alpha_0 U_0^2 + \alpha_1 U_1^2 + \alpha_2 U_2^2 + \alpha_3 U_3^2 + 2\alpha_4 U_2 U_3 &= 0, \\
\beta_0 U_0^2 + \beta_1 U_1^2 + \beta_2 U_2^2 + \beta_3 U_3^2 + 2\beta_4 U_2 U_3 &= 0,
\end{aligned}\right\} \tag{67}$$

where the α_i, β_i are in $\mathfrak{o}$. We have just seen that any 2-covering of an elliptic curve with one rational 2-division point can be put in this form, and we shall now prove the converse. Write $d_{ij} = \alpha_i\beta_j - \alpha_j\beta_i$; then the curve (67) takes the more convenient form

$$\left.\begin{aligned}
d_{10}U_0^2 + d_{12}U_2^2 + 2d_{14}U_2 U_3 + d_{13}U_3^2 &= 0, \\
d_{01}U_1^2 + d_{02}U_2^2 + 2d_{04}U_2 U_3 + d_{03}U_3^2 &= 0.
\end{aligned}\right\} \tag{68}$$

If we write $U_0 = 2Z_0(d_{14}^2 - d_{12}d_{13})$ and $U_1 = Z_1/4d_{34}(d_{14}^2 - d_{12}d_{13})$, this last pair of equations can be identified with (66) provided that

$$\begin{aligned}
a = -2(2d_{14}d_{34} + d_{13}d_{23})(d_{14}^2 - d_{12}d_{13}), \quad b &= d_{01}^{-1}d_{13}(d_{14}^2 - d_{12}d_{13}), \\
c = 4d_{04}d_{14} - 2d_{02}d_{13} - 2d_{03}d_{12}, \quad d &= 4d_{01}^2(d_{23}^2 + 4d_{24}d_{34});
\end{aligned}$$

it also follows from these that

$$c^2 - d = 16(d_{04}^2 - d_{02}d_{03})(d_{14}^2 - d_{12}d_{13}),$$
$$m'' = a^2 - db^2 = 16d_{34}^2(d_{14}^2 - d_{12}d_{13})^3.$$

We assume that $d(c^2 - d) \neq 0$, so that (67) defines a nonsingular curve of genus 1.

Now let $\mathscr{S}$ be a finite set of places which contains the infinite places, the primes which divide 2, the odd primes of bad reduction for E' (or E'') and a set of generators for the ideal class group of k. For any v in $\mathscr{S}$ we write

$$V'_v = \mathrm{H}^1(k_v, \{O', P'\}) \sim k_v^* / k_v^{*2}$$

and similarly for V''_v; and we denote by W'_v the image of $E''(k_v)/\phi'E'(k_v)$ in V'_v and similarly for W''_v. Thus m' lies in W'_v if and only if Γ' is soluble over k_v, and similarly for W''_v. There is a non-degenerate canonical pairing

$$V'_v \times V''_v \to \{\pm 1\} \tag{69}$$

induced by the Hilbert symbol, under which the orthogonal complement of W'_v is W''_v. As in Sect. 10, we write

$$V'_{\mathscr{S}} = \oplus_{v \in \mathscr{S}} V'_v, \quad W'_{\mathscr{S}} = \oplus_{v \in \mathscr{S}} W'_v$$

and similarly for V'' and W''. The machinery in the first half of Sect. 10 needs to be modified to take account of the changed circumstances, but the proofs involve no new ideas.

Lemma 13.1. *Let $\mathscr{S}_0$ consist of the infinite places, the even primes, and a set of generators for the ideal class group of k. For each v in $\mathscr{S}$ there exist subspaces $K'_v \subset V'_v$ and $K''_v \subset V''_v$ such that*

(i) *K''_v is the orthogonal complement of K'_v under the pairing (69);*
(ii) *$V'_{\mathscr{S}} = U'_{\mathscr{S}} \oplus K'_{\mathscr{S}}$ and $V''_{\mathscr{S}} = U''_{\mathscr{S}} \oplus K''_{\mathscr{S}}$ where $U'_{\mathscr{S}}, U''_{\mathscr{S}}$ are the images of $X_{\mathscr{S}} \times X_{\mathscr{S}} = (\mathfrak{o}_{\mathscr{S}}^*/\mathfrak{o}_{\mathscr{S}}^{*2})^2$ in $V'_{\mathscr{S}}$ and $V''_{\mathscr{S}}$ respectively;*
(iii) *If v is not in $\mathscr{S}_0$ we can take K'_v and K''_v to be the images of $(\mathfrak{o}_v^*/\mathfrak{o}_v^{*2})^2$.*

It follows from (69) that there is a non-degenerate canonical pairing

$$V'_{\mathscr{S}} \times V''_{\mathscr{S}} \to \{\pm 1\} \tag{70}$$

and from (i) that $K''_{\mathscr{S}} = \oplus_{v \in \mathscr{S}} K''_v$ is the orthogonal complement of $K'_{\mathscr{S}}$ under this pairing.

Lemma 13.2. *If $\mathscr{S} \supset \mathscr{S}_0$ then S'_ϕ is isomorphic to each of $U'_{\mathscr{S}} \cap W'_{\mathscr{S}}$, the left kernel of the map $U'_{\mathscr{S}} \times W''_{\mathscr{S}} \to \{\pm 1\}$ induced by (70), and the left kernel of the map $W'_{\mathscr{S}} \times U''_{\mathscr{S}} \to \{\pm 1\}$ induced by (70). A similar result holds for S''_ϕ.*

Let $t'_{\mathscr{S}} : V'_{\mathscr{S}} \to U'_{\mathscr{S}}$ be the projection along $K'_{\mathscr{S}}$ and similarly for $t''_{\mathscr{S}}$. We now diverge from the notation of Sect. 10, writing

$$\mathbf{U}'_{\mathscr{S}} = U'_{\mathscr{S}} \cap (W'_{\mathscr{S}} + K'_{\mathscr{S}}), \quad \mathbf{W}'_{\mathscr{S}} = W'_{\mathscr{S}} / (W'_{\mathscr{S}} \cap K'_{\mathscr{S}})$$

and similarly for $\mathbf{U}''_{\mathscr{S}}$ and $\mathbf{W}''_{\mathscr{S}}$; as in Sect. 10, the map $t'_{\mathscr{S}}$ induces an isomorphism $\tau'_{\mathscr{S}} : \mathbf{W}'_{\mathscr{S}} \to \mathbf{U}'_{\mathscr{S}}$, and there is an analogous isomorphism $\tau''_{\mathscr{S}} : \mathbf{W}''_{\mathscr{S}} \to \mathbf{U}''_{\mathscr{S}}$. The pairing (70) induces pairings

$$\mathbf{U}'_{\mathscr{S}} \times \mathbf{W}''_{\mathscr{S}} \to \{\pm 1\}, \quad \mathbf{W}'_{\mathscr{S}} \times \mathbf{U}''_{\mathscr{S}} \to \{\pm 1\} \tag{71}$$

and the action of $\tau'_{\mathscr{S}} \times (\tau''_{\mathscr{S}})^{-1}$ takes the first pairing into the second. The left kernel of either of these pairings is isomorphic to S'_ϕ and the right kernel to S''_ϕ. The action of $\tau'_{\mathscr{S}} \times 1$ takes the first pairing into the pairing .

$$\mathbf{W}'_{\mathscr{S}} \times \mathbf{W}''_{\mathscr{S}} \to \{\pm 1\}.$$

Our objective is to prove the solubility in k of pencils of curves (67) under suitable conditions. The appropriate modification of Lemma 11.1 is as follows.

Lemma 13.3. *Suppose that P' is the only primitive 2-division point of E' defined over k and similarly for P'' on E''. If the orders of S'_ϕ and S''_ϕ are 2 and 4 respectively then the order of S'_2 is at most 4.*

Proof. Let Γ' be a 2-covering of E' and denote by C'' the quotient of Γ' by the action of the group $\{O', P'\}$; then C'' is a ϕ''-covering of E' and we have a commutative diagram

$$
\begin{array}{ccccc}
E' & \xrightarrow{\phi'} & E'' & \xrightarrow{\phi''} & E' \\
\| & & \| & & \| \\
\Gamma' & \longrightarrow & C'' & \longrightarrow & E'
\end{array}
$$

where the first two vertical double lines mean that Γ' and C'' are principal homogeneous spaces for E' and E'' respectively. If Γ' is identified with the element f of $\mathrm{H}^1(k, E'[2])$ then C'' is identified with $\phi' \circ f$ as an element of $\mathrm{H}^1(k, E''[\phi''])$. If Γ' is in S'_2 then C'' is in S''_ϕ; so we can construct all the elements of S'_2 by lifting back the elements of S''_ϕ. But by hypothesis P'' is not in $\phi' E'(k)$, so the two elements of S'_ϕ must correspond to the points O'' and P'' as members of $E''(k)/\phi' E'(k)$; hence regarded as elements of S'_2 they are equivalent. In other words, E'' regarded as an element of S''_ϕ lifts back to only one element of S'_2; so the same is true of each element of S''_ϕ. $\square$

We now have to study simultaneously the ϕ'-descent on E'' and the ϕ''-descent on E'. As in Sect. 11, by introducing a sequence of well-chosen primes we reduce S'_ϕ and S''_ϕ until we can apply Lemma 13.3; but the process is more complicated than in Sect. 11. The strongest version of the argument is due to Wittenberg [14]; assuming Schinzel's Hypothesis and the finiteness of III, to prove solubility he needs little more than the triviality of the Brauer-Manin obstruction.

14 Del Pezzo Surfaces of Degree 4

Let V be a Del Pezzo surface of degree 4 (that is, the smooth intersection of two quadrics in $\mathbf{P}^4$) defined over an algebraic number field k. Salberger and Skoroboga-tov [60] have shown that the only obstruction to weak approximation on V is the Brauer-Manin obstruction, and a more elementary proof can be found in [61].

Theorem 14.1. *Suppose that $V(k)$ is not empty. Let $\mathscr{A}$ be the subset of the adelic space $V(\mathbf{A})$ consisting of the points $\prod P_v$ such that*

$$\sum \mathrm{inv}_v(A(P_v)) = 0 \text{ in } \mathbf{Q}/\mathbf{Z}$$

for all A in the Brauer group $\mathrm{Br}(V)$. *Then the image of $V(k)$ is dense in $\mathscr{A}$.*

The idea behind the proof in [61] is that we can use the existence of a point of $V(k)$ to fibre V by conics. Theorem 9.4 now allows us to find a positive 0-cycle of degree 8 on V defined over k satisfying pre-assigned approximation conditions; and the proof is then completed by a modification of an argument of Coray [62]. Coray's result is Theorem 14.3 below; it will probably turn out to be a fundamental tool in the Diophantine theory of Del Pezzo equations of degree 4. Lemma 14.2 is weaker than Theorem 14.3, but appears to be a necessary step in the proof of the latter.

Lemma 14.2. *Let V be a Del Pezzo surface of degree 4, defined over a field L of characteristic 0. If V contains a positive 0-cycle of degree 2 and a positive 0-cycle of odd degree n, both defined over L, then $V(L)$ is not empty.*

Proof. We can suppose V embedded in $\mathbf{P}^4$ as the intersection of two quadrics. We proceed by induction on n. If the given 0-cycle of degree 2 consists of the two points P' and P'' then we can suppose that they are conjugate over L and distinct, because otherwise the lemma would be trivial. By a standard result, there are infinitely many points on V defined over $L(P')$ and hence infinitely many positive 0-cycles of degree 2 defined over L. Choose d so that

$$2d(d+1) > n > 2d(d-1)$$

and let $\{P'_i, P''_i\}$ be $\frac{1}{2}\{2d(d+1) - n - 1\}$ distinct pairs of points of V, the points of each pair being conjugate over L. The hypersurfaces of degree d cut out on V a system of curves of dimension $2d(d+1)$; hence there is at least a pencil of such curves passing through the P'_i and P''_i and the points of the given 0-cycle of degree n, and this pencil is defined over L. We have accounted for $2d(d+1) - 1$ of the $4d^2$ base points of the pencil; so the remaining ones form a positive 0-cycle of degree $2d(d-1) + 1$ defined over L. This completes the induction step unless $n = 2d(d-1) + 1$.

In this latter case we must have $d > 1$ because if $d = 1$ then $n = 1$ and the lemma is already proved; hence $2d(d+1) - n - 1 = 4d - 2 \geq 6$. Instead of the previous construction we now choose our pencil of curves to have double points at P'_0 and P''_0 and to pass through $\frac{1}{2}\{2d(d+1) - n - 7\}$ other pairs P'_i, P''_i as well as through the points of the given 0-cycle of degree n. In this case each of P_0' and P_0'' is a base point of the pencil with multiplicity 4; so we have accounted for $2d(d+1)+1$ of the base points of the pencil, and the remaining ones form a positive 0-cycle of degree $2d(d-1)-1$ defined over L. This completes the induction step in this case. $\square$

Theorem 14.3. *Let V be a del Pezzo surface of degree 4, defined over a field L of characteristic 0. If V contains a 0-cycle of odd degree defined over L then $V(L)$ is not empty.*

Proof. By decomposing the 0-cycle into its irreducible components, we can assume that V contains a positive 0-cycle $\mathfrak{a}$ of odd degree defined over L. We can write V as the intersection of two quadrics, each defined over L; let W be one of them. We can find a field $L_1 \supset L$ with $[L_1 : L] \leq 2$ and a point P on W defined over L_1. The lines on W through P are parametrised by the points of a conic, so we can find a field $L_2 \supset L_1$ with $[L_2 : L_1] \leq 2$ and a line ℓ on W, passing through P and defined over L_2. The intersection of this line with another quadric containing V cuts out on V a positive 0-cycle of degree 2 defined over L_2. Applying Lemma 14.2 to $\mathfrak{a}$ and this 0-cycle, we obtain a point P_2 on V defined over L_2. Repeating this argument for $\mathfrak{a}$ and the positive 0-cycle of degree 2 consisting of P_2 and its conjugate over L_1, we obtain a point P_1 on V defined over L_1; and one further repetition of the argument gives us a point on V defined over L. $\square$

To use and then collapse a field extension in this way is a device which probably has a number of uses. For such a collapse step to be feasible, the degree of the field extension needs to be prime to the degree of the variety; and this leads one to phrase the same property somewhat differently.

Question 14.4. Let V be a variety defined over a field K, not necessarily of a number-theoretic kind. For what families of V is it true that if V contains a 0-cycle of degree 1 defined over K then it contains a point defined over K?

As stated above, this is true for Del Pezzo surfaces of degree 4. For pencils of conics it is in general false, even for algebraic number fields K, as was shown at the end of Sect. 9. For Del Pezzo surfaces of degree 3 the question is open: I expect it to be true for algebraic number fields K but false for general fields.

The methods of Sect. 13 have enabled Wittenberg [14] to prove the solubility of almost all Del Pezzo surfaces of degree 4 on which there is no Brauer-Manin obstruction. His starting point is as follows. Let V be a nonsingular Del Pezzo surface of degree 4, defined over an algebraic number field k and everywhere locally soluble. Then, after a field extension of odd degree, we can exhibit a family of hyperplane sections of V which is of the form considered in Sect. 13. This family is parametrised by the points of $\mathbf{P}^3$ blown up along a certain curve and at four other points. The construction, which was first sketched in [15], is as follows.

The surface V is the base locus of a pencil of quadrics; because V is nonsingular, the pencil contains exactly 5 cones defined over $\bar{k}$ and these are all distinct. Hence one at least of them is defined over a field k_1 which is of odd degree over k; and by Theorem 14.3 it is enough to ask whether V contains points defined over k_1. Henceforth we work over k_1. After a change of variables, we can assume that the singular quadric just described has vertex $(1,0,0,0,0)$ and therefore an equation of the form $f(X_1,X_2,X_3,X_4) = 0$. By absorbing multiples of the other X_i into X_0, we can write V in the form

$$f(X_1,X_2,X_3,X_4) = 0, \quad aX_0^2 + g(X_1,X_2,X_3,X_4) = 0 \tag{72}$$

with $a \neq 0$. Now let P be any point on $X_0 = 0$, let Q be the quadric of the pencil (72) which passes through P, and let Π be the tangent hyperplane to Q at P. For general P the curve of genus 1 in which Π meets V can be put in the form (68) and hence is of the type considered in Sect. 13. For provided that P does not lie on $f = 0$, by a further change of variables we can take P to be $(0,1,0,0,0)$ and require

$$f(X_1,X_2,X_3,X_4) = bX_1^2 + f_1(X_2,X_3,X_4).$$

The equation of Q has no term in X_1^2, so by a further change of variables we can take it to have the form

$$aX_0^2 + cX_1X_4 + h(X_2,X_3,X_4) = 0 \tag{73}$$

with $c \neq 0$; this is equivalent to requiring the equation of Π to be $X_4 = 0$. Since V is given by $f = 0$ and (73), its intersection with $X_4 = 0$ has the required form.

This construction breaks down if P lies on V or is the vertex of one of the other singular quadrics of the pencil, because then Π is no longer well-defined. To remedy this, what we do is to choose a point P on $X_0 = 0$ together with a hyperplane Π which touches at P some quadric of the pencil (72). Thus P should be considered as a point of the variety W obtained by blowing up $X_0 = 0$ (which can be identified with $\mathbf{P}^3$) along the curve $V \cap \{X_0 = 0\}$ and at the vertices of the other four singular quadrics of the pencil.

Denote by U the variety over W whose fibres are the curves $V \cap \Pi$ in the construction above; then what we have obtained is a diagram

$$W \longleftarrow U \longrightarrow V$$

in which the left hand map is a fibration. The right hand map here is not a fibration, and it seems unlikely that there is even a subvariety of U on which the restriction of the map is a fibration. But this is not important. What matters is the existence of a section – that is, a map $V \to U$ such that the composite map $V \to U \to V$ is the identity; and for this we only need the map $V \to U$ to be rational rather than everywhere defined. In the notation of (72) let $P_0 = (x_0, \ldots, x_4)$ be a point of V with $x_0 \neq 0$, and choose $P = (0, x_1, x_2, x_3, x_4)$. The equation of Π has no term in X_0; hence since P lies on Π so does P_0. This defines the rational map $V \to U$. Provided V is everywhere locally soluble, so is U. If we can find a field extension k_2/k_1 of odd degree such that U is soluble in k_2, then V will also be soluble in k_2 and two applications of Theorem 14.3 will show that V is soluble in k.

We cannot apply Wittenberg's results cited in Sect. 13 to W directly, because W is too big; but it is simple enough to find a line L defined over k_1 in the $\mathbf{P}^3$ which underlies W such that

- L is in sufficiently general position, and
- The inverse image of L in U is everywhere locally soluble.

To do this, we choose any P_1 on $X_0 = 0$ and defined over k_1. The fibre above P_1 is locally soluble except at a finite set $\mathscr{S}$ of places. For each of these places there is a point of U in the corresponding local field, and this maps down to a point of $\mathbf{P}^3$. Using weak approximation on $\mathbf{P}^3$ we can therefore find a point P_2 in $\mathbf{P}^3$ such that the fibre above P_2 is locally soluble at each place in $\mathscr{S}$. We can now take L to be the line $P_1 P_2$.

References

1. Silverberg, A.: Open questions in arithmetic algebraic geometry. In: Arithmetic algebraic geometry, pp. 85–142. AMS, RI (2002)
2. Vaughan, R.C.: The Hardy-Littlewood method, 2nd edn. Cambridge (1997)
3. Davis, M., Matijasevič, Y., Robinson, J.: Hilbert's tenth problem: Diophantine equations: positive aspects of a negative solution. In: Mathematical developments arising from Hilbert problems (Proc. Sympos. Pure Math. Vol. XXVIII, Northern Illinois Univ., De Kalb, Ill., 1974), pp. 323–378. (loose erratum) Amer. Math. Soc., Providence, RI (1976)
4. Matiyasevich, Y.: Hilbert's tenth problem: what was done and what is to be done. In: Hilbert's tenth problem: relations with arithmetic and algebraic geometry (Ghent, 1999), Contemp. Math., vol. 270, pp. 1–47, Amer. Math. Soc., Providence, RI (2000)
5. Sir Swinnerton-Dyer P.: Weak approximation and R-equivalence on Cubic Surfaces. In: Peyre, E., Tschinkel, Y. (eds.) Rational points on algebraic varieties, Progress in Mathematics, vol. 199, pp. 357–404. Birkhäuser (2001)
6. Skorobogatov, A.N.: Beyond the Manin obstruction. Invent. Math. **135**, 399–424 (1999)
7. Harari, D.: Obstructions de Manin transcendantes. In: Number theory (Paris, 1993–1994). London Mathematical Society Lecture Note Series, vol. 235, pp. 75–87. Cambridge University Press, Cambridge (1996)
8. Bright, M., Sir Peter Swinnerton-Dyer: Computing the Brauer-Manin obstructions. Math. Proc. Camb. Phil. Soc. **137**, 1–16 (2004)
9. Manin, Yu.I.: Cubic forms. North-Holland, Amsterdam (1974)
10. Colliot-Thélène, J.-L., Sansuc, J.-J., Sir Peter Swinnerton-Dyer: Intersections of two quadrics and Châtelet surfaces. J. reine angew. Math. **373**, 37–107 (1987); **374**, 72–168 (1987)
11. Colliot-Thélène, J.-L., Sir Peter Swinnerton-Dyer: Hasse principle and weak approximation for pencils of Severi-Brauer and similar varieties. J. reine angew. Math. **453**, 49–112 (1994)
12. Sir Peter Swinnerton-Dyer: Rational points on pencils of conics and on pencils of quadrics. J. Lond. Math. Soc. **50**(2), 231–242 (1994)
13. Skorobogatov, A.: Torsors and rational points. Cambridge Tracts in Mathematics, vol. 144. Cambridge University Press, Cambridge (2001)
14. Wittenberg, O.: Intersections de deux quadriques et pinceaux de courbes de genre 1. Springer, Heidelberg (2007)
15. Bender, A.O., Sir Peter Swinnerton-Dyer: Solubility of certain pencils of curves of genus 1, and of the intersection of two quadrics in $\mathbf{P}^4$. Proc. Lond. Math. Soc. **83**(3), 299–329 (2001)

16. Colliot-Thélène, J.-L., Skorobogatov, A.N., Sir Peter Swinnerton-Dyer: Hasse principle for pencils of curves of genus one whose Jacobians have rational 2-division points. Invent. Math. **134**, 579–650 (1998)

17. Sir Peter Swinnerton-Dyer: Arithmetic of diagonal quartic surfaces II. Proc. Lond. Math. Soc. **80**(3), 513–544 (2000)

18. Sir Peter Swinnerton-Dyer: The solubility of diagonal cubic surfaces. Ann. Scient. Éc. Norm. Sup. **34**(4), 891–912 (2001)

19. Kleiman, S.L.: Algebraic cycles and the Weil conjectures. In: Grothendieck, A., Kuiper, N.H. (eds.) Dix exposés sur la cohomologie des schémas, pp. 359–386. North-Holland, Amsterdam (1968)

20. Serre, J.-P.: Facteurs locaux des fonctions zêta des variétés algébriques (définitions et conjectures). Séminaire Delange-Pisot-Poitou 1969/70, exp. 19

21. Borel, A.: Cohomologie de SL_n et valeurs de fonctions zeta aux points entiers. Ann. Scuola Norm. Sup. Pisa Cl. Sci. (4), **4**(4), 613–636 (1977)

22. Tate, J.T.: On the conjectures of Birch and Swinnerton-Dyer and a geometric analog. Sém. Bourbaki **306** (1966)

23. Swinnerton-Dyer, P.: The conjectures of Birch and Swinnerton-Dyer, and of Tate. In: Springer, T.A. (ed.) Proceedings of a Conference on Local Fields, Driebergen, 1966, pp. 132–157. Springer, NY (1967)

24. Rapaport, M., Schappacher, N., Schneider, P. (eds.): Beilinson's conjectures on special values of L-functions. Academic, NY (1988)

25. Hulsbergen, W.W.J.: Conjectures in arithmetic algebraic geometry. Vieweg, Braunschweig (1992)

26. Bloch, S.J.: Higher regulators, algebraic K-theory and zeta-functions of elliptic curves. CRM Monograph Series 11. AMS, RI (2000)

27. Bloch, S.J., Kato, K.: L-functions and Tamagawa numbers of motives. In: The Grothendieck Festschrift, vol. I, pp. 333–400. Birkhauser, Boston (1990)

28. Siegel, C.L.: Normen algebraischer Zahlen, Werke, Band IV, pp. 250–268

29. Raghavan, S.: Bounds for minimal solutions of Diophantine equations. Nachr. Akad. Wiss. Göttingen Math. -Phys. Kl. II, (9):109–114 (1975)

30. Faltings, G., Wüstholz, G.(eds.): Rational points, 3rd edn. Vieweg, Braunschweig (1992)

31. Mazur, B.: Rational isogenies of prime degree. Invent. Math. **44**, 129–162 (1978)

32. Merel, L.: Bornes pour la torsion des courbes elliptiques sur les corps de nombres. Invent. Math. **124**, 437–449 (1996)

33. Rubin, K., Silverberg, A.: Ranks of elliptic curves. Bull. Amer. Math. Soc. **39**, 455–474 (2002)

34. Gebel, J., Zimmer, H.G.: Computing the Mordell-Weil group of an elliptic curve over **Q**. In: Kisilevsky, H., Ram Murthy, M. (eds.) Elliptic curves and related topics. CRM Proceedings and Lecture Notes, vol. 4, pp. 61–83. AMS, RI (1994)

35. Gross, B., Kohnen, W., Zagier, D.: Heegner points and derivatives of L-series II. Math. Ann. **278**, 497–562 (1987)

36. Gross, B., Zagier, D.: Heegner points and derivatives of L-series. Invent. Math. **84**, 225–320 (1986)

37. Gross, B.: Kolyvagin's work for modular elliptic curves. In: Coates, J., Taylor, M.J. (eds.) L-functions and arithmetic, pp. 235–256. Cambridge (1991)

38. Kolyvagin, V.A.: Finiteness of $E(\mathbf{Q})$ and $III(E/\mathbf{Q})$ for a class of Weil curves. Izv. Akad. Nauk SSSR **52**, 522–540 (1988); translation in Math. USSR-Izv. **33**, 523–541 (1989)

39. Rubin, K.: Elliptic curves with complex multiplication and the conjecture of Birch and Swinnerton-Dyer. In: Viola, C. (ed.) Arithmetic theory of elliptic curves, pp. 167–234. Springer Lecture Notes 1716 (1999)

40. Beauville, A.: Complex algebraic surfaces, 2nd edn. London Mathematical Society Student Texts, 34, Cambridge University Press, Cambridge (1996)

41. Birch, B.J.: Homogeneous forms of odd degree in a large number of variables. Mathematika **4**, 102–105 (1957)

42. Hooley, C.: On nonary cubic forms. J. Reine Angew. Math. **386**, 32–98 (1988); **415**, 95–165 (1991); **456**, 53–63 (1994)
43. Browning, T.D.: An overview of Manin's conjecture for Del Pezzo surfaces. In: Duke, W., Tschinkel, Y. (eds.) Analytic number theory: A tribute to Gauss and Dirichlet. AMS, RI (2007)
44. Peyre, E.: Hauteurs et mesures de Tamagawa sur les variétés de Fano. Duke J. Math. **79**, 101–218 (1995)
45. Slater, J.B., Sir Peter Swinnerton-Dyer: Counting points on cubic surfaces I. Astérisque **251**, 1–11 (1998)
46. de la Brèteche, R., Browning, T.D.: On Manin's conjecture for singular Del Pezzo surfaces of degree four, I. Mich. Math. J (to appear)
47. de la Brèteche, R., Browning, T.D.: On Manin's conjecture for singular del Pezzo surfaces of degree four. II. Math. Proc. Cambridge Philos. Soc., **143**(3), 579–605 (2007)
48. de la Brèteche, R., Browning, T.D., Derenthal, U.: On Manin's conjecture for a certain singular cubic surface. Ann. Sci. École Norm. Sup. (4), **40**(1), 1–50 (2007)
49. de la Brèteche, R.: Sur le nombre de points de hauteur bornée d'une certaine surface cubique singulière. Astérisque **251**, 51–77 (1998)
50. de la Brèteche, R., Swinnerton-Dyer, P.: Fonction zêta des hauteurs associée à une certaine surface cubique. Bull. Soc. Math. France, **135**(1), (2007)
51. Swinnerton-Dyer, P.: A canonical height on $X_0^3 = X_1X_2X_3$. In: Diophantine geometry. CRM Series, vol. 4, pp. 309–322. Ed. Norm., Pisa (2007)
52. Colliot-Thélène, J.-L., Skorobogatov, A.N. Sir Peter Swinnerton-Dyer: Rational points and zero-cycles on fibred varieties: Schinzel's Hypothesis and Salberger's device. J. reine angew. Math. **495**, 1–28 (1998)
53. Sir Peter Swinnerton-Dyer: Some applications of Schinzel's hypothesis to diophantine equations. In: Györy, K., Iwaniec, H., Urbanowicz, J. (eds.) Number theory in progress, vol. 1, pp. 503–530. de Gruyter, Berlin (1999)
54. Milne, J.S.: Arithmetic duality theorems. Perspectives in Mathematics 1, Academic Press Inc., Boston (1986)
55. Skorobogatov, A.N., Swinnerton-Dyer, P.: 2-descent on elliptic curves and rational points on certain Kummer surfaces. Adv. Math. **198**, 448–483 (2005)
56. Colliot-Thélène, J.-L., Skorobogatov, A.N., Sir Peter Swinnerton-Dyer: Double fibres and double covers: paucity of rational points. Acta Arith. **79**, 113–135 (1997)
57. Bright, M.: Ph.D. dissertation. Cambridge (2002)
58. Cassels, J.W.S.: Second descents for elliptic curves. J. reine angew. Math. **494**, 101–127 (1998)
59. Colliot-Thélène, J.-L.: Hasse principle for pencils of curves of genus one whose Jacobians have a rational 2-division point (close variation on a paper of Bender and Swinnerton-Dyer). In: Rational points on algebraic varieties. Progr. Math. **199**, 117–161 (2001)
60. Salberger, P., Skorobogatov, A.N.: Weak approximation for surfaces defined by two quadratic forms. Duke J. Math. **63**, 517–536 (1991)
61. Swinnerton-Dyer, P.: Weak approximation on Del Pezzo surfaces of degree 4. In: Arithmetic of higher-dimensional algebraic varieties; Progr. Math. **226**, 235–257 (2004)
62. Coray, D.: Points algébriques sur les surfaces de Del Pezzo. C. R. Acad. Sci. Paris **284**, 1531–1534 (1977)

Received: 19th October 2007, revised: 13th March 2008

Diophantine Approximation
and Nevanlinna Theory

Paul Vojta

1 Introduction

Beginning with the work of Osgood [65], it has been known that the branch of complex analysis known as *Nevanlinna theory* (also called *value distribution theory*) has many similarities with Roth's theorem on diophantine approximation. This was extended by the author [87] to include an explicit dictionary and to include geometric results as well, such as Picard's theorem and Mordell's conjecture (Faltings' theorem). The latter analogy ties in with Lang's conjecture that a projective variety should have only finitely many rational points over any given number field (i.e., is *Mordellic*) if and only if it is Kobayashi hyperbolic.

This circle of ideas has developed further in the last 20 years: Lang's conjecture on sharpening the error term in Roth's theorem was carried over to a conjecture in Nevanlinna theory which was proved in many cases. In the other direction, Bloch's conjectures on holomorphic curves in abelian varieties (later proved; see Sect. 15 for details) led to proofs of the corresponding results in number theory (again, see Sect. 15). More recently, work in number theory using Schmidt's Subspace Theorem has led to corresponding results in Nevanlinna theory.

This relation with Nevanlinna theory is in some sense similar to the (much older) relation with function fields, in that one often looks to function fields or Nevanlinna theory for ideas that might translate over to the number field case, and that work over function fields or in Nevanlinna theory is often easier than work in the number field case. On the other hand, both function fields and Nevanlinna theory have concepts that (so far) have no counterpart in the number field case. This is especially true of derivatives, which exist in both the function field case and in Nevanlinna theory. In the number field case, however, one would want the "derivative with respect to p," which remains as a major stumbling block, although (separate) work of

P. Vojta (✉)
Department of Mathematics, University of California, 970 Evans Hall #3840, Berkeley, CA 94720-3840, USA
e-mail: vojta@math.berkeley.edu

P. Corvaja and C. Gasbarri (eds.), *Arithmetic Geometry*, Lecture Notes in Mathematics 2009, DOI 10.1007/978-3-642-15945-9_3,
© Springer-Verlag Berlin Heidelberg 2011

Buium and of Minhyong Kim may ultimately provide some answers. The search for such a derivative is also addressed in these notes, using a potential approach using successive minima.

It is important to note, however, that the relation with Nevanlinna theory does not "go through" the function field case. Although it is possible to look at the function field case over $\mathbb{C}$ and apply Nevanlinna theory to the functions representing the rational points, this is not the analogy being described here. Instead, in the analogy presented here, *one* holomorphic function corresponds to *infinitely many* rational or algebraic points (whether over a number field or over a function field). Thus, the analogy with Nevanlinna theory is less concrete, and may be regarded as a more distant analogy than function fields.

These notes describe some of the work in this area, including much of the necessary background in diophantine geometry. Specifically, Sects. 2–4 recall basic definitions of number theory and the theory of heights of elements of number fields, culminating in the statement of Roth's theorem and some equivalent formulations of that theorem. This part assumes that the reader knows the basics of algebraic number theory and algebraic geometry at the level of Lang [45] and Hartshorne [36], respectively. Some proofs are omitted, however; for those the interested reader may refer to Lang [46].

Sections 5–7 briefly introduce Nevanlinna theory and the analogy between Roth's theorem and the classical work of Nevanlinna. Again, many proofs are omitted; references include Shabat [75], Nevanlinna [60], and Goldberg and Ostrovskii [29] for pure Nevanlinna theory, and Vojta [87] and Ru [69] for the analogy.

Sections 8–16 generalize the content of the earlier sections, in the more geometric context of varieties over the appropriate fields (number fields, function fields, or $\mathbb{C}$). Again, proofs are often omitted; most may be found in the references given above.

Section 15 in particular introduces the main conjectures being discussed here: Conjecture 15.2 in Nevanlinna theory ("Griffiths' conjecture") and its counterpart in number theory, the author's Conjecture 15.6 on rational points on varieties.

Sections 17 and 18 round out the first part of these notes, by discussing the function field case and the subject of the exceptional sets that come up in the study of higher dimensional varieties.

In both Nevanlinna theory and number theory, these conjectures have been proved only in very special cases, mostly involving subvarieties of semiabelian varieties. This includes the case of projective space minus a collection of hyperplanes in general position (Cartan's theorem and Schmidt's Subspace Theorem). Recent work of Corvaja, Zannier, Evertse, Ferretti, and Ru has shown, however, that using geometric constructions one can use Schmidt's Subspace Theorem and Cartan's theorem to derive other weak special cases of the conjectures mentioned above. This is the subject of Sects. 19–23.

Sections 24–28 present generalizations of Conjectures 15.2 and 15.6. Conjecture 15.2, in Nevanlinna theory, can be generalized to involve truncated counting functions (as was done by Nevanlinna in the 1-dimensional case), and can also be posed in the context of finite ramified coverings. In number theory, Conjecture 15.6 can also be generalized to involve truncated counting functions. The simplest nontrivial case of this conjecture, involving the divisor $[0] + [1] + [\infty]$ on $\mathbb{P}^1$, is the celebrated "abc conjecture" of Masser and Oesterlé. Thus, Conjecture 23.5 can be regarded as a generalization of the abc conjecture as well as of Conjecture 15.6. One can also generalize Conjecture 15.6 to treat algebraic points; this corresponds to finite ramified coverings in Nevanlinna theory. This is Conjecture 25.1, which can also be posed using truncated counting functions (Conjecture 25.3).

Sections 29 and 30 briefly discuss the question of derivatives in Nevanlinna theory, and Nevanlinna's "Lemma on the Logarithmic Derivative." A geometric form of this lemma, due to Kobayashi, McQuillan, and Wong, is given, and it is shown how this form leads to an inequality in Nevanlinna theory, due to McQuillan, called the "tautological inequality." This inequality then leads to a conjecture in number theory (Conjecture 30.1), which of course should then be called the "tautological conjecture." This conjecture is discussed briefly; it is of interest since it may shed some light on how one might take "derivatives" in number theory.

The abc conjecture infuses much of this theory, not only because a Nevanlinna-like conjecture with truncated counting functions contains the abc conjecture as a special case, but also because other conjectures also imply the abc conjecture, and therefore are "at least as hard" as abc. Specifically, Conjecture 25.1, on algebraic points, implies the abc conjecture, even if known only in dimension 1, and Conjecture 15.6, on rational points, also implies abc if known in high dimensions. This latter implication is the subject of Sect. 31. Finally, implications in the other direction are explored in Sect. 32.

2 Notation and Basic Results: Number Theory

We assume that the reader has an understanding of the fundamental basic facts of number theory (and algebraic geometry), up through the definitions of (Weil) heights. References for these topics include [46] and [87]. We do, however, recall some of the basic conventions here since they often differ from author to author.

Throughout these notes, k will usually denote a number field; if so, then $\mathcal{O}_k$ will denote its ring of integers and M_k its set of places. This latter set is in one-to-one correspondence with the disjoint union of the set of nonzero prime ideals of $\mathcal{O}_k$, the set of real embeddings $\sigma : k \hookrightarrow \mathbb{R}$, and the set of unordered complex conjugate pairs $(\sigma, \bar{\sigma})$ of complex embeddings $\sigma : k \hookrightarrow \mathbb{C}$ with image not contained in $\mathbb{R}$. Such elements of M_k are called **non-archimedean** or **finite** places, **real** places, and **complex** places, respectively.

The real and complex places are collectively referred to as **archimedean** or **infinite** places. The set of these places is denoted S_∞. It is a finite set.

To each place $v \in M_k$ we associate a **norm** $\| \cdot \|_v$, defined for $x \in k$ by $\|x\|_v = 0$ if $x = 0$ and

$$\|x\|_v = \begin{cases} (\mathcal{O}_k : \mathfrak{p})^{\mathrm{ord}_\mathfrak{p}(x)} & \text{if } v \text{ corresponds to } \mathfrak{p} \subseteq \mathcal{O}_k; \\ |\sigma(x)| & \text{if } v \text{ corresponds to } \sigma : k \hookrightarrow \mathbb{R}; \text{ and} \\ |\sigma(x)|^2 & \text{if } v \text{ is a complex place, corresponding to } \sigma : k \hookrightarrow \mathbb{C} \end{cases} \tag{1}$$

if $x \neq 0$. Here $\mathrm{ord}_\mathfrak{p}(x)$ means the exponent of $\mathfrak{p}$ in the factorization of the fractional ideal (x). If we use the convention that $\mathrm{ord}_\mathfrak{p}(0) = \infty$, then (1) is also valid when $x = 0$.

We refer to $\| \cdot \|_v$ as a norm instead of an absolute value, because $\| \cdot \|_v$ does not satisfy the triangle inequality when v is a complex place. However, let

$$N_v = \begin{cases} 0 & \text{if } v \text{ is non-archimedean;} \\ 1 & \text{if } v \text{ is real; and} \\ 2 & \text{if } v \text{ is complex.} \end{cases} \tag{2}$$

Then the norm associated to a place v of k satisfies the axioms

(3.1) $\|x\|_v \geq 0$, with equality if and only if $x = 0$;
(3.2) $\|xy\|_v = \|x\|_v \|y\|_v$ for all $x, y \in k$; and
(3.3) $\|x_1 + \cdots + x_n\|_v \leq n^{N_v} \max\{\|x_1\|_v, \ldots, \|x_n\|_v\}$ for all $x_1, \ldots, x_n \in k$, $n \in \mathbb{N}$.

(In these notes, $\mathbb{N} = \{0, 1, 2, \ldots\}$.)

Some authors treat complex conjugate embeddings as distinct places. We do not do so here, because they give rise to the same norms.

Note that, if $x \in k$, then x lies in the ring of integers if and only if $\|x\|_v \leq 1$ for all non-archimedean places v. Indeed, if $x \neq 0$ then both conditions are equivalent to the fractional ideal (x) being a genuine ideal.

Let L be a finite extension of a number field k, and let w be a place of L. If w is non-archimedean, corresponding to a nonzero prime ideal $\mathfrak{q} \subseteq \mathscr{O}_L$, then $\mathfrak{p} := \mathfrak{q} \cap \mathscr{O}_k$ is a nonzero prime of $\mathscr{O}_k$, and gives rise to a non-archimedean place $v \in M_k$. If v arises from w in this way, then we say that w **lies over** v, and write $w \mid v$. Likewise, if w is archimedean, then it corresponds to an embedding $\tau \colon L \hookrightarrow \mathbb{C}$, and its restriction $\tau|_k \colon k \hookrightarrow \mathbb{C}$ gives rise to a unique archimedean place $v \in M_k$, and again we say that w lies over v and write $w \mid v$.

For each $v \in M_k$, the set of $w \in M_L$ lying over it is nonempty and finite. If $w \mid v$ then we also say that v **lies under** w.

If S is a subset of M_k, then we say $w \mid S$ if w lies over some place in S; otherwise we write $w \nmid S$.

If $w \mid v$, then we have

$$\|x\|_w = \|x\|_v^{[L_w : k_v]} \qquad \text{for all } x \in k, \tag{4}$$

where L_w and k_v denote the completions of L and k at w and v, respectively. We also have

$$\prod_{\substack{w \in M_L \\ w \mid v}} \|y\|_w = \|N_k^L y\|_v \qquad \text{for all } v \in M_k \text{ and all } y \in L. \tag{5}$$

This is proved by using the isomorphism $L \otimes_k k_v \cong \prod_{w \mid v} L_w$; see for example [59, Chap. II, Cor. 8.4].

Let $L/K/k$ be a tower of number fields, and let w' and v be places of L and k, respectively. Then $w' \mid v$ if and only if there is a place w of K satisfying $w' \mid w$ and $w \mid v$.

The field $k = \mathbb{Q}$ has no complex places, one real place corresponding to the inclusion $\mathbb{Q} \subseteq \mathbb{R}$, and infinitely many non-archimedean places, corresponding to prime rational integers. Thus, we write

$$M_{\mathbb{Q}} = \{\infty, 2, 3, 5, \ldots\} .$$

Places of a number field satisfy a **Product Formula**

$$\prod_{v \in M_k} \|x\|_v = 1 \qquad \text{for all } x \in k^* . \tag{6}$$

This formula plays a key role in number theory: it is used to show that certain expressions for the height are well defined, and it also implies that if $\prod_v \|x\|_v < 1$ then $x = 0$.

The Product Formula is proved first by showing that it is true when $k = \mathbb{Q}$ (by direct verification) and then using (5) to pass to an arbitrary number field.

3 Heights

The height of a number, or of a point on a variety, is a measure of the complexity of that number. For example, $100/201$ and $1/2$ are very close to each other (using the norm at the infinite place, at least), but the latter is a much "simpler" number since it can be written down using fewer digits.

We define the **height** (also called the **Weil height**) of an element $x \in k$ by the formula

$$H_k(x) = \prod_{v \in M_k} \max\{\|x\|_v, 1\} . \tag{7}$$

As an example, consider the special case in which $k = \mathbb{Q}$. Write $x = a/b$ with $a, b \in \mathbb{Z}$ relatively prime. For all (finite) rational primes p, if p^i is the largest power of p dividing a, and p^j is the largest power dividing b, then $\|a\|_p = p^{-i}$ and $\|b\|_p = p^{-j}$, and therefore $\max\{\|x\|_p, 1\} = p^b$. Therefore the product of all terms in (7) over all finite places v is just $|b|$. At the infinite place, we have $\|x\|_\infty = |a/b|$, so this gives

$$H_{\mathbb{Q}}(x) = \max\{|a|, |b|\} . \tag{8}$$

Similarly, if $P \in \mathbb{P}^n(k)$ for some $n \in \mathbb{N}$, we define the Weil height $h_k(P)$ as follows. Let $[x_0 : \ldots : x_n]$ be homogeneous coordinates for P (with the x_i always assumed to lie in k). Then we define

$$H_k(P) = \prod_{v \in M_k} \max\{\|x_0\|_v, \ldots, \|x_n\|_v\} . \tag{9}$$

By the Product Formula (6), this quantity is independent of the choice of homogeneous coordinates.

If we identify k with $\mathbb{A}^1(k)$ and identify the latter with a subset of $\mathbb{P}^1(k)$ via the standard injection $i: \mathbb{A}^1 \hookrightarrow \mathbb{P}^1$, then we note that $H_k(x) = H_k(i(x))$ for all $x \in k$. Similarly, we can identify k^n with $\mathbb{A}^n(k)$, and the standard embedding of $\mathbb{A}^n$ into $\mathbb{P}^n$ gives us a height

$$H_k(x_1,\ldots,x_n) = \prod_{v \in M_k} \max\{\|x_1\|_v, \ldots, \|x_n\|_v, 1\} \, .$$

The height functions defined so far, all using capital "H," are called **multiplicative heights**. Usually it is more convenient to take their logarithms and define **logarithmic heights**:

$$h_k(x) = \log H_k(x) = \sum_{v \in M_k} \log^+ \|x\|_v \tag{10}$$

and

$$h_k([x_0 : \ldots : x_n]) = \log H_k([x_0 : \ldots : x_n]) = \sum_{v \in M_k} \log \max\{\|x_0\|_v, \ldots, \|x_n\|_v\}.$$

Here

$$\log^+(x) = \max\{\log x, 0\}.$$

Equation (5) tells us how heights change when the number field k is extended to a larger number field L:

$$h_L(x) = [L : k]h_k(x) \tag{11}$$

and

$$h_L([x_0 : \ldots : x_n]) = [L : k]h_k([x_0 : \ldots : x_n]) \tag{12}$$

for all $x \in k$ and all $[x_0, \ldots, x_n] \in \mathbb{P}^n(k)$, respectively.

Then, given $x \in \overline{\mathbb{Q}}$, we define

$$h_k(x) = \frac{1}{[L : k]}h_L(x)$$

for any number field $L \supseteq k(x)$, and similarly given any $[x_0 : \ldots : x_n] \in \mathbb{P}^n(\overline{\mathbb{Q}})$, we define

$$h_k([x_0 : \ldots : x_n]) = \frac{1}{[L : k]}h_L([x_0 : \ldots : x_n])$$

for any number field $L \supseteq k(x_0,\ldots,x_n)$. These expressions are independent of the choice of L by (11) and (12), respectively.

Following EGA, if x is a point on $\mathbb{P}^n_k$, then $\kappa(x)$ will denote the residue field of the local ring at x. If x is a closed point then the homogeneous coordinates can be chosen such that $k(x_0,\ldots,x_n) = \kappa(x)$.

With these definitions, (11) and (12) remain valid without the conditions $x \in k$ and $[x_0 : \ldots : x_n] \in \mathbb{P}^n(k)$, respectively.

It is common to assume $k = \mathbb{Q}$ and omit the subscript k. The resulting heights are called **absolute heights**.

It is obvious from (7) that $h_k(x) \geq 0$ for all $x \in k$, and that equality holds if $x = 0$ or if x is a root of unity. Conversely, $h_k(x) = 0$ implies $\|x\|_v \leq 1$ for all v; if $x \neq 0$ then the Product Formula implies $\|x\|_v = 1$ for all v. Thus x must be a unit, and the known structure of the unit group then leads to the fact that x must be a root of unity.

Therefore, there are infinitely many elements of $\overline{\mathbb{Q}}$ with height 0. If one bounds the degree of such elements over $\mathbb{Q}$, then there are only finitely many; more generally, we have:

Theorem 3.1. (Northcott's finiteness theorem) *For any $r \in \mathbb{Z}_{>0}$ and any $C \in \mathbb{R}$, there are only finitely many $x \in \overline{\mathbb{Q}}$ such that $[\mathbb{Q}(x) : \mathbb{Q}] \leq r$ and $h(x) \leq C$. Moreover, given any $n \in \mathbb{N}$ there are only finitely many $x \in \mathbb{P}^n(\overline{\mathbb{Q}})$ such that $[\kappa(x) : \mathbb{Q}] \leq r$ and $h(x) \leq C$.*

The first assertion is proved using the fact that, for any $x \in \overline{\mathbb{Q}}$, if one lets $k = \mathbb{Q}(x)$, then $H_k(x)$ is within a constant factor of the largest absolute value of the largest coefficient of the irreducible polynomial of x over $\mathbb{Q}$, when that polynomial is multiplied by a rational number so that its coefficients are relatively prime integers. The second assertion then follows as a consequence of the first. For details, see [48, Chap. II, Thm. 2.2].

This result plays a central role in number theory, since (for example) proving an upper bound on the heights of rational points is equivalent to proving finiteness.

4 Roth's Theorem

Roth [67] proved a key and much-anticipated theorem on how well an algebraic number can be approximated by rational numbers. Of course rational numbers are dense in the reals, but if one limits the size of the denominator then one can ask meaningful and nontrivial questions.

Theorem 4.1. (Roth) *Fix $\alpha \in \overline{\mathbb{Q}}$, $\varepsilon > 0$, and $C > 0$. Then there are only finitely many $a/b \in \mathbb{Q}$, where a and b are relatively prime integers, such that*

$$\left| \frac{a}{b} - \alpha \right| \leq \frac{C}{|b|^{2+\varepsilon}}. \tag{13}$$

Example 4.2. As a diophantine application of Roth's theorem, consider the diophantine equation

$$x^3 - 2y^3 = 11, \qquad x, y \in \mathbb{Z}. \tag{14}$$

If (x, y) is a solution, then x/y must be close to $\sqrt[3]{2}$ (assuming $|x|$ or $|y|$ is large, which would imply both are large):

$$\left|\frac{x}{y} - \sqrt[3]{2}\right| = \left|\frac{11}{y(x^2 + xy\sqrt[3]{2} + y^2\sqrt[3]{4})}\right| \ll \frac{1}{|y|^3}.$$

Thus Roth's theorem implies that (14) has only finitely many solutions.

More generally, if $f \in \mathbb{Z}[x,y]$ is homogeneous of degree ≥ 3 and has no repeated factors, then for any $a \in \mathbb{Z}$ $f(x,y) = a$ has only finitely many integral solutions. This is called the **Thue equation** and historically was the driving force behind the development of Roth's theorem (which is sometimes called the Thue-Siegel-Roth theorem, sometimes also mentioning Schneider, Dyson, and Mahler).

The inequality (13) is best possible, in the sense that the 2 in the exponent on the right-hand side cannot be replaced by a smaller number. This can be shown using continued fractions. Of course one can conjecture a sharper error term [49, Intro. to Chap. I].

If a/b is close to α, then after adjusting C one can replace $|b|$ in the right-hand side of (13) with $H_{\mathbb{Q}}(a/b)$ (see (8)). Moreover, the theorem has been generalized to allow a finite set of places (possibly non-archimedean) and to work over a number field:

Theorem 4.3. *Let k be a number field, let S be a finite set of places of k containing all archimedean places, fix $\alpha_v \in \overline{\mathbb{Q}}$ for each $v \in S$, let $\varepsilon > 0$, and let $C > 0$. Then only finitely many $x \in k$ satisfy the inequality*

$$\prod_{v \in S} \min\{1, \|x - \alpha_v\|_v\} \leq \frac{C}{H_k(x)^{2+\varepsilon}}. \tag{15}$$

(Strictly speaking, S can be any finite set of places at this point, but requiring S to contain all archimedean places does not weaken the theorem, and this assumption will be necessary in Sect. 6. See, for example, (29).)

Taking $-\log$ of both sides of (15), dividing by $[k : \mathbb{Q}]$, and rephrasing the logic, the above theorem is equivalent to the assertion that for all $c \in \mathbb{R}$ the inequality

$$\frac{1}{[k : \mathbb{Q}]} \sum_{v \in S} \log^+ \left\|\frac{1}{x - \alpha_v}\right\|_v \leq (2 + \varepsilon)h(x) + c \tag{16}$$

holds for all but finitely many $x \in k$.

In writing (15), we assume that one has chosen an embedding $i_v : \bar{k} \hookrightarrow \overline{k_v}$ over k for each $v \in S$. Otherwise the expression $\|x - \alpha_v\|_v$ may not make sense.

This is mostly a moot point, however, since we may restrict to $\alpha_v \in k$ for all v. Clearly this restricted theorem is implied by the theorem without the additional restriction, but in fact it also implies the original theorem. To see this, suppose k, S, ε, and c are as above, and assume that $\alpha_v \in \overline{\mathbb{Q}}$ are given for all $v \in S$. Let L be the Galois closure over k of $k(\alpha_v : v \in S)$, and let T be the set of all places of L lying over places in S. We assume that L is a subfield of $\bar{k}$, so that $\alpha_v \in L$ for all $v \in S$. Then $(i_v)|_L : L \to \overline{k_v}$ induces a place w_0 of L over v, and all other places w of L over v are conjugates by elements $\sigma_w \in \mathrm{Gal}(L/k)$:

$\|x\|_w = \|\sigma_w^{-1}(x)\|_{w_0}$ for all $x \in L$. Letting $\alpha_w = \sigma_w(\alpha_v)$ for all $w \mid v$, we then have
$\|x - \alpha_w\|_w = \|\sigma_w^{-1}(x - \alpha_w)\|_{w_0} = \|x - \alpha_v\|_v^{[L_{w_0}:k_v]}$ for all $x \in k$ by (4), and therefore

$$\sum_{w \mid v} \log^+ \left\| \frac{1}{x - \alpha_w} \right\|_w = \sum_{w \mid v} [L_{w_0} : k_v] \log^+ \left\| \frac{1}{x - \alpha_v} \right\|_v = [L : k] \log^+ \left\| \frac{1}{x - \alpha_v} \right\|_v$$

since L/k is Galois. Thus

$$\frac{1}{[k : \mathbb{Q}]} \sum_{v \in S} \log^+ \left\| \frac{1}{x - \alpha_v} \right\|_v = \frac{1}{[L : \mathbb{Q}]} \sum_{w \in T} \log^+ \left\| \frac{1}{x - \alpha_w} \right\|_w$$

for all $x \in k$. Applying Roth's theorem over the field L (where now $\alpha_w \in L$ for all $w \in T$) then gives (16) for almost all $x \in k$.

Finally, we note that Roth's theorem (as now rephrased) is equivalent to the following statement.

Theorem 4.4. *Let k be a number field, let $S \supseteq S_\infty$ be a finite set of places of k, fix distinct $\alpha_1, \dots, \alpha_q \in k$, let $\varepsilon > 0$, and let $c \in \mathbb{R}$. Then the inequality*

$$\frac{1}{[k : \mathbb{Q}]} \sum_{v \in S} \sum_{i=1}^q \log^+ \left\| \frac{1}{x - \alpha_i} \right\|_v \leq (2 + \varepsilon) h(x) + c \tag{17}$$

holds for almost all $x \in k$.

Indeed, given $\alpha_v \in k$ for all $v \in S$, let $\alpha_1, \dots, \alpha_q$ be the distinct elements of the set $\{\alpha_v : v \in S\}$. Then

$$\frac{1}{[k : \mathbb{Q}]} \sum_{v \in S} \log^+ \left\| \frac{1}{x - \alpha_v} \right\|_v \leq \frac{1}{[k : \mathbb{Q}]} \sum_{v \in S} \sum_{i=1}^q \log^+ \left\| \frac{1}{x - \alpha_i} \right\|_v ,$$

so Theorem 4.4 implies the earlier form of Roth's theorem (as modified).

Conversely, given distinct $\alpha_1, \dots, \alpha_q \in k$, we note that any given $x \in k$ can be close to only one of the α_i at each place v (where the value of i may depend on v). Therefore, for each v,

$$\sum_{i=1}^q \log^+ \left\| \frac{1}{x - \alpha_i} \right\|_v \leq \log^+ \left\| \frac{1}{x - \alpha_v} \right\|_v + c_v$$

for some constant c_v independent of x and some $\alpha_v \in \{\alpha_1, \dots, \alpha_q\}$ depending on x and v. Thus, for each $x \in k$, there is a choice of α_v for each $v \in S$ such that

$$\frac{1}{[k : \mathbb{Q}]} \sum_{v \in S} \sum_{i=1}^q \log^+ \left\| \frac{1}{x - \alpha_i} \right\|_v \leq \frac{1}{[k : \mathbb{Q}]} \sum_{v \in S} \log^+ \left\| \frac{1}{x - \alpha_v} \right\|_v + c',$$

where c' is independent of x. Since there are only finitely many choices of the system $\{\alpha_v : v \in S\}$, finitely many applications of the earlier version of Roth's theorem suffice to imply Theorem 4.4.

5 Basics of Nevanlinna Theory

Nevanlinna theory, developed by R. and F. Nevanlinna in the 1920s, concerns the distribution of values of holomorphic and meromorphic functions, in much the same way that Roth's theorem concerns approximation of elements of a number field.

One can think of it as a generalization of a theorem of Picard, which says that a nonconstant holomorphic function from $\mathbb{C}$ to $\mathbb{P}^1$ can omit at most two points. This, in turn, generalizes Liouville's theorem.

An example relevant to Picard's theorem is the exponential function e^z, which omits the values 0 and ∞. When r is large, the circle $|z| = r$ is mapped to many values close to ∞ (when $\mathrm{Re}\,z$ is large) and many values close to 0 (when $\mathrm{Re}\,z$ is highly negative).

So even though e^z omits these two values, it spends a lot of time very close to them. This observation can be made precise, in what is called Nevanlinna's *First Main Theorem*. In order to state this theorem, we need some definitions.

First we recall that $\log^+ x = \max\{\log x, 0\}$, and similarly define

$$\mathrm{ord}_z^+ f = \max\{\mathrm{ord}_z f, 0\}$$

if f is a meromorphic function and $z \in \mathbb{C}$.

Definition 5.1. Let f be a meromorphic function on $\mathbb{C}$. We define the **proximity function** of f by

$$m_f(r) = \int_0^{2\pi} \log^+ |f(re^{i\theta})| \frac{d\theta}{2\pi} \tag{18}$$

for all $r > 0$. We also define

$$m_f(\infty, r) = m_f(r) \qquad \text{and} \qquad m_f(a, r) = m_{1/(f-a)}(r)$$

when $a \in \mathbb{C}$.

The integral in (18) converges when f has a zero or pole on the circle $|z| = r$, so it is defined everywhere. The proximity function $m_f(a, r)$ is large to the extent that the values of f on $|z| = r$ are close to a.

Definition 5.2. Let f be a meromorphic function on $\mathbb{C}$. For $r > 0$ let $n_f(r)$ be the number of poles of f in the open disc $|z| < r$ of radius r (counted with multiplicity), and let $n_f(0)$ be the order of the pole (if any) at $z = 0$. We then define the **counting function** of f by

$$N_f(r) = \int_0^r (n_f(s) - n_f(0)) \frac{ds}{s} + n_f(0) \log r. \tag{19}$$

As before, we also define

$$N_f(\infty, r) = N_f(r) \qquad \text{and} \qquad N_f(a, r) = N_{1/(f-a)}(r)$$

when $a \in \mathbb{C}$.

The counting function can also be written

$$N_f(a, r) = \sum_{0 < |z| < r} \operatorname{ord}_z^+ (f - a) \cdot \log \frac{r}{|z|} + \operatorname{ord}_0^+ (f - a) \cdot \log r. \tag{20}$$

Thus, the expression $N_f(a, r)$ is a weighted count, with multiplicity, of the number of times f takes on the value a in the disc $|z| < r$.

Definition 5.3. Let f be as in Definition 5.1. Then the **height function** of f is the function $T_f \colon (0, \infty) \to \mathbb{R}$ given by

$$T_f(r) = m_f(r) + N_f(r). \tag{21}$$

Classically, the above function is called the characteristic function, but here we will use the term height function, since this is more in parallel with terminology in the number field case. The height function T_f does, in fact, measure the complexity of the meromorphic function f.

In particular, if f is constant then so is $T_f(r)$; otherwise,

$$\liminf_{r \to \infty} \frac{T_f(r)}{\log r} > 0. \tag{22}$$

Moreover, it is known that $T_f(r) = O(\log r)$ if and only if f is a rational function. Although this is a well-known fact, I was unable to find a convenient reference, so a proof is sketched here. If f is rational, then direct computation gives $T_f(r) = O(\log r)$. Conversely, if $T_f(r) = O(\log r)$ then f can have only finitely many poles; clearing these by multiplying f by a polynomial changes T_f by at most $O(\log r)$, so we may assume that f is entire. We may also assume that f is nonconstant. By [37, Thm. 1.8], if f is entire and nonconstant and $K > 1$, then

$$\liminf_{r \to \infty} \frac{\log \max_{|z|=r} |f(z)|}{T_f(r) (\log T_f(r))^K} = 0.$$

This implies that $f(z)/z^n$ has a removable singularity at ∞ for sufficiently large n, hence is a polynomial.

The following theorem relates the height function to the proximity and counting functions at points other than ∞.

Theorem 5.4. (First Main Theorem) *Let f be a meromorphic function on $\mathbb{C}$, and let $a \in \mathbb{C}$. Then*

$$T_f(r) = m_f(a,r) + N_f(a,r) + O(1),$$

where the constant in $O(1)$ depends only on f and a.

This theorem is a straightforward consequence of **Jensen's formula**

$$\log|c_f| = \int_0^{2\pi} \log|f(re^{i\theta})| \frac{d\theta}{2\pi} + N_f(\infty,r) - N_f(0,r),$$

where c_f is the leading coefficient in the Laurent expansion of f at $z = 0$. For details, see [60, Chap. VI, (1.2')] or [69, Cor. A1.1.3].

As an example, let $f(z) = e^z$. This function is entire, so $N_f(\infty,r) = 0$ for all r. We also have

$$m_f(\infty,r) = \int_0^{2\pi} \log^+ e^{r\cos\theta} \frac{d\theta}{2\pi} = r \int_{-\pi/2}^{\pi/2} \cos\theta \frac{d\theta}{2\pi} = \frac{r}{\pi}.$$

Thus

$$T_f(r) = \frac{r}{\pi}.$$

Similarly, we have $N_f(0,r) = 0$ and $m_f(0,r) = r/\pi$ for all r, confirming the First Main Theorem in the case $a = 0$.

The situation with $a = -1$ is more difficult. The integral in the proximity function seems to be beyond any hope of computing exactly. Since $e^z = -1$ if and only if z is an odd integral multiple of πi, we have

$$N_f(-1,r) = 2 \int_0^r \left[\frac{s}{2\pi} + \frac{1}{2} \right] \frac{ds}{s} \approx 2 \int_0^r \frac{s}{2\pi} \frac{ds}{s} = \frac{r}{\pi},$$

where $[\cdot]$ denotes the greatest integer function. The error in the above approximation should be $o(r)$, which would give $m(-1,r) = o(r)$. Judging from the general shape of the exponential function, similar estimates should hold for all nonzero $a \in \mathbb{C}$.

In one way of thinking, the First Main Theorem gives an upper bound on the counting function. As the above example illustrates, there is no lower bound for an individual counting function (other than 0), but it is known that there cannot be many values of a for which $N_f(a,r)$ is much smaller than the height. This is what the Second Main Theorem shows.

Theorem 5.5. (Second Main Theorem) *Let f be a meromorphic function on $\mathbb{C}$, and let $a_1, \ldots, a_q \in \mathbb{C}$ be distinct numbers. Then*

$$\sum_{j=1}^q m_f(a_j,r) \leq_{\text{exc}} 2T_f(r) + O(\log^+ T_f(r)) + o(\log r), \tag{23}$$

where the implicit constants depend only on f and $a_1, \ldots, a_q$.

Here the notation $\leq_{\mathrm{exc}}$ means that the inequality holds for all $r > 0$ outside of a set of finite Lebesgue measure.

By the First Main Theorem, (23) can be rewritten as a lower bound on the counting functions:

$$\sum_{j=1}^{q} N_f(a_j, r) \geq_{\mathrm{exc}} (q-2)T_f(r) - O(\log^+ T_f(r)) - o(\log r). \qquad (24)$$

As another variation, (23) can be written with a weaker error term:

$$\sum_{j=1}^{q} m_f(a_j, r) \leq_{\mathrm{exc}} (2+\varepsilon)T_f(r) + c \qquad (25)$$

for all $\varepsilon > 0$ and any constant c. The next section will show that this correponds to Roth's theorem.

Corollary 5.6. (Picard's "little" theorem) *If $a_1, a_2, a_3 \in \mathbb{P}^1(\mathbb{C})$ are distinct, then any holomorphic function $f \colon \mathbb{C} \to \mathbb{P}^1(\mathbb{C}) \setminus \{a_1, a_2, a_3\}$ must be constant.*

Proof. Assume that $f \colon \mathbb{C} \to \mathbb{P}^1(\mathbb{C}) \setminus \{a_1, a_2, a_3\}$ is a nonconstant holomorphic function. After applying an automorphism of $\mathbb{P}^1$ if necessary, we may assume that all a_j are finite. We may regard f as a meromorphic function on $\mathbb{C}$.

Since f never takes on the values a_1, a_2, or a_3, the left-hand side of (24) vanishes. Since f is nonconstant, the right-hand side approaches $+\infty$ by (22). This is a contradiction. $\square$

As we have seen, (24) has some advantages over (23). Other advantages include the fact that $q - 2$ on the right-hand side is the Euler characteristic of $\mathbb{P}^1$ minus q points, and it will become clear later that the dependence on a metric is restricted to the height term. It is also the preferred form when comparing with the abc conjecture.

6 Roth's Theorem and Nevanlinna Theory

We now claim that Nevanlinna's Second Main Theorem corresponds very closely to Roth's theorem. To see this, we make the following definitions in number theory.

Definition 6.1. Let k be a number field and $S \supseteq S_\infty$ a finite set of places of k. For $x \in k$ we define the **proximity function** to be

$$m_S(x) = \sum_{v \in S} \log^+ \|x\|_v$$

and, for $a \in k$ with $a \neq x$,

$$m_S(a,x) = m_S\left(\frac{1}{x-a}\right) = \sum_{v \in S} \log^+ \left\|\frac{1}{x-a}\right\|_v. \tag{26}$$

Similarly, for distinct $a,x \in k$ the **counting function** is defined as

$$N_S(x) = \sum_{v \notin S} \log^+ \|x\|_v$$

and

$$N_S(a,x) = N_S\left(\frac{1}{x-a}\right) = \sum_{v \notin S} \log^+ \left\|\frac{1}{x-a}\right\|_v. \tag{27}$$

By (10) it then follows that

$$m_S(x) + N_S(x) = \sum_{v \in M_k} \log^+ \|x\|_v = h_k(x) \tag{28}$$

for all $x \in k$. This corresponds to (21).

Note that k does not appear in the notation for the proximity and counting functions, since it is implied by S.

We also note that all places outside of S are non-archimedean, hence correspond to nonzero prime ideals $\mathfrak{p} \subseteq \mathcal{O}_k$. Thus, by (1), (27) can be rewritten

$$N_S(a,x) = \sum_{v \notin S} \operatorname{ord}_{\mathfrak{p}}^+ (x-a) \cdot \log(\mathcal{O}_k : \mathfrak{p}), \tag{29}$$

where $\mathfrak{p}$ in the summand is the prime ideal corresponding to v. This corresponds to (20).

The number field case has an analogue of the First Main Theorem, which we prove as follows.

Lemma 6.2. *Let v be a place of a number field k, and let $a,x \in k$. Then*

$$\left| \log^+ \|x\|_v - \log^+ \|x-a\|_v \right| \leq \log^+ \|a\|_v + N_v \log 2. \tag{30}$$

Proof. **Case I**: v is archimedean.

We first claim that

$$\log^+(s+t) \leq \log^+ s + \log^+ t + \log 2 \tag{31}$$

for all real $s,t \geq 0$. Indeed, let $f(s,t) = \log^+(s+t) - \log^+ s - \log^+ t$. By considering partial derivatives, for each fixed s the function has a global maximum at $t = 1$, and for each fixed t it has a global maximum at $s = 1$. Therefore all s and t satisfy $f(s,t) \leq f(1,1) = \log 2$.

Now let $z, b \in \mathbb{C}$. Since $|z| \le |z - b| + |b|$, (31) with $s = |z - b|$ and $t = |b|$ gives

$$\log^+ |z| - \log^+ |z - b| \le \log^+ (|z - b| + |b|) - \log^+ |z - b| \le \log^+ |b| + \log 2.$$

Similarly, since $|z - b| \le |z| + |b|$, we have

$$\log^+ |z - b| - \log^+ |z| \le \log^+ (|z| + |b|) - \log^+ |z| \le \log^+ |b| + \log 2.$$

These two inequalities together imply (30).

Case II: v is non-archimedean.

In this case $N_v = 0$, so the last term vanishes. Also, since v is non-archimedean, at least two of $\|x\|_v$, $\|x - a\|_v$, and $\|a\|_v$ are equal, and the third (if different) is smaller. If $\|x\|_v = \|x - a\|_v$, then the result is obvious, so we may assume that $\|a\|_v$ is equal to one of the other two. If $\|a\|_v = \|x\|_v$, then

$$\left| \log^+ \|x\|_v - \log^+ \|x - a\|_v \right| = \log^+ \|x\|_v - \log^+ \|x - a\|_v \le \log^+ \|x\|_v = \log^+ \|a\|_v$$

since $0 \le \log^+ \|x - a\|_v \le \log^+ \|x\|_v$. If $\|a\|_v = \|x - a\|_v$ then (30) follows by a similar argument. $\qquad\square$

Corresponding to Theorem 5.4, we then have:

Theorem 6.3. *Let k be a number field, let $S \supseteq S_\infty$ be a finite set of places of k, and fix $a \in k$. Then*

$$h_k(x) = m_S(a, x) + N_S(a, x) + O(1),$$

where the constant in $O(1)$ depends only on k and a. In fact, the constant can be taken to be $h_k(a) + [k : \mathbb{Q}] \log 2$.

Proof. First, we note that

$$m_S(a, x) + N_S(a, x) = m_S \left(\frac{1}{x - a} \right) + N_S \left(\frac{1}{x - a} \right) = h_k \left(\frac{1}{x - a} \right).$$

Next, by comparing with the height on $\mathbb{P}^1$, we have

$$h_k \left(\frac{1}{x - a} \right) = h_k([x - a : 1]) = h_k([1 : x - a]) = h_k(x - a).$$

Therefore, it suffices to show that

$$h_k(x - a) = h_k(x) + O(1),$$

with the constant in $O(1)$ equal to $[k : \mathbb{Q}] \log 2$. This follows immediately by applying Lemma 6.2 termwise to the sums in the two height functions. $\qquad\square$

It will later be clear that this theorem is a well-known geometric property of heights.

We now consider the Second Main Theorem. With the notation of Definition 6.1, Roth's theorem can be made to look very similar to Nevanlinna's Second Main Theorem. Indeed, multiplying (17) by $[k : \mathbb{Q}]$ and substituting the definition (26) of the proximity function gives the inequality

$$\sum_{j=1}^{q} m_S(a_j, x) \leq (2 + \varepsilon) h_k(x) + c,$$

which corresponds to (25). As has been mentioned earlier, it has been conjectured that Roth's theorem should hold with sharper error terms, corresponding to (23). Such conjectures predated the emergence of the correspondence between number theory and Nevanlinna theory, but the latter spurred renewed work in the area. See, for example, [101, 49, 12].

Unfortunately, the correspondence between the statements of Roth's theorem and Nevanlinna's Second Main Theorem does not extend to the proofs of these theorems. Roth's theorem is proved by taking sufficiently many $x \in k$ not satisfying the inequality, using them to construct an auxiliary polynomial, and then deriving a contradiction from the vanishing properties of that polynomial. Nevanlinna's Second Main Theorem has a number of proofs; for example, one proof uses curvature arguments, one follows from Nevanlinna's "lemma on the logarithmic derivative," and one uses Ahlfors' theory of covering spaces. All of these proofs make essential use of the derivative of the meromorphic function, and it is a major unsolved question in the field to find some analogue of this in number theory.

A detailed discussion of these proofs would be beyond the scope of these notes.

Beyond Roth's theorem and the Second Main Theorem, one can define the *defect* of an element of $\mathbb{C}$ or of an element $a \in k$, as follows.

Definition 6.4. Let f be a meromorphic function on $\mathbb{C}$, and let $a \in \mathbb{C} \cup \{\infty\}$. Then the **defect** of a is

$$\delta_f(a) = \liminf_{r \to \infty} \frac{m_f(a, r)}{T_f(r)}.$$

Similarly, let $S \supseteq S_\infty$ be a finite set of places of a number field k, let $a \in k$, and let Σ be an infinite subset of k. Then the defect is defined as

$$\delta_S(a) = \liminf_{x \in \Sigma} \frac{m_S(a, x)}{h_k(x)}.$$

By the First Main Theorem (Theorems 5.4 and 6.3), we then have

$$0 \leq \delta_f(a) \leq 1 \qquad \text{and} \qquad 0 \leq \delta_S(a) \leq 1,$$

respectively. The Second Main Theorems (Theorems 5.5 and 4.4) then give

$$\sum_{a\in\mathbb{C}} \delta_f(a) \le 2 \qquad \text{and} \qquad \sum_{j=1}^{q} \delta_S(a_j) \le 2,$$

respectively. This is just an equivalent formulation of the Second Main Theorem, with a weaker error term in the case of Nevanlinna theory, so it is usually better to work directly with the inequality of the Second Main Theorem.

The defect gets it name because it measures the extent to which $N_f(a,r)$ or $N_S(a,x)$ is smaller than the maximum indicated by the First Main Theorem.

We conclude this section by noting that Definition 6.1 can be extended to $x \in \bar{k}$. Indeed, let k and S be as in Definition 6.1, and let $x \in \bar{k}$. Let L be a number field containing $k(x)$, and let T be the set of places of L lying over places in S. If $L' \supseteq L$ is another number field, and if T' is the set of places of L' lying over places of k, then (4) gives

$$m_{T'}(x) = [L' : L]m_T(x) \qquad \text{and} \qquad N_{T'}(x) = [L' : L]N_T(x). \tag{32}$$

This allows us to make the following definition.

Definition 6.5. Let k, S, x, L, and T be as above. Then we define

$$m_S(x) = \frac{1}{[L:k]}m_T(x) \qquad \text{and} \qquad N_S(x) = \frac{1}{[L:k]}N_T(x).$$

These expressions are independent of $L \supseteq k(x)$ by (32). As in (26) and (27), we also let

$$m_S(a,x) = m_S\left(\frac{1}{x-a}\right) \qquad \text{and} \qquad N_S(a,x) = N_S\left(\frac{1}{x-a}\right).$$

Likewise, Theorem 6.3 (the number-theoretic First Main Theorem) extends to $x \in \bar{k}$, by (11), (32), and (6.5). The expression (28) for the height also extends. Roth's theorem, however, does not extend in this manner, and questions of extending Roth's theorem even to algebraic numbers of bounded degree are quite deep and unresolved.

7 The Dictionary (Non-Geometric Case)

The discussion in the preceding section suggests that there should be an analogy between the fields of Nevanlinna theory and number theory. This section describes this dictionary in more detail.

The existence of an analogy between number theory and Nevanlinna theory was first observed by Osgood [65, 66], but he did not provide an explicit dictionary for comparing the two theories. This was provided by Vojta [87]. An updated version of that dictionary is provided here as Table 1.

The first and most important thing to realize about the dictionary is that the analogue of a holomorphic (or meromorphic) function is an *infinite sequence* of rational numbers. While it is tempting to compare number theory with Nevanlinna theory by way of function fields – by viewing a single rational point as being analogous to a rational point over a function field over $\mathbb{C}$ and then applying Nevanlinna theory to

Table 1 The dictionary in the one-dimensional case

Nevanlinna theory	Number theory
$f\colon \mathbb{C} \to \mathbb{C}, \quad$ non-constant	$\{x\} \subseteq k, \quad$ infinite
r	x
θ	$v \in S$
$\|f(re^{i\theta})\|$	$\|x\|_v, \quad v \in S$
$\mathrm{ord}_z f$	$\mathrm{ord}_v x, \quad v \notin S$
$\log \dfrac{r}{\|z\|}$	$\log(\mathscr{O}_k : \mathfrak{p})$
Height function	Logarithmic height
$T_f(r) = \displaystyle\int_0^{2\pi} \log^+ \|f(re^{i\theta})\|\, \dfrac{d\theta}{2\pi} + N_f(\infty, r)$	$h_k(x) = \displaystyle\sum_{v \in M_k} \log^+ \|x\|_v$
Proximity function	
$m_f(a,r) = \displaystyle\int_0^{2\pi} \log^+ \left\| \dfrac{1}{f(re^{i\theta}) - a} \right\| \dfrac{d\theta}{2\pi}$	$m_S(a,x) = \displaystyle\sum_{v \in S} \log^+ \left\| \dfrac{1}{x - a} \right\|_v$
Counting function	
$N_f(a,r) = \displaystyle\sum_{\|z\| < r} \mathrm{ord}_z^+ (f - a) \log \dfrac{r}{\|z\|}$	$N_S(a,x) = \displaystyle\sum_{v \notin S} \mathrm{ord}_{\mathfrak{p}}^+ (x - a) \log(\mathscr{O}_k : \mathfrak{p})$
First main theorem	Property of heights
$N_f(a,r) + m_f(a,r) = T_f(r) + O(1)$	$N_S(a,x) + m_S(a,x) = h_k(x) + O(1)$
Second main theorem	Conjectured refinement of Roth
$\displaystyle\sum_{i=1}^m m_f(a_i, r) \leq_{\mathrm{exc}} 2T_f(r) - N_{1,f}(r)$ $\qquad\qquad\qquad + O(r \log T_f(r))$	$\displaystyle\sum_{i=1}^m m_S(a_i, x) \leq 2h_k(x) + O(\log h_k(x))$
Defect	
$\delta(a) = \displaystyle\liminf_{r \to \infty} \dfrac{m_f(a,r)}{T_f(r)}$	$\delta(a) = \displaystyle\liminf_x \dfrac{m_S(a,x)}{h_k(x)}$
Defect relation	Roth's theorem
$\displaystyle\sum_{a \in \mathbb{C}} \delta(a) \leq 2$	$\displaystyle\sum_{a \in k} \delta(a) \leq 2$
Jensen's formula	Artin-Whaples product formula
$\log \|c_f\| = \displaystyle\int_0^{2\pi} \log \|f(re^{i\theta})\| \dfrac{d\theta}{2\pi}$ $\qquad\qquad + N_f(\infty, r) - N_f(0, r)$	$\displaystyle\sum_{v \in M_k} \log \|x\|_v = 0$

the corresponding section map – this is not what is being compared here. Note that the Second Main Theorem posits the non-existence of a meromorphic function violating the inequality for too many r, and Roth's theorem claims the non-existence of an infinite sequence of rational numbers not satisfying its main inequality.

We shall now describe Table 1 in more detail. Much of it (below the top six rows) has already been described in Sect. 6, with the exception of the last line. This is left to the reader.

The bottom two-thirds of the table can be broken down further, leading to the top six rows. The first row has been described above. One can say more, though. The analogue of a single rational number can be viewed as the restriction of f to the closed disc $\overline{\mathbb{D}}_r$ of radius r. Of course $f\big|_{\overline{\mathbb{D}}_r}$ for varying r are strongly related, in the sense that if one knows one of them then all of them are uniquely determined. This is not true of the number field case (as far as is known); thus the analogy is not perfect.

However, when comparing $f\big|_{\overline{\mathbb{D}}_r}$ to a given element of k, there are further similarities between the respective proximity functions and counting functions. As far as the proximity functions are concerned, in Nevanlinna theory $m_f(a,r)$ depends only on the values of f on the circle $|z| = r$, whereas in number theory $m_S(a,r)$ involves only the places in S. So places in S correspond to $\partial\mathbb{D}_r$, and both types of proximity functions involve the absolute values at those places. Moreover, in Nevanlinna theory the proximity function is an integral over a set of finite measure, while in number theory the proximity function is a finite sum.

As for counting functions, they involve the open disc $\mathbb{D}_r$ in Nevanlinna theory, and places outside of S (all of which are non-archimedean) in number theory. Both types of counting functions involve an infinite weighted sum of orders of vanishing at those places, and the sixth line of Table 1 compares these weights.

It should also be mentioned that many of these theorems in Nevanlinna theory have been extended to holomorphic functions with domains other than $\mathbb{C}$. In one direction, one can replace the domain with $\mathbb{C}^m$ for some $m > 0$. While this is useful from the point of view of pure Nevanlinna theory, it is less interesting from the point of view of the analogy with number theory, since number rings are one-dimensional. Moreover, in Nevanlinna theory, the proofs that correspond most closely to proofs in number theory concern maps with domain $\mathbb{C}$.

There is one other way to change the domain of a holomorphic function, though, which is highly relevant to comparisons with number theory. Namely, one can replace the domain $\mathbb{C}$ with a ramified cover. Let B be a connected Riemann surface, let $\pi\colon B \to \mathbb{C}$ be a proper surjective holomorphic map, and let $f\colon B \to \mathbb{C}$ be a meromorphic function. In place of $\overline{\mathbb{D}}_r$ in the above discussion, one can work with $\pi^{-1}(\overline{\mathbb{D}}_r)$ and define the proximity, counting, and height functions accordingly. For detailed definitions, see Sect. 27.

When working with a finite ramified covering, though, the Second Main Theorem requires an additional term $N_{\mathrm{Ram}(\pi)}(r)$, which is a counting function for ramification points of π (Definition 27.3c). The main inequality (23) of the Second Main Theorem then becomes

$$\sum_{j=1}^{q} m_f(a_j,r) \leq_{\text{exc}} 2T_f(r) + N_{\text{Ram}(\pi)}(r) + O(\log^+ T_f(r)) + o(\log r)$$

in this context.

In number theory, the corresponding situation involves algebraic numbers of bounded degree over k instead of elements of k itself. Again, the inequality in the Second Main Theorem becomes weaker in this case, conjecturally by adding the following term.

Definition 7.1. Let D_k denote the discriminant of a number field k, and for number fields $L \supseteq k$ define

$$d_k(L) = \frac{1}{[L:k]} \log|D_L| - \log|D_k|.$$

For $x \in \bar{k}$ we then define

$$d_k(x) = d_k(k(x)).$$

It is then conjectured that Roth's theorem for $x \in \bar{k}$ of bounded degree over k still holds, with inequality

$$\sum_{j=1}^{q} m_S(a_j,x) \leq (2+\varepsilon)h_k(x) + d_k(x) + C. \tag{33}$$

For further discussion of this situation, including its relation to the abc conjecture, see Sects. 25–26.

8 Cartan's Theorem and Schmidt's Subspace Theorem

In both Nevanlinna theory and number theory, the first extensions of the Second Main Theorem and its counterpart to higher dimensions were theorems involving approximation to hyperplanes in projective space.

We start with a definition needed for both theorems.

Definition 8.1. A collection of hyperplanes in $\mathbb{P}^n$ is in **general position** if for all $j \leq n$ the intersection of any j of them has dimension $n - j$, and if the intersection of any $n + 1$ of them is empty.

The Second Main Theorem for approximation to hyperplanes in $\mathbb{P}^n$ was first proved by Cartan [11]. Before stating it, we need to define the proximity, counting, and height functions.

Definition 8.2. Let H be a hyperplane in $\mathbb{P}^n(\mathbb{C})$ $(n > 0)$, and let $a_0 x_0 + \cdots + a_n x_n$ be a linear form defining it. Let $P \in \mathbb{P}^n \setminus H$ be a point, and let $[x_0 : \ldots : x_n]$ be homogeneous coordinates for P. We then define

$$\lambda_H(P) = -\frac{1}{2}\log\frac{|a_0x_0 + \cdots + a_nx_n|^2}{|x_0|^2 + \cdots + |x_n|^2} \qquad (34)$$

(this depends on $a_0, \ldots, a_n$, but only up to an additive constant). It is independent of the choice of homogeneous coordinates for P.

If $n = 1$ and H is a finite number $a \in \mathbb{C}$ (via the usual identification of $\mathbb{C}$ as a subset of $\mathbb{P}^1(\mathbb{C})$), then

$$\lambda_H(x) = \log^+\left|\frac{1}{x-a}\right| + O(1). \qquad (35)$$

Recall that a **holomorphic curve** in a complex variety X is a holomorphic function from $\mathbb{C}$ to $X(\mathbb{C})$.

Definition 8.3. Let n, H, and λ_H be as in Definition 8.2, and let $f\colon \mathbb{C} \to \mathbb{P}^n$ be a holomorphic curve whose image is not contained in H. Then the **proximity function** for H is

$$m_f(H,r) = \int_0^{2\pi} \lambda_H(f(re^{i\theta}))\,\frac{d\theta}{2\pi}. \qquad (36)$$

For the following, recall that an **analytic divisor** on $\mathbb{C}$ is a formal sum

$$\sum_{z\in\mathbb{C}} n_z \cdot z,$$

where $n_z \in \mathbb{Z}$ for all z and the set $\{z \in \mathbb{C} : n_z \neq 0\}$ is a discrete set (which may be infinite).

Definition 8.4. Let n, H, and f be as above. Then f^*H is an analytic divisor on $\mathbb{C}$, and for $z \in \mathbb{C}$ we let $\mathrm{ord}_z f^*H$ denote its multiplicity at the point z. Then the **counting function** for H is defined to be

$$N_f(H,r) = \sum_{0<|z|<r} \mathrm{ord}_z f^*H \cdot \log\frac{r}{|z|} + \mathrm{ord}_0 f^*H \cdot \log r \qquad (37)$$

Definition 8.5. Let $f\colon \mathbb{C} \to \mathbb{P}^n(\mathbb{C})$ be a holomorphic curve ($n > 0$). We then define the **height** of f to be

$$T_f(r) = m_f(H,r) + N_f(H,r)$$

for any hyperplane H not containing the image of f. The First Main Theorem can be shown to hold in the context of hyperplanes in projective space, so the height depends on H only up to $O(1)$.

We may now state Cartan's theorem.

Theorem 8.6. (Cartan) *Let $n > 0$ and let $H_1, \ldots, H_q$ be hyperplanes in $\mathbb{P}^n$ in general position. Let $f\colon \mathbb{C} \to \mathbb{P}^n(\mathbb{C})$ be a holomorphic curve whose image is not contained in any hyperplane. Then*

$$\sum_{j=1}^{q} m_f(H_j, r) \leq_{\text{exc}} (n+1)T_f(r) + O(\log^+ T_f(r)) + o(\log r). \qquad (38)$$

If $n = 1$ then by (35) this reduces to the classical Second Main Theorem (Theorem 5.5).

Inequality (38) can also be expressed using counting functions as

$$\sum_{j=1}^{q} N_f(H_j, r) \geq_{\text{exc}} (q-n-1)T_f(r) - O(\log^+ T_f(r)) - o(\log r) \qquad (39)$$

(cf. (24)).

The corresponding definitions and theorem in number theory are as follows. These will all assume that k is a number field, that $S \supseteq S_\infty$ is a finite set of places of k, and that $n > 0$.

Definition 8.7. Let H be a hyperplane in $\mathbb{P}_k^n$ and let $a_0 x_0 + \cdots + a_n x_n = 0$ be a linear form defining it. (Since $\mathbb{P}_k^n$ is a scheme over k, this implies that $a_0, \ldots, a_n \in k$.) For all places v of k and all $P \in \mathbb{P}^n(k)$ not lying on H we then define

$$\lambda_{H,v}(P) = -\log \frac{\|a_0 x_0 + \cdots + a_n x_n\|_v}{\max\{\|x_0\|_v, \ldots, \|x_n\|_v\}}, \qquad (40)$$

where $[x_0 : \ldots : x_n]$ are homogeneous coordinates for P. Again, this is independent of the choice of homogeneous coordinates $[x_0 : \ldots : x_n]$ and depends on the choice of $a_0, \ldots, a_n$ only up to a bounded function which is zero for almost all v.

These functions are special cases of *Weil functions* (Definition 9.6), with domain restricted to $\mathbb{P}^n(k) \setminus H$.

Definition 8.8. For H and P as above, the **proximity function** for H is defined to be

$$m_S(H, P) = \sum_{v \in S} \lambda_{H,v}(P), \qquad (41)$$

and the **counting function** is defined by

$$N_S(H, P) = \sum_{v \notin S} \lambda_{H,v}(P). \qquad (42)$$

We then note that

$$\begin{aligned}
m_S(H,P) + N_S(H,P) &= \sum_{v \in M_k} -\log \frac{\|a_0 x_0 + \cdots + a_n x_n\|_v}{\max\{\|x_0\|_v, \ldots, \|x_n\|_v\}} \\
&= \sum_{v \in M_k} \log \max\{\|x_0\|_v, \ldots, \|x_n\|_v\} \\
&= h_k(P)
\end{aligned}$$

by the Product Formula.

Although this equality holds exactly, the proximity and counting functions depend (up to $O(1)$) on the choice of linear form $a_0 x_0 + \cdots + a_n x_n$ describing D, so we regard them as being defined only up to $O(1)$.

The counterpart to Theorem 8.6 (with, of course, a weaker error term) is a slightly weaker form of Schmidt's Subspace Theorem.

Theorem 8.9. (Schmidt) *Let k, S, and n be as above, let $H_1, \ldots, H_q$ be hyperplanes in $\mathbb{P}^n_k$ in general position, let $\varepsilon > 0$, and let $c \in \mathbb{R}$. Then*

$$\sum_{j=1}^{q} m_S(H_j, x) \leq (n+1+\varepsilon) h_k(x) + c \qquad (43)$$

for all $x \in \mathbb{P}^n(k)$ outside of a finite union of proper linear subspaces of $\mathbb{P}^n_k$. This latter set depends on k, S, $H_1, \ldots, H_q$, ε, c, and the choices used in defining the $m_S(H_j, x)$, but not on x.

When $n = 1$ this reduces to Roth's theorem (in the form of Theorem 4.4).

Note, in particular, that the H_i are hyperplanes in the k-scheme $\mathbb{P}^n_k$. This automatically implies that they can be defined by linear forms with coefficients in k. This corresponds to requiring the α_j to lie in k in the case of Roth's theorem. Schmidt's original formulation of his theorem allowed hyperplanes with algebraic coefficients; the reduction to hyperplanes in $\mathbb{P}^n_k$ is similar to the reduction for Roth's theorem and is omitted here. Also, Schmidt's original formulation was stated in terms of hyperplanes in $\mathbb{A}^{n+1}_k$ passing through the origin and points in $\mathbb{A}^{n+1}_k$ with integral coefficients. He also used the *size* instead of the height. For details on the equivalence of his original formulation and the form given here, see [87, Chap. 2, Sect. 2].

Theorem 8.9 is described as a slight weakening of Schmidt's Subspace Theorem because Schmidt actually allowed the set of hyperplanes to vary with v. Thus, to get a statement that was fully equivalent to Schmidt's original theorem, (43) would need to be replaced by

$$\sum_{v \in S} \sum_{j=1}^{q_v} m_S(H_{v,j}, x) \leq (n+1+\varepsilon) h_k(x) + c,$$

where for each $v \in S$, $H_{v,1}, \ldots, H_{v,q_v}$ are hyperplanes in general position (but in totality the set $\{H_{v,j} : v \in S, 1 \leq j \leq q_v\}$ need not be in general position, even after eliminating duplicates). Of course, at a given place v a point can be close to at most n of the $H_{v,j}$, so we may assume $q_v = n$ for all v (or actually $n+1$ is somewhat easier to work with).

Thus, a full statement of Schmidt's Subspace Theorem, rendered using the notation of Sect. 6, is as follows. It has been stated in a form that most readily carries over to Nevanlinna theory.

Theorem 8.10. (Schmidt's Subspace Theorem [71, Chap. VIII, Thm. 7A]) *Let k, S, and n be as above, and let $H_1, \ldots, H_q$ be distinct hyperplanes in $\mathbb{P}^n_k$. Then for all $\varepsilon > 0$ and all $c \in \mathbb{R}$ the inequality*

$$\sum_{v \in S} \max_J \sum_{j \in J} \lambda_{H_j, v}(x) \leq (n+1+\varepsilon) h_k(x) + c \tag{44}$$

holds for all $x \in \mathbb{P}^n(k)$ outside of a finite union of proper linear subspaces depending only on k, S, $H_1, \ldots, H_q$, ε, c, and the choices used in defining the $\lambda_{H_j, v}$. The max in this inequality is taken over all subsets J of $\{1, \ldots, q\}$ corresponding to subsets of $\{H_1, \ldots, H_q\}$ in general position.

In Nevanlinna theory there are infinitely many angles θ, so if one allowed the collection of hyperplanes to vary with θ without additional restriction, then the resulting statement could involve infinitely many hyperplanes, and would therefore likely be false (although this has not been proved). Therefore an overall restriction on the set of hyperplanes is needed in the case of Cartan's theorem, and is why Theorem 8.10 was stated in the way that it was.

Cartan's theorem itself can be generalized as follows.

Theorem 8.11. [90] *Let $n \in \mathbb{Z}_{>0}$, let $H_1, \ldots, H_q$ be hyperplanes in $\mathbb{P}^n_{\mathbb{C}}$, and let $f \colon \mathbb{C} \to \mathbb{P}^n(\mathbb{C})$ be a holomorphic curve whose image is not contained in a hyperplane. Then*

$$\int_0^{2\pi} \max_J \sum_{j \in J} \lambda_{H_j}(f(re^{i\theta})) \frac{d\theta}{2\pi} \leq_{\text{exc}} (n+1) T_f(r) + O(\log^+ T_f(r)) + o(\log r),$$

where J varies over the same collection of sets as in Theorem 8.10.

This has proved to be a useful formulation for applications; see [90] and [68]. The latter reference also improves the error term in Theorem 8.11.

Remark 8.12. It has been further shown that in Theorem 8.10, the finite set of linear subspaces can be taken to be the union of a finite number of points (depending on the same data as given in the theorem), together with a finite union of linear subspaces (of higher dimension) depending only on the collection of hyperplanes [88]. In other words, the higher-dimensional part of the exceptional set depends only on the geometric data. Correspondingly, Theorem 8.11 holds for all nonconstant holomorphic curves whose image is not contained in the union of this latter set [90]. For an example of the collection of higher dimensional subspaces for a specific set of lines in $\mathbb{P}^2$, see Example 14.3.

9 Varieties and Weil Functions

The goal of this section and the next is to carry over the definitions of the proximity, counting, and height functions to the context of varieties.

First it is necessary to define variety. Generally speaking, varieties and other algebro-geometric objects are as defined in [36], except that varieties (when discussing number theory at least) may be defined over a field that is not necessarily algebraically closed.

Definition 9.1. A **variety** over a field k, or a k-variety, is an integral separated scheme of finite type over k (i.e., over $\mathrm{Spec}\,k$). A **curve** over k is a variety over k of dimension 1. A **morphism** of varieties over k is a morphism of k-schemes. Finally, a **subvariety** of a variety (resp. closed subvariety, open subvariety) is an integral subscheme (resp. closed integral subscheme, open integral subscheme) of that variety (with induced map to $\mathrm{Spec}\,k$).

As an example, $X := \mathrm{Spec}\,\mathbb{Q}[x,y]/(y^2 - 2x^2)$ is a variety over $\mathbb{Q}$. Indeed, it is an integral scheme because the ring $\mathbb{Q}[x,y]/(y^2 - 2x^2)$ is entire. However, $X \times_{\mathbb{Q}} \mathbb{Q}(\sqrt{2})$ is not a variety over $\mathbb{Q}(\sqrt{2})$, since $\mathbb{Q}[x,y]/(y^2 - 2x^2) \otimes_{\mathbb{Q}} \mathbb{Q}(\sqrt{2})$ is not entire (the polynomial $y^2 - 2x^2$ is not irreducible over $\mathbb{Q}(\sqrt{2})$). Therefore, some authors require a variety to be *geometrically integral,* but we do not do so here. The advantage of not requiring geometric integrality is that every reduced closed subset is a finite union of closed subvarieties, without requiring base change to a larger field.

Many people would be tempted to say that the variety $X := \mathrm{Spec}\,\mathbb{Q}[x,y]/(y^2 - 2x^2)$ is not *defined over* $\mathbb{Q}$. Such wording does not make sense in this context (the variety is, after all, a $\mathbb{Q}$-variety). This wording usually comes about because the variety (in this instance) is associated to the line $y = \sqrt{2}x$ in $\mathbb{A}_{\mathbb{Q}}^2$, which does not come from any subvariety of $\mathbb{A}_{\mathbb{Q}}^2$ (without also obtaining the conjugate $y = -\sqrt{2}x$). The correct way to express this situation is to say that X is not *geometrically irreducible* (or not geometrically integral).

Strictly speaking, if $k \subseteq L$ are distinct fields, then $X(k)$ and $X(L)$ are disjoint sets. However, we will at times identify $X(k)$ with a subset of $X(L)$ in the obvious way. Following EGA, if $x \in X$ is a point, then $\kappa(x)$ will denote the residue field of the local ring at x. If $x \in X(L)$, then it is technically a morphism, but by abuse of notation $\kappa(x)$ will refer to the corresponding point on X (so $\kappa(x)$ may be smaller than L).

We also recall that the function field of a variety X is denoted $K(X)$. If ξ is the generic point of X, then $K(X) = \kappa(\xi)$.

The next goal of this section is to introduce Weil functions. These functions were introduced in Weil's thesis [98] and further developed in a later paper [99]. Weil functions give a way to write the height as a sum over places of a number field, and are exactly what is needed in order to generalize the proximity and counting functions to the geometric setting.

The description provided here will be somewhat brief; for a fuller treatment, see [46, Chap. 10].

We start with the very easy setting used in Nevanlinna theory.

Weil functions are best described using Cartier divisors.

Definition 9.2. Let D be a Cartier divisor on a complex variety X. Then a **Weil function** for D is a function $\lambda_D \colon (X \setminus \mathrm{Supp}\,D)(\mathbb{C}) \to \mathbb{R}$ such that for all $x \in X$

there is an open neighborhood U of x in X, a nonzero function $f \in K(X)$ such that $D|_U = (f)$, and a continuous function $\alpha \colon U(\mathbb{C}) \to \mathbb{R}$ such that

$$\lambda_D(x) = -\log|f(x)| + \alpha(x) \tag{45}$$

for all $x \in (U \setminus \mathrm{Supp}\, D)(\mathbb{C})$. Here the topology on $U(\mathbb{C})$ is the complex topology.

It is fairly easy to show that if λ_D is a Weil function, then the above condition is satisfied for any open set U and any nonzero $f \in \mathscr{O}_U$ satisfying $D|_U = (f)$.

Recall that linear equivalence classes of Cartier divisors on a variety are in natural one-to-one correspondence with isomorphism classes of line sheaves (invertible sheaves) on that variety. Moreover, for each divisor D on a variety X, if $\mathscr{L}$ is the corresponding line sheaf, then there is a nonzero rational section s of $\mathscr{L}$ whose vanishing describes D: $D = (s)$. As was noted by Néron, Weil functions on D correspond to metrics on $\mathscr{L}$.

Recall that if X is a complex variety and $\mathscr{L}$ is a line sheaf on X, then a **metric** on $\mathscr{L}$ is a collection of norms on the fibers of the complex line bundle corresponding to the sheaf $\mathscr{L}$, varying smoothly or continuously with the point on X. Such a metric is called a **smooth metric** or **continuous metric**, respectively. In these notes, **smooth** means C^∞. If X is singular, then we say that a function $f \colon X(\mathbb{C}) \to \mathbb{C}$ is C^∞ at a point $P \in X(\mathbb{C})$ if there is an open neighborhood U of P in $X(\mathbb{C})$ in the complex topology, a holomorphic function $\phi \colon U \to \mathbb{C}^n$ for some n, and a C^∞ function $g \colon \mathbb{C}^n \to \mathbb{C}$ such that $f = g \circ \phi$. This reduces to the usual concept of C^∞ function at smooth points of X.

To describe a metric on $\mathscr{L}$ in concrete terms, let U be an open subset of X and let $\phi_U \colon \mathscr{O}_U \xrightarrow{\sim} \mathscr{L}|_U$ be a local trivialization. Then the function $\rho_U \colon U(\mathbb{C}) \to \mathbb{R}_{>0}$ given by $\rho_U(x) = |\phi_U(1)(x)|$ is smooth (resp. continuous), and for any section $s \in \mathscr{L}(U)$ and any $x \in U(\mathbb{C})$, we have $|s(x)| = \rho_U(x) \cdot |\phi_U^{-1}(s)(x)|$. Moreover, if V is another open set in X and $\phi_V \colon \mathscr{O}_V \xrightarrow{\sim} \mathscr{L}|_V$ is a local trivialization on V, then $\phi_U^{-1} \circ \phi_V$ (appropriately restricted) is an automorphism of $\mathscr{O}_{U \cap V}$ corresponding to multiplication by a function $\alpha_{UV} \in \mathscr{O}_{U \cap V}^*$. Again letting $\rho_V(x) = |\phi_V(1)(x)|$, we see that ρ_U and ρ_V are related by $\rho_V(x) = |\alpha_{UV}(x)| \rho_U(x)$ for all $x \in (U \cap V)(\mathbb{C})$.

Conversely, an isomorphism class of line sheaves on X can be uniquely specified by giving an open cover $\mathscr{U}$ of X and $\alpha_{UV} \in \mathscr{O}_{U \cap V}^*$ for all $U, V \in \mathscr{U}$ satisfying $\alpha_{UU} = 1$ and $\alpha_{UW} = \alpha_{UV} \alpha_{VW}$ on $U \cap V \cap W$ for all $U, V, W \in \mathscr{U}$. Moreover, with these data, one can specify a metric on the associated line sheaf by giving smooth or continuous functions $\rho_U \colon U(\mathbb{C}) \to \mathbb{R}_{>0}$ for each $U \in \mathscr{U}$ that satisfy $\rho_V = |\alpha_{UV}| \rho_U$ on $U \cap V$ for all $U, V \in \mathscr{U}$.

A continuous metric on a line sheaf $\mathscr{L}$ determines a Weil function for any associated Cartier divisor D. Indeed, if s is a nonzero rational section of $\mathscr{L}$ such that $D = (s)$, then $\lambda_D(x) = -\log|s(x)|$ is a Weil function for D. Conversely, a Weil function for D determines a continuous metric on $\mathscr{L}$.

In Nevanlinna theory it is customary to work only with smooth metrics, and hence it is often better to work with Weil functions associated to smooth metrics (equivalently, to Weil functions for which the functions α in (45) are all smooth).

An example of a Weil function in Nevanlinna theory (and perhaps the primary example) is the function λ_H of Definition 8.2 used in Cartan's theorem.

Likewise, the function $\lambda_{H,v}$ of Definition 8.7 is an example of a Weil function in number theory. In this case, it is no longer sufficient to say that two Weil functions agree up to $O(1)$: the implied constant also has to vanish for almost all v. For example, Lemma 6.2 compares the difference of two Weil functions, and shows that the difference is bounded by a bound that vanishes for almost all v. A plain bound of $O(1)$ would not suffice to give a finite bound in Theorem 6.3.

Before defining Weil functions in the number theory case, we first give some definitions relevant to the domains of Weil functions.

Definition 9.3. Let v be a place of a number field k. Then $\mathbb{C}_v$ is the completion of the algebraic closure $\bar{k}_v$ of the completion k_v of k at v.

Recall [42, Chap. III, Sects. 3 and 4] that if v is non-archimedean then $\bar{k}_v$ is not complete, but its completion $\mathbb{C}_v$ is algebraically closed. If v is archimedean, then $\mathbb{C}_v$ is isomorphic to the field of complex numbers (as is $\bar{k}_v$). The norm $\|\cdot\|_v$ on k extends uniquely to norms on k_v, on $\bar{k}_v$, and on $\mathbb{C}_v$. If X is a variety, then the norm on $\mathbb{C}_v$ defines a topology on $X(\mathbb{C}_v)$, called the v-topology. It is defined to be the coarsest topology such that for all open $U \subseteq X$ and all $f \in \mathcal{O}(U)$, $U(\mathbb{C}_v)$ is open and $f: U(\mathbb{C}_v) \to \mathbb{C}_v$ is continuous.

One can also work just with the algebraic closure $\bar{k}_v$ when defining Weil functions, without any essential difference.

Definition 9.4. Let X be a variety over a number field k. Then $X(M_k)$ is the disjoint union

$$X(M_k) = \coprod_{v \in M_k} X(\mathbb{C}_v).$$

This set is given a topology defined by the condition that $A \subseteq X(M_k)$ is open if and only if $A \cap X(\mathbb{C}_v)$ is open in the v-topology for all v.

Definition 9.5. Let k be a number field. Then an M_k-**constant** is a collection (c_v) of constants $c_v \in \mathbb{R}$ for each $v \in M_k$, such that $c_v = 0$ for almost all v. If X is a variety over k, then a function $\alpha: X(M_k) \to \mathbb{R}$ is said to be $O_{M_k}(1)$ if there is an M_k-constant (c_v) such that $|\alpha(x)| \leq c_v$ for all $x \in X(\mathbb{C}_v)$ and all $v \in M_k$.

We may then define Weil functions as follows.

Definition 9.6. Let X be a variety over a number field k, and let D be a Cartier divisor on X. Then a **Weil function** for D is a function $\lambda_D: (X \setminus \operatorname{Supp} D)(M_k) \to \mathbb{R}$ that satisfies the following condition. For each $x \in X$ there is an open neighborhood U of x, a nonzero function $f \in \mathcal{O}(U)$ such that $D|_U = (f)$, and a continuous locally M_k-bounded function $\alpha: U(M_k) \to \mathbb{R}$ satisfying

$$\lambda_D(x) = -\log \|f(x)\|_v + \alpha(x) \tag{46}$$

for all $v \in M_k$ and all $x \in (U \setminus \operatorname{Supp} D)(\mathbb{C}_v)$.

For the definition of locally M_k-bounded function, see [46, Chap. 10, Sect. 1]. The definition is more complicated than one would naively expect, stemming from the fact that $\mathbb{C}_v$ is totally disconnected, and not locally compact. For our purposes, though, it suffices to note that if X is a complete variety then such a function is $O_{M_k}(1)$. (In other contexts, these problems are dealt with by using Berkovich spaces, but Weil's work does not use them, not least because it came much earlier.)

As with Definition 9.2, if λ is a Weil function for D, then it can be shown that the above condition is true for all open $U \subseteq X$ and all $f \in \mathscr{O}(U)$ for which $D|_U = (f)$.

If λ_D is a Weil function for D, then we write

$$\lambda_{D,v} = \lambda_D\big|_{(X\setminus \mathrm{Supp}\, D)(\mathbb{C}_v)}$$

for all places v of k. If v is an archimedean place, then $\mathbb{C}_v \cong \mathbb{C}$, and $\lambda_{D,v}$ is a Weil function for D in the sense of Definition 9.2(up to a factor $1/2$ if v is a complex place).

In the future, if $x \in X(M_k)$ and f is a function on X, then $\|f(x)\|$ will mean $\|f(x)\|_v$ for the (unique) place v such that $x \in X(\mathbb{C}_v)$. Thus, (46) could be shortened to $\lambda(x) = -\log\|f(x)\| + \alpha(x)$ for all $x \in (U \setminus \mathrm{Supp}\, D)(M_k)$.

Of course, this discussion would be academic without the following theorem.

Theorem 9.7. *Let k be a number field, let X be a projective variety over k, and let D be a Cartier divisor on X. Then there exists a Weil function for D.*

For the proof, see [46, Chap. 10]. This is also true for complete varieties, using Nagata's embedding theorem to construct a model for X and then using Arakelov theory to define the Weil function. But, again, the details are beyond the scope of these notes.

Weil functions have the following properties.

Theorem 9.8. *Let X be a complete variety over a number field k. Then*

(a) **Additivity:** *If λ_1 and λ_2 are Weil functions for Cartier divisors D_1 and D_2 on X, respectively, then $\lambda_1 + \lambda_2$ extends uniquely to a Weil function for $D_1 + D_2$.*

(b) **Functoriality:** *If λ is a Weil function for a Cartier divisor D on X, and if $f: X' \to X$ is a morphism of k-varieties such that $f(X') \not\subseteq \mathrm{Supp}\, D$, then $x \mapsto \lambda(f(x))$ is a Weil function for the Cartier divisor f^*D on X'.*

(c) **Normalization:** *If $X = \mathbb{P}^n_k$, and if D is the hyperplane at infinity, then the function*

$$\lambda_{D,v}([x_0 : \ldots : x_n]) := -\log \frac{\|x_0\|_v}{\max\{\|x_0\|_v, \ldots, \|x_n\|_v\}} \tag{47}$$

is a Weil function for D.

(d) **Uniqueness:** *If both λ_1 and λ_2 are Weil functions for a Cartier divisor D on X, then $\lambda_1 = \lambda_2 + O_{M_k}(1)$.*

(e) **Boundedness from below:** *If D is an effective Cartier divisor and λ is a Weil function for D, then λ is bounded from below by an M_k-constant.*

(f) **Principal divisors:** *If D is a principal divisor (f), then $-\log\|f\|$ is a Weil function for D.*

The proofs of these properties are left to the reader (modulo the properties of locally M_k-bounded functions).

Parts (b) and (c) of the above theorem combine to give a way of computing Weil functions for effective very ample divisors. This, in turn, gives rise to the "max-min" method for computing Weil functions for arbitrary Cartier divisors on projective varieties.

Lemma 9.9. *Let $\lambda_1, \dots, \lambda_n$ be Weil functions for Cartier divisors $D_1, \dots, D_n$, respectively, on a complete variety X over a number field k. Assume that the divisors D_i are of the form $D_i = D_0 + E_i$, where D_0 is a fixed Cartier divisor and E_i are effective for all i. Assume also that $\operatorname{Supp} E_1 \cap \cdots \cap \operatorname{Supp} E_n = \emptyset$. Then the function*

$$\lambda(x) = \min\{\lambda_i(x) : x \notin \operatorname{Supp} E_i\}$$

is defined everywhere on $(X \setminus \operatorname{Supp} D_0)(M_k)$, and is a Weil function for D_0.

Proof. See [46, Chap. 10, Prop. 3.2]. □

Theorem 9.10. (Max-min) *Let X be a projective variety over a number field k, and let D be a Cartier divisor on X. Then there are positive integers m and n, and nonzero rational functions f_{ij} on X, $1 = 1, \dots, n$, $j = 1, \dots, m$, such that*

$$\lambda(x) := \max_{1 \le i \le n} \min_{1 \le j \le m} \left(-\log \|f_{ij}\| \right)$$

defines a Weil function for D.

Proof. We may write D as a difference $E - F$ of very ample divisors. Let $E_1, \dots, E_n$ be effective Cartier divisors linearly equivalent to E such that $\bigcap \operatorname{Supp} E_i = \emptyset$ (for example, pull-backs of hyperplane sections via a projective embedding associated to E). Likewise, let $F_1, \dots, F_m$ be effective Cartier divisors linearly equivalent to F with $\bigcap \operatorname{Supp} F_j = \emptyset$. Then $D - E_i + F_j$ is a principal divisor for all i and j; hence

$$D - E_i + F_j = (f_{ij})$$

for some $f_{ij} \in K(X)^*$ and all i and j. Applying Lemma 9.9 to $-\log \|f_{ij}\|$ then implies that $\min_{1 \le j \le m}(-\log \|f_{ij}\|)$ is a Weil function for $D - E_i$ for all i. Applying Lemma 9.9 again to the negatives of these Weil functions then gives the theorem. □

To conclude the section, we give some notation that will be useful for working with rational and algebraic points.

Definition 9.11. Let X be a variety over a number field k, let D be a Cartier divisor on X, and let λ_D be a Weil function for D. If L is a number field containing k, and if w is a place of L lying over a place v of k, then we identify $\mathbb{C}_w$ with $\mathbb{C}_v$ in the obvious manner, and write

$$\lambda_{D,w} = [L_w : k_v]\lambda_{D,v}. \tag{48}$$

(Recall that $\|x\|_w = \|x\|_v^{[L_w:k_v]}$ for all $x \in \mathbb{C}_v$, by (4).) Finally, each point $x \in X(L)$ gives rise to points $x_w \in X(\mathbb{C}_w)$ for all $w \in M_L$, and we define

$$\lambda_{D,w}(x) = \lambda_{D,w}(x_w) \tag{49}$$

if $x \notin \operatorname{Supp} D$.

Note that, if $x \in (X \setminus \operatorname{Supp} D)(L)$, if L' is a number field containing L, if w is a place of L, and if w' is a place of L' lying over w, then

$$\lambda_{D,w'}(x) = [L'_{w'} : L_w]\lambda_{D,w}(x), \tag{50}$$

regardless of whether the left-hand side is defined using (48) or (49) (by regarding $X(L)$ as a subset of $X(L')$ for the latter).

If (c_v) is an M_k-constant, if w is a place of a number field L containing k, and if v is the place of k lying under w, then we write

$$c_w = [L_w : k_v]c_v, \tag{51}$$

so that the condition $\lambda_{D,w} \leq c_w$ will be equivalent to $\lambda_{D,v} \leq c_v$, by (48).

10 Height Functions on Varieties in Number Theory

Weil functions can be used to generalize the height h_k (defined in Sect. 3) to arbitrary complete varieties over k. This can also be done by working directly with heights; see [46, Chap. 3] or [77].

> Throughout this section, k is a number field, X is a complete variety over k, and D is a Cartier divisor on X, unless otherwise specified.

Definition 10.1. Let λ be a Weil function for D, and let $x \in X(\bar{k})$ be an algebraic point with $x \notin \operatorname{Supp} D$. Then the **height** of x relative to λ and k is defined as

$$h_{\lambda,k}(x) = \frac{1}{[L:k]} \sum_{w \in M_L} \lambda_w(x) \tag{52}$$

for any number field $L \supseteq \kappa(x)$. It is independent of the choice of L by (50).

In particular, if $x \in X(k)$, then

$$h_{\lambda,k}(x) = \sum_{v \in M_k} \lambda_v(x).$$

Specializing in a different direction, if $X = \mathbb{P}^n_k$, if D is the hyperplane at infinity, and if λ is the Weil function (47), then

$$
\begin{aligned}
h_{\lambda,k}([x_0 : \ldots : x_n]) &= -\frac{1}{[L:k]} \sum_{w \in M_L} \log \frac{\|x_0\|_v}{\max\{\|x_0\|_v, \ldots, \|x_n\|_v\}} \\
&= \frac{1}{[L:k]} \sum_{w \in M_L} \log \max\{\|x_0\|_v, \ldots, \|x_n\|_v\} \\
&= h_k([x_0 : \ldots : x_n])
\end{aligned}
\tag{53}
$$

for all $[x_0 : \ldots : x_n] \in \mathbb{P}^n(\bar{k})$ with $x_0 \neq 0$, where L is any number field containing the field of definition of this point.

The restriction $x \notin \operatorname{Supp} D$ can be eliminated as follows.

Let D' be another Cartier divisor on X linearly equivalent to D, say $D' = D + (f)$; then $\lambda' := \lambda - \log\|f\|$ is a Weil function for D'. If $x \in X(\bar{k})$ does not lie on $\operatorname{Supp} D \cup \operatorname{Supp} D'$, and if L is a number field containing $\kappa(x)$, then

$$
\begin{aligned}
h_{\lambda',k}(x) &= \frac{1}{[L:k]} \sum_{w \in M_L} \lambda'_v(x) \\
&= \frac{1}{[L:k]} \sum_{w \in M_L} \lambda_v(x) - \frac{1}{[L:k]} \sum_{w \in M_L} \log\|f(x)\|_w \\
&= h_{\lambda,k}(x)
\end{aligned}
\tag{54}
$$

by the Product Formula (6). Thus we have:

Definition 10.2. Let λ be a Weil function for D, and let $x \in X(\bar{k})$. Then, for any $f \in K(X)^*$ such that the support of $D + (f)$ does not contain x, we define

$$
h_{\lambda,k}(x) = h_{\lambda - \log\|f\|,k}(x),
$$

where $h_{\lambda - \log\|f\|,k}$ on the right-hand side is defined using Definition 10.1. This expression is independent of the choice of f by (54), and agrees with Definition 10.1 when $x \notin \operatorname{Supp} D$ since we can take $f = 1$ in that case.

With this definition, (53) holds without the restriction $x_0 \neq 0$.

Proposition 10.3. *If both λ and λ' are Weil functions for D, then*

$$
h_{\lambda',k} = h_{\lambda,k} + O(1).
$$

Proof. Indeed, this is immediate from Theorem 9.8d. □

Thus, the height function defined above depends only on the divisor; moreover, by (54) it depends only on the linear equivalence class of the divisor.

Definition 10.4. The height $h_{D,k}(x)$ for points $x \in X(\bar{k})$ is defined, up to $O(1)$, as

$$h_{D,k}(x) = h_{\lambda,k}(x)$$

for any Weil function λ for D. If $\mathscr{L}$ is a line sheaf on X, then we define

$$h_{\mathscr{L},k}(x) = h_{D,k}(x)$$

for points $x \in X(\bar{k})$, where D is any Cartier divisor for which $\mathscr{O}(D) \cong \mathscr{L}$. Again, it is only defined up to $O(1)$.

By (53), we then have

$$h_{\mathscr{O}(1),k} = h_k + O(1)$$

on $\mathbb{P}_k^n$ for all $n > 0$. Since the automorphism group of $\mathbb{P}_k^n$ is transitive on the set of rational points, and since automorphisms preserve the line sheaf $\mathscr{O}(1)$, the term $O(1)$ in the above formula cannot be eliminated without additional structure. Thus, Definition 10.4 cannot give an exact definition for the height without additional structure. (This additional structure can be given using Arakelov theory.)

Theorem 9.8 and (50) also immediately imply the following properties of heights:

Theorem 10.5. *(a)* ***Functoriality:*** *If $f : X' \to X$ is a morphism of k-varieties, and if $\mathscr{L}$ is a line sheaf on X, then*

$$h_{f^*\mathscr{L},k}(x) = h_{\mathscr{L},k}(f(x)) + O(1)$$

for all $x \in X'(\bar{k})$, where the implied constant depends only on f, $\mathscr{L}$, and the choices of the height functions.

(b) ***Additivity:*** *If $\mathscr{L}_1$ and $\mathscr{L}_2$ are line sheaves on X, then*

$$h_{\mathscr{L}_1 \otimes \mathscr{L}_2,k}(x) = h_{\mathscr{L}_1,k}(x) + h_{\mathscr{L}_2,k}(x) + O(1)$$

for all $x \in X(\bar{k})$, where the implied constant depends only on $\mathscr{L}_1$, $\mathscr{L}_2$, and the choices of the height functions.

(c) ***Base locus:*** *If $h_{D,k}$ is a height function for D, then it is bounded from below outside of the base locus of the complete linear system $|D|$.*

(d) ***Globally generated line sheaves:*** *If $\mathscr{L}$ is a line sheaf on X, and is generated by its global sections, then $h_{\mathscr{L},k}(x)$ is bounded from below for all $x \in X(\bar{k})$, by a bound depending only on $\mathscr{L}$ and the choice of height function.*

(e) ***Change of number field:*** *If $L \supseteq k$ then*

$$h_{\mathscr{L},L}(x) = [L : k] h_{\mathscr{L},k}(x)$$

for all line sheaves $\mathscr{L}$ on X and all $x \in X(\bar{k})$. (Strictly speaking, the left-hand side should be $h_{\mathscr{L}',L}(x')$, where $\mathscr{L}'$ is the pull-back of $\mathscr{L}$ to $X_L := X \times_k L$ and x' is the point in $X_L(\bar{k})$ corresponding to $x \in X(\bar{k})$.)

Corollary 10.6. *If $\mathscr{L}$ is an ample line sheaf on X, then $h_{\mathscr{L},k}(x)$ is bounded from below for all $x \in X(\bar{k})$, by a bound depending only on $\mathscr{L}$ and the choice of height function.*

Proof. By Theorem 10.5b, we may replace $\mathscr{L}$ with $\mathscr{L}^{\otimes n}$ for any positive integer n, and therefore may assume that $\mathscr{L}$ is very ample. Then the result follows from Theorem 10.5d.   □

The following result shows that heights relative to ample line sheaves are the largest possible heights, up to a constant multiple.

Proposition 10.7. *Let $\mathscr{L}$ and $\mathscr{M}$ be line sheaves on X, with $\mathscr{L}$ ample. Then there is a constant C, depending only on $\mathscr{L}$ and $\mathscr{M}$, such that*

$$h_{\mathscr{M},k}(x) \leq C h_{\mathscr{L},k}(x) + O(1)$$

for all $x \in X(\bar{k})$, where the implied constant depends only on $\mathscr{L}$, $\mathscr{M}$, and the choices of height functions.

Proof. By the definition of ampleness, there is an integer n such that the line sheaf $\mathscr{L}^{\otimes n} \otimes \mathscr{M}^{\vee}$ is generated by global sections. Therefore an associated height function

$$h_{\mathscr{L}^{\otimes n} \otimes \mathscr{M}^{\vee},k} = n h_{\mathscr{L},k} - h_{\mathscr{M},k} + O(1)$$

is bounded from below, giving the result with $C = n$.   □

For projective varieties, Northcott's finiteness theorem can be carried over.

Theorem 10.8. (Northcott) *Assume that X is projective, and let $\mathscr{L}$ be an ample line sheaf on X. Then, for all integers d and all $c \in \mathbb{R}$, the set*

$$\{x \in X(\bar{k}) : [\kappa(x) : k] \leq d \text{ and } h_{\mathscr{L},k}(x) \leq c\} \tag{55}$$

is finite.

Proof. First, if $X = \mathbb{P}_k^n$ and $\mathscr{L} = \mathscr{O}(1)$, then the result follows by bounding the heights of $[x_i : x_0]$ (if $x_0 \neq 0$, which we assume without loss of generality), and applying Theorem 3.1 to these points for each i. The general case then follows by replacing $\mathscr{L}$ with a very ample positive multiple and using an associated projective embedding and functoriality of heights.   □

Of course, if X is not projective then it has no ample divisors, making the above two statements vacuous. Complete varieties have a notion that is almost as good, though.

Definition 10.9. Let X be a complete variety over an arbitrary field. A line sheaf $\mathscr{L}$ on X is **big** if there is a constant $c > 0$ such that

$$h^0(X, \mathscr{L}^{\otimes n}) \geq cn^{\dim X}$$

for all sufficiently large and divisible n. A Cartier divisor D on X is big if $\mathscr{O}(D)$ is big.

If X is a complete variety over an arbitrary field, then by Chow's lemma there is a projective variety X' and a proper birational morphism $\pi\colon X' \to X$. If $\mathscr{L}$ is a big line sheaf on X, then $\pi^*\mathscr{L}$ is big on X'. Therefore, it makes some sense to compare big line sheaves with ample ones.

Proposition 10.10. (Kodaira's lemma) *Let X be a projective variety over an arbitrary field, and let $\mathscr{L}$ and $\mathscr{A}$ be line sheaves on X, with $\mathscr{A}$ ample. Then $\mathscr{L}$ is big if and only if there is a positive integer n such that $H^0(X, \mathscr{L}^{\otimes n} \otimes \mathscr{A}^\vee) \neq 0$. Equivalently, if D and A are Cartier divisors on X, with A ample, then D is big if and only if some positive multiple of it is linearly equivalent to the sum of A and an effective divisor.*

Proof. See [87, Prop. 1.2.7]. □

The above allows us to show that heights relative to big line sheaves are also, well, big.

Proposition 10.11. *Let X be a complete variety over a number field. Let $\mathscr{L}$ and $\mathscr{M}$ be line sheaves on X, with $\mathscr{L}$ big. Then there is a constant C and a proper Zariski-closed subset Z of X, depending only on $\mathscr{L}$ and $\mathscr{M}$, such that*

$$h_{\mathscr{M},k}(x) \leq Ch_{\mathscr{L},k}(x) + O(1)$$

for all $x \in X(\bar{k})$ outside of Z, where the implied constant depends only on $\mathscr{L}$, $\mathscr{M}$, and the choices of height functions.

Proof. After applying Chow's lemma and pulling back $\mathscr{L}$ and $\mathscr{M}$, we may assume that X is projective. We may also replace $\mathscr{L}$ with a positive multiple, and hence may assume that $\mathscr{L}$ is isomorphic to $\mathscr{A} \otimes \mathscr{O}(D)$, where $\mathscr{A}$ is an ample line sheaf and D is an effective Cartier divisor. Then the result follows from Proposition 10.7, with $Z = \operatorname{Supp} D$, by Theorem 10.5. □

Unfortunately, it is still not true that an arbitrary complete variety must have a big line sheaf. But it is true if the variety is nonsingular, since one can then take the complement of any open affine subset.

For general complete varieties, we can do the following.

Remark 10.12. For a general complete variety X over k, we can define a **big height** to be a function $h\colon X(\bar{k}) \to \mathbb{R}$ for which there exist disjoint subvarieties $U_1, \ldots, U_n$ of X (not necessarily open or closed), with $\bigcup U_i = X$; and for each $i = 1, \ldots, n$ a projective embedding $U_i \hookrightarrow \overline{U}_i$, an ample line sheaf $\mathscr{L}_i$ on $\overline{U}_i$, and real constants $c_i > 0$ and C_i such that $h(x) \geq c_i h_{\mathscr{L}_i,k}(x) + C_i$ for all $x \in U_i(\bar{k})$. One can then show:

- Every complete variety over k has a big height;
- Any two big heights on a given complete variety are bounded from above by linear functions of each other;
- If X is a projective variety and $\mathscr{L}$ is an ample line sheaf on X then $h_{\mathscr{L},k}$ is a big height on X;
- If $\mathscr{L}$ is a line sheaf on X then there are real constants c and C such that $h_{\mathscr{L},k}(x) \leq c\,h(x) + C$ for all $x \in X(\bar{k})$;
- The restriction of a big height to a closed subvariety is a big height on that subvariety; and
- A counterpart to Proposition 10.13 (below) holds for big heights on complete varieties.

Details of these assertions are left to the reader.

Big heights are useful for error terms: the conjectures and theorems that follow are generally stated for projective varieties, with an error term involving a height relative to an ample divisor. However, they can also be stated more generally for complete varieties if the height is changed to a big height. For concreteness, though, the more restricted setting of projective varieties is used.

Finally, we note a case in which Z can be bounded explicitly. This will be used in the proof of Proposition 30.3.

Proposition 10.13. *Let $f\colon X_1 \to X_2$ be a morphism of projective varieties over a number field, and let $\mathscr{A}_1$ and $\mathscr{A}_2$ be ample line sheaves on X_1 and X_2, respectively. Then there is a constant C, depending only on f, $\mathscr{A}_1$, and $\mathscr{A}_2$ such that*

$$h_{\mathscr{A}_1,k}(x) \leq C h_{\mathscr{A}_2,k}(f(x)) + O(1) \tag{56}$$

for all points $x \in X_1(\bar{k})$ that are isolated in their fibers of f, where the implied constant depends only on f, $\mathscr{A}_1$, $\mathscr{A}_2$, and the choices of height functions.

Proof. If no closed points x of X_1 are isolated in their fibers of f, then there is nothing to prove. If there is at least one such point x, then $\dim f(X_1) = \dim X_1$, so $f^*\mathscr{A}_2$ is big. The result then follows by Proposition 10.11 and noetherian induction applied to the irreducible components of the set Z in that proposition $\square$

Note that if any fiber component of f has dimension >0, then it contains algebraic points of arbitrarily large height, so (56) cannot possibly hold for all such points.

11   Proximity and Counting Functions on Varieties in Number Theory

The definitions of proximity and counting functions given in Sects. 6 and 8 also generalize readily to points on varieties.

Throughout this section, k is a number field, S is a finite set of places of k containing S_∞, and X is a complete variety over k.

Definition 11.1. Let D be a Cartier divisor on X, let λ_D be a Weil function for D, let $x \in X(\bar{k})$ with $x \notin \mathrm{Supp}\,D$, let $L \supseteq k$ be a number field such that $x \in X(L)$, and let T be the set of places of L lying over places in S. Then the **proximity function** and **counting function** in this setting are defined up to $O(1)$ by

$$m_S(D,x) = \frac{1}{[L:k]} \sum_{w \in T} \lambda_{D,w}(x) \qquad \text{and} \qquad N_S(D,x) = \frac{1}{[L:k]} \sum_{w \notin T} \lambda_{D,w}(x).$$

These expressions are independent of the choice of L, by (48). They depend on the choice of λ_D only up to $O(1)$.

Unlike the height, the proximity and counting functions depend on D, even within a linear equivalence class. Therefore the restriction $x \notin \mathrm{Supp}\,D$ cannot be eliminated.

By (52) and Definition 10.4, we have

$$h_{D,k}(x) = m_S(D,x) + N_S(D,x)$$

for all $x \in X(\bar{k})$ outside of the support of D. This is, basically, the First Main Theorem. The Second Main Theorem in this context is still a conjecture (Conjecture 15.6).

Theorem 9.8 immediately implies the following properties of proximity and counting functions.

Proposition 11.2. *In number theory, proximity and counting functions have the following properties.*

(a) **Additivity:** *If D_1 and D_2 are Cartier divisors on X, then*

$$m_S(D_1 + D_2, x) = m_S(D_1, x) + m_S(D_2, x) + O(1)$$

and

$$N_S(D_1 + D_2, x) = N_S(D_1, x) + N_S(D_2, x) + O(1)$$

for all $x \in X(\bar{k})$ outside of the supports of D_1 and D_2.

(b) **Functoriality:** *If $f : X' \to X$ is a morphism of complete k-varieties and D is a divisor on X whose support does not contain the image of f, then*

$$m_S(f^*D, x) = m_S(D, f(x)) + O(1)$$

and

$$N_S(f^*D,x) = N_S(D,f(x)) + O(1)$$

for all $x \in X'(\bar{k})$ outside of the support of f^*D.

(c) **Effective divisors:** *If D is an effective Cartier divisor on X, then $m_S(D,x)$ and $N_S(D,x)$ are bounded from below for all $x \in X(\bar{k})$ outside of the support of D.*

(d) **Change of number field:** *If L is a number field containing k and if T is the set of places of L lying over places in S, then*

$$m_T(D,x) = [L:k]m_S(D,x) + O(1)$$

and

$$N_T(D,x) = [L:k]N_S(D,x) + O(1)$$

for all $x \in X(\bar{k})$ outside of the support of D (with the same abuse of notation as in Theorem 10.5e).

In each of the above cases, the implied constant in $O(1)$ depends on the varieties, divisors, and morphisms, but not on x.

When working with proximity and height functions, the divisor D is almost always assumed to be effective.

12 Height, Proximity, and Counting Functions in Nevanlinna Theory

The height, proximity, and counting functions of Nevanlinna theory can also be generalized to the context of a divisor on a complete complex variety.

In this section, X is a complete complex variety, D is a Cartier divisor on X, and $f: \mathbb{C} \to X$ is a holomorphic curve whose image is not contained in the support of D. Throughout these notes, we will often implicitly think of a complex variety X as a complex analytic space [36, App. B].

We begin with the proximity and counting functions.

Definition 12.1. Let λ be a Weil function for D. Then the **proximity function** for f relative to D is the function

$$m_f(D,r) = \int_0^{2\pi} \lambda(f(re^{i\theta})) \frac{d\theta}{2\pi}.$$

It is defined only up to $O(1)$.

If λ' is another Weil function for D, then $|\lambda - \lambda'|$ is bounded, so the proximity function is independent of D (up to $O(1)$).

Definition 12.2. The **counting function** for f relative to D is the function

$$N_f(D,r) = \sum_{0<|z|<r} \mathrm{ord}_z f^*D \cdot \log\frac{r}{|z|} + \mathrm{ord}_0 f^*D \cdot \log r.$$

Unlike the proximity function and the counting function in Nevanlinna theory, this function is defined exactly.

Corresponding to Proposition 11.2, we then have

Proposition 12.3. *In Nevanlinna theory, proximity and counting functions have the following properties.*

(a) *Additivity: If D_1 and D_2 are Cartier divisors on X, then*

$$m_f(D_1+D_2,r) = m_f(D_1,r) + m_f(D_2,r) + O(1)$$

and

$$N_f(D_1+D_2,r) = N_f(D_1,r) + N_f(D_2,r).$$

(b) *Functoriality: If $\phi: X \to X'$ is a morphism of complete complex varieties and D' is a Cartier divisor on X' whose support does not contain the image of $\phi \circ f$, then*

$$m_f(\phi^*D',r) = m_{\phi\circ f}(D',r) + O(1) \qquad and \qquad N_f(\phi^*D',r) = N_{\phi\circ f}(D',r).$$

(c) *Effective divisors: If D is effective, then $m_f(D,r)$ is bounded from below and $N_f(D,r)$ is nonnegative.*

In each of the above cases, the implied constant in $O(1)$ depends on the varieties, divisors, and morphisms, but not on f or r.

We can now define the height.

Definition 12.4. The **height function** relative to D is defined, up to $O(1)$, as

$$T_{D,f}(r) = m_f(D,r) + N_f(D,r).$$

Proposition 12.5. *The height function $T_{D,f}$ is additive in D, is functorial, and is bounded from below if D is effective, as in Proposition 12.3.*

Proof. Immediate from Proposition 12.3.   □

Proposition 12.6. (First Main Theorem) *Let D' be another Cartier divisor on X whose support does not contain the image of f, and assume that D' is linearly equivalent to D. Then*

$$T_{D',f}(r) = T_{D,f}(r) + O(1). \tag{57}$$

Proof. We first consider the special case $X = \mathbb{P}^1_{\mathbb{C}}$, $D = [0]$ (the image of the point 0 under the injection $\mathbb{A}^1 \hookrightarrow \mathbb{P}^1$), and $D' = [\infty]$ (the point at infinity, with multiplicity one). Then $T_{D',f}(r) = T_f(r) + O(1)$, $m_f(D,r) = m_f(0,r) + O(1)$, and $N_f(D,r) = N_f(0,r)$ (where $T_f(r)$, $m_f(0,r)$, and $N_f(0,r)$ are as defined in Sect. 5). The result then follows by Theorem 5.4 (the First Main Theorem for meromorphic functions).

In the general case, write $D - D' = (g)$ for some $g \in K(X)^*$. Then g defines a rational map $X \dashrightarrow \mathbb{P}^1_{\mathbb{C}}$. Let X' be the closure of the graph, with projections $p \colon X' \to X$ and $q \colon X' \to \mathbb{P}^1_{\mathbb{C}}$. By the additivity property of heights, (57) is equivalent to $T_{D-D',f}(r)$ being bounded. By the special case proved already, $T_{[0]-[\infty],g \circ f}(r)$ is bounded. The holomorphic curve $f \colon \mathbb{C} \to X$ lifts to a function $f' \colon \mathbb{C} \to X'$ that satisfies $p \circ f' = f$ and $q \circ f' = g \circ f$. By functoriality, we then have

$$\begin{aligned}
T_{D-D',f}(r) &= T_{p^*(D-D'),f'}(r) + O(1) \\
&= T_{q^*([0]-[\infty]),f'}(r) + O(1) \\
&= T_{[0]-[\infty],q \circ f'}(r) + O(1) \\
&= T_{[0]-[\infty],g \circ f}(r) + O(1) \\
&= O(1),
\end{aligned}$$

which implies the proposition.   □

Definition 12.7. The **height function** of f relative to a line sheaf $\mathscr{L}$ on X is defined to be $T_{\mathscr{L},f}(r) = T_{D,f}(r) + O(1)$ for any divisor D such that $\mathscr{O}(D) \cong \mathscr{L}$ and such that the support of D does not contain the image of f. It is defined only up to $O(1)$.

One can obtain a precise height function (defined exactly, not up to $O(1)$), by fixing a Weil function for any such D, or by choosing a metric on $\mathscr{L}$. It is also possible to use a $(1,1)$-form associated to such a metric (the **Ahlfors-Shimizu height**), but this will not be used in these notes.

Continuing on with the development of the height, we have the following counterpart to Theorem 10.5.

Theorem 12.8. (a) *Functoriality: If $\phi \colon X \to X'$ is a morphism of complete complex varieties and if $\mathscr{L}$ is a line sheaf on X', then*

$$T_{\phi^*\mathscr{L},f}(r) = T_{\mathscr{L},\phi \circ f}(r) + O(1).$$

(b) *Additivity: If $\mathscr{L}_1$ and $\mathscr{L}_2$ are line sheaves on X, then*

$$T_{\mathcal{L}_1 \otimes \mathcal{L}_2, f}(r) = T_{\mathcal{L}_1, f}(r) + T_{\mathcal{L}_2, f}(r) + O(1).$$

(c) **Base locus:** *If the image of f is not contained in the base locus of the complete linear system $|D|$, then $T_{D,f}(r)$ is bounded from below.*

(d) **Globally generated line sheaves:** *If $\mathcal{L}$ is a line sheaf on X, and is generated by its global sections, then $T_{\mathcal{L},f}(r)$ is bounded from below.*

The implicit constants can probably also be made to depend only on the geometric data and the choice of height functions (and not on f), but this is not very important since it is the independence of r that is useful.

The following three results correspond to similar results in the end of Sect. 10.

Corollary 12.9. *If $\mathcal{L}$ is an ample line sheaf on X, then $T_{\mathcal{L},f}(r)$ is bounded from below, is bounded from above if and only if f is constant, and is $O(\log r)$ if and only if f is algebraic.*

Proof. When $X = \mathbb{P}^1$, see [29, Chap. 1, Thm. 6.4] for the second assertion. The general case is left as an exercise for the reader.   □

Proposition 12.10. *Let $\mathcal{L}$ and $\mathcal{M}$ be line sheaves on X, with $\mathcal{L}$ ample. Then there is a constant C, depending only on $\mathcal{L}$ and $\mathcal{M}$, such that*

$$T_{\mathcal{M},f}(r) \leq C T_{\mathcal{L},f}(r) + O(1).$$

Proof. This is true for essentially the same reasons as Proposition 10.7. The details are left to the reader.   □

Proposition 12.11. *Let $\mathcal{L}$ and $\mathcal{M}$ be line sheaves on X, with $\mathcal{L}$ big. Then there is a constant C and a proper Zariski-closed subset Z of X, depending only on $\mathcal{L}$ and $\mathcal{M}$, such that*

$$T_{\mathcal{M},f}(r) \leq C T_{\mathcal{L},f}(r) + O(1),$$

provided that the image of f is not contained in Z.

Proof. Similar to the proof of Proposition 10.11; details are again left to the reader.   □

Remark 12.12. For an arbitrary complete variety X over $\mathbb{C}$ and a holomorphic curve $f \colon \mathbb{C} \to X$, one can define a **big height** to be a real-valued function $T_{\mathrm{big},f}(r)$ with the property that if Z is the Zariski closure of the image of f, if $\widetilde{Z} \to Z$ is a proper birational morphism with $\widetilde{Z}$ projective, if $\mathcal{L}$ is an ample line sheaf on $\widetilde{Z}$, and if $\tilde{f} \colon \mathbb{C} \to \widetilde{Z}$ is a lifting of f, then there are constants $c > 0$ and C such that $T_{\mathrm{big},f}(r) \geq c T_{\mathcal{L},f}(r) + C$ for all $r > 0$. This condition is independent of the choices of $\mathcal{L}$ and $\widetilde{Z}$. This height satisfies the same properties as in Remark 10.12. (There is no list of subvarieties in this case since in Nevanlinna theory f is usually fixed; however, one could define the big height instead by using the same $U_1, \ldots, U_n$, $\overline{U}_1, \ldots, \overline{U}_n$, and $\mathcal{L}_1, \ldots, \mathcal{L}_n$ as in Remark 10.12; then extend $f^{-1}(U_i) \to U_i$ to a map $\mathbb{C} \to \overline{U}_i$ for i such that U_i contains the generic point of Z.)

13 Integral Points

Weil functions can be used to study integral points on varieties. This includes not only affine varieties, but also non-affine varieties. Integral points on non-affine varieties come up in some important applications, such as moduli spaces of abelian varieties.

To begin, let k be a number field and recall that a point $(x_1, \ldots, x_n) \in \mathbb{A}^n(k)$ is an **integral point** if all x_i lie in $\mathscr{O}_k$. More generally, if $S \supseteq S_\infty$ is a finite set of places of k, then $(x_1, \ldots, x_n)$ as above is an S-**integral point** if all x_i lie in the ring

$$\mathscr{O}_{k,S} := \{x \in k : \|x\|_v \leq 1 \text{ for all } v \notin S\} \tag{58}$$

of S-integers. Algebraic points $(x_1, \ldots, x_n) \in \mathbb{A}^n(\bar{k})$ are **integral** (resp. S-**integral**) if all of the x_i are integral over $\mathscr{O}_k$ (resp. $\mathscr{O}_{k,S}$). (Of course, $\mathscr{O}_k = \mathscr{O}_{k,S_\infty}$, so only one definition is really needed.) These definitions are inherited by points on a closed subvariety X of $\mathbb{A}^n_k$.

Given an abstract affine variety X over k, however, the situation becomes a little more complicated. Indeed, for any rational point $x \in X(k)$, there is a closed embedding into $\mathbb{A}^n_k$ for some n that takes x to an integral point. The same is true for algebraic points.

Instead, therefore, we refer to integrality of a *set* of points [74, § 1.3]: Let X be an affine variety over k. Then a set $\Sigma \subseteq X(k)$ (resp. $\Sigma \subseteq X(\bar{k})$) is S-**integral** if there is a closed immersion $i: X \hookrightarrow \mathbb{A}^n_k$ for some n and a nonzero element $a \in k$ such that, for all $x \in \Sigma$, all coordinates of $i(x)$ lie in $(1/a)\mathscr{O}_{k,S}$ (resp. a times all coordinates are integral over $\mathscr{O}_{k,S}$).

As noted above, this definition is meaningful only if Σ is an infinite set.

This definition can be phrased in geometric terms using Weil functions. Indeed, we identify $\mathbb{A}^n_k$ with the complement of the hyperplane $x_0 = 0$ in $\mathbb{P}^n_k$, by identifying $(x_1, \ldots, x_n) \in \mathbb{A}^n_k$ with the point $[1 : x_1 : \ldots : x_n] \in \mathbb{P}^n_k$. Let H denote the hyperplane $x_0 = 0$ at infinity, and let λ_H be the Weil function (47):

$$\begin{aligned}
\lambda_{H,v}([1 : x_1 : \ldots : x_n]) &= -\log \frac{\|1\|_v}{\max\{\|1\|_v, \|x_1\|_v, \ldots, \|x_n\|_v\}} \\
&= \log \max\{\|1\|_v, \|x_1\|_v, \ldots, \|x_n\|_v\}.
\end{aligned} \tag{59}$$

Now let a be a nonzero element of k, let $x \in \bar{k}$, and let L be a number field containing $k(x)$. Then ax is integral over $\mathscr{O}_{k,S}$ if and only if $\|ax\|_w \leq 1$ for all places w of M_L lying over places in $M_k \setminus S$, which holds if and only if $\|x\|_w \leq \|a\|_w^{-1}$ for all such w. Thus, by (59), $\Sigma \subseteq X(\bar{k})$ is S-integral if and only if there is a closed immersion $i: X \hookrightarrow \mathbb{A}^n_k$ for some n and an M_k-constant (c_v) with the following property. For all $x \in \Sigma$, $\lambda_{H,w}(x) \leq c_w$ for all places w of $M_{k(x)}$ lying over places not in S. (Here, as above, we identify $\mathbb{A}^n_k$ with $\mathbb{P}^n_k \setminus H$.)

By functoriality of Weil functions (Theorem 9.8b), this leads to the following definition.

Definition 13.1. Let k be a number field, let $S \supseteq S_\infty$ be a finite set of places of k, let X be a complete variety over k, and let D be an effective Cartier divisor on X. Then a set $\Sigma \subseteq X(\bar{k})$ is a (D,S)**-integral set of points** if (i) no point $x \in \Sigma$ lies in the support of D, and (ii) there is a Weil function λ_D for D and an M_k-constant (c_v) such that

$$\lambda_{D,w}(x) \le c_w$$

for all $x \in \Sigma$ and all places w of $M_{k(x)}$ not lying over places in S.

We may eliminate S from the notation if it is clear from the context, and refer to a D-integral set of points.

From the above discussion, it follows that the condition in the earlier definition of integrality holds for some closed immersion into $\mathbb{A}_k^n$, then it holds for all such closed immersions (with varying n).

Similarly, by Theorem 9.8d, one can use a fixed Weil function λ_D in Definition 13.1 (after adjusting (c_v)). One can also vary the divisor, as follows.

Proposition 13.2. *If k, S, and X are as above, and if D_1 and D_2 are effective Cartier divisors on X with the same support, then a set $\Sigma \subseteq X(\bar{k})$ is D_1-integral if and only if it is D_2-integral.*

Proof. This follows from Theorem 9.8a, e (additivity and boundedness of Weil functions). Details are left to the reader.   □

Thus, D-integrality depends only on the support of D. In fact, one can go further: It depends only on the open subvariety $X \setminus \operatorname{Supp} D$:

Proposition 13.3. *Let k and S be as above, let X_1 and X_2 be complete k-varieties, and let D_1 and D_2 be effective Cartier divisors on X_1 and X_2, respectively. Assume that*

$$\phi \colon X_1 \setminus \operatorname{Supp} D_1 \xrightarrow{\sim} X_2 \setminus \operatorname{Supp} D_2$$

is an isomorphism. Then a set $\Sigma \subseteq X_1(\bar{k})$ is a D_1-integral set on X_1 if and only if

$$\phi(\Sigma) := \{\phi(x) : x \in \Sigma\}$$

is a D_2-integral set on X_2.

Proof. By working with the closure of the graph, we may assume that ϕ extends to a morphism from X_1 to X_2. In that case, it follows from Theorem 9.8a, e.   □

Definition 13.4. Let k and S be as above, and let U be a variety over k. A set $\Sigma \subseteq U(\bar{k})$ is **integral** if there is an open immersion $i \colon U \to X$ of U into a complete variety X over k and an effective Cartier divisor D on X such that $i(U) = X \setminus \operatorname{Supp} D$ and $i(\Sigma)$ is a D-integral set on X.

Proposition 13.5. *Let $\phi \colon X_1 \to X_2$ be a morphism of complete k-varieties, and let D_1 and D_2 be effective Cartier divisors on X_1 and X_2, respectively. Assume that the* *support of D_2 does not contain the image of ϕ, and that the support of D_1 contains the support of $\phi^* D_2$. If Σ is a D_1-integral set on X_1, then*

$$\phi(\Sigma) := \{\phi(x) : x \in \Sigma\}$$

is a D_2-integral set on X_2.

Proof. By Proposition 13.2, we may assume that $D_1 - \phi^* D_2$ is effective, and then the result follows by Theorem 9.8a, e.   □

If we let $U_1 = X_1 \setminus \operatorname{Supp} D_1$ and $U_2 = X_2 \setminus \operatorname{Supp} D_2$, then the above conditions on the supports of D_1 and D_2 are equivalent to $\phi(U_1) \subseteq U_2$. Therefore Proposition 13.5 says that integral sets of points on varieties are preserved by morphisms of those varieties. This phenomenon is more obvious when using models over $\operatorname{Spec} \mathcal{O}_k$ to work with integral points, but this will not be explored in these notes.

We also note that Definition 13.4 does not require U to be affine. Indeed, many moduli spaces are neither affine nor projective, and it is often useful to work with integral points on those moduli spaces (although this is usually done using models). In an extreme case, U can be a complete variety. This corresponds to taking $D = 0$ in Definition 13.1, and the integrality condition is therefore vacuous in that case.

When working with rational points, Definition 13.1 can be stated using counting functions instead: $\Sigma \subseteq X(k)$ is integral if and only if $N_S(D, x)$ is bounded for $x \in \Sigma$. This is no longer equivalent when working with algebraic points, or when working over function fields, though.

The discussion of the corresponding notion in Nevanlinna theory is quite short: an (infinite) D-integral set of rational points on a complete k-variety X corresponds to a holomorphic curve f in a complete complex variety X whose image is disjoint from the support of a given Cartier divisor D on X. (In other words, $N_f(D, r) = 0$ for all r.) The next section will discuss an example of this comparison.

Of course, holomorphic curves omitting divisors also behave as in Proposition 13.3: Let $\phi \colon X_1 \to X_2$ be a morphism of complete complex varieties, let D_1 and D_2 be effective Cartier divisors on X_1 and X_2, respectively, with

$$\phi^{-1}(\operatorname{Supp} D_2) \subseteq \operatorname{Supp} D_1,$$

and let $f \colon \mathbb{C} \to X_1$ be a holomorphic curve which omits D_1. Then $\phi \circ f \colon \mathbb{C} \to X_2$ omits D_2, for trivial reasons.

Now consider the situation where $\phi \colon X' \to X$ is a morphism of complete complex varieties, D is an effective Cartier divisor on X, and D' is an effective Cartier divisor on X' with $\operatorname{Supp} D' = \phi^{-1}(\operatorname{Supp} D)$. Assume that ϕ is étale outside of $\operatorname{Supp} D'$. Then any holomorphic curve $f \colon \mathbb{C} \to X \setminus \operatorname{Supp} D$ lifts to a holomorphic curve $f' \colon \mathbb{C} \to X' \setminus \operatorname{Supp} D'$ such that $\phi \circ f' = f$, essentially for topological reasons.

What is surprising is that this situation carries over to the number field case. Indeed, let $\phi \colon X' \to X$ be a morphism of complete k-varieties, and let D and D' be as above, with ϕ étale outside of $\operatorname{Supp} D'$. If Σ is a set of D-integral points in $X(k)$, then $\phi^{-1}(\Sigma)$ is a set of integral points in $X'(\bar{k})$. The Chevalley-Weil theorem extends to integral points by Serre [74], Sect. 4.2 or Vojta [87], Sect. 5.1, and implies that although the points of Σ' may not lie in $X'(k)$, the ramification of the fields $k(x)$ over k is bounded uniformly for all $x \in X'(k)$. Combining this with the Hermite-Minkowski theorem, it then follows that there is a number field $L \supseteq k$ such that $\Sigma' \subseteq X'(L)$.

14 Units and the Borel Lemma

Units in a number field k can be related to integral points on the affine variety $xy = 1$ in $\mathbb{P}^2_k$: u is a unit if and only if there is a point (u, v) on this variety with $u, v \in \mathcal{O}_k$. This variety is isomorphic to $\mathbb{P}^1$ minus two points, which we may take to be 0 and ∞. More generally, a set of rational points on $\mathbb{P}^1 \setminus \{0, \infty\}$ is integral if and only if it is contained in finitely many cosets of the units in the group k^*.

Units therefore correspond to entire functions that never vanish. An entire function f never vanishes if and only if it can be written as e^g for an entire function g. This leads to what is called the "Borel lemma" in Nevanlinna theory.

Theorem 14.1. [7] *If* $g_1, \ldots, g_n$ *are entire functions such that*

$$e^{g_1} + \cdots + e^{g_n} = 1, \tag{60}$$

then some g_j *is constant.*

Proof. We may assume that $n \geq 2$. The homogeneous coordinates $[e^{g_1} : \ldots : e^{g_n}]$ define a holomorphic curve $f \colon \mathbb{C} \to \mathbb{P}^{n-1}(\mathbb{C})$. The image of this map omits the n coordinate hyperplanes, and also omits the hyperplane $x_1 + \cdots + x_n = 0$ (expressed in homogeneous coordinates $[x_1 : \ldots : x_n]$). Therefore $N_f(H_j, r) = 0$, as H_j varies over these $n + 1$ hyperplanes. This contradicts (39) unless the image of f is contained in a hyperplane (note that n is different in (39)). One can then use the linear relation between the coordinates of f to eliminate one of the terms e^{g_j} and then conclude by induction. $\square$

In fact, by induction, it can be shown that some nontrivial subsum of the terms on the left-hand side of (60) must vanish.

To find the counterpart of this result in number theory, change the e^{g_j} to units. This theorem is due to van der Poorten and Schlickewei [84], and independently to Evertse [19].

Theorem 14.2. (Unit Theorem) *Let k be a number field and let $S \supseteq S_\infty$ be a finite set of places of k. Let $\mathcal{U}$ be a collection of n-tuples $(u_1, \ldots, u_n)$ of S-units in k that satisfy the equation*

$$a_1 u_1 + \cdots + a_n u_n = 1, \tag{61}$$

where $a_1, \ldots, a_n$ are fixed nonzero elements of k. Then all but finitely many elements of $\mathcal{U}$ have the property that there is a nonempty proper subset J of $\{1, 2, \ldots, n\}$ such that $\sum_{j \in J} a_j u_j = 0$.

Proof. Assume that the theorem is false, and let $\mathscr{U}'$ be the set of all $(u_1, \ldots, u_n)$ for which there is no such J as above. Then $\mathscr{U}'$ is infinite.

If we regard each $(u_1, \ldots, u_n) \in \mathscr{U}'$ as a point $[u_1 : \ldots : u_n] \in \mathbb{P}_k^{n-1}$, then by looking directly at the formula (40) for Weil functions, we see that $N_S(H_j, x)$ is bounded as x varies over $\mathscr{U}'$, for the same set of $n+1$ hyperplanes as in the previous proof. This gives $m_S(H_j, x) = h_k(x) + O(1)$ for all $x \in U'$ and all j, contradicting Theorem 8.9 unless all points in $\mathbb{P}^{n-1}$ corresponding to points in $\mathscr{U}'$ lie in a finite union of proper linear subspaces.

Consider one of those linear subspaces containing infinitely many points of $\mathscr{U}'$. Combining the equation of some hyperplane containing that subspace with (61) allows one to eliminate one or more of the u_j, since by assumption there is no set J as in the statement of the theorem. We then proceed by induction on n (the base case $n = 1$ is trivial).   □

Example 14.3. The condition with the set J is essential because, for example, the unit equation (61) with $n = 3$ and $a_1 = a_2 = a_3 = 1$ has solutions $u + (-u) + 1 = 1$ for infinitely many units u (if k or S is large enough). Geometrically, if $H_1, \ldots, H_4$ are the hyperplanes in $\mathbb{P}_k^n$ occurring in the proofs of Theorems 14.1 and 14.2, then the possible sets $J = \{1, 2\}, J = \{1, 3\}$, and $J = \{2, 3\}$ correspond to the line joining the points $H_1 \cap H_2$ and $H_3 \cap H_4$, the line joining the points $H_1 \cap H_3$ and $H_2 \cap H_4$ and the line joining the points $H_1 \cap H_4$ and $H_2 \cap H_3$. Each such line meets the divisor $D := \sum H_j$ in only two points, so if we map $\mathbb{P}^1$ to that line in such a way that 0 and ∞ are taken to those two points, then integral points on $\mathbb{P}_k^1 \setminus \{0, \infty\}$ (i.e., units) are taken to integral points on $\mathbb{P}_k^2 \setminus D$.

Finally, we note that theorems on exponentials of entire functions that can be reduced to Theorem 14.1 by elementary geometric arguments can be readily translated to theorems on units, by replacing the use of Theorem 14.1 with Theorem 14.2. For example, Theorem 14.4 below leads directly to Theorem 14.5.

Theorem 14.4. [16, Théorème XVI]; see also [26] and [30]. *Let $f : \mathbb{C} \to \mathbb{P}^n$ be a holomorphic curve that omits $n + m$ hyperplanes in general position, $m \geq 1$. Then the image of f is contained in a linear subspace of dimension $\leq [n/m]$, where $[\cdot]$ denotes the greatest integer function.*

Theorem 14.5. [50, Cor. 3] *Let $\Sigma \subseteq \mathbb{P}^n(k)$ be a set of D-integral points, where D is the sum of $n + m$ hyperplanes in general position, $m \geq 1$. Then Σ is contained in a finite union of linear subspaces of dimension $\leq [n/m]$.*

15   Conjectures in Nevanlinna Theory and Number Theory

Since the canonical line sheaf $\mathscr{K}$ of $\mathbb{P}^1$ is $\mathscr{O}(-2)$, the main inequality of Theorem 5.5 can be stated in the form

$$m_f(D, r) + T_{\mathscr{K}, f}(r) \leq_{\text{exc}} O(\log^+ T_f(r)) + o(\log r),$$

leading to a general conjecture in Nevanlinna theory. This first requires a definition.

Definition 15.1. A subset Z of a smooth complex variety X is said to have **normal crossings** if each $P \in X(\mathbb{C})$ has an open neighborhood U and holomorphic local coordinates $z_1, \ldots, z_n$ in U such that $Z \cap U$ is given by $z_1 = \cdots = z_r = 0$ for some r ($0 \leq r \leq n$). A divisor on X is **reduced** if all multiplicities occurring in it are either 0 or 1. Finally, a **normal crossings divisor** on X is a reduced divisor whose support has normal crossings.

(Note that not all authors assume that a normal crossings divisor is reduced.)

Conjecture 15.2. Let X be a smooth complex projective variety, let D be a normal crossings divisor on X, let $\mathscr{K}$ be the canonical line sheaf on X, and let $\mathscr{A}$ be an ample line sheaf on X. Then:

(a) The inequality

$$m_f(D,r) + T_{\mathscr{K},f}(r) \leq_{\text{exc}} O(\log^+ T_{\mathscr{A},f}(r)) + o(\log r) \tag{62}$$

holds for all holomorphic curves $f \colon \mathbb{C} \to X$ with Zariski-dense image.
(b) For any $\varepsilon > 0$ there is a proper Zariski-closed subset Z of X, depending only on $X, D, \mathscr{A}$, and ε, such that the inequality

$$m_f(D,r) + T_{\mathscr{K},f}(r) \leq_{\text{exc}} \varepsilon T_{\mathscr{A},f}(r) + C \tag{63}$$

holds for all nonconstant holomorphic curves $f \colon \mathbb{C} \to X$ whose image is not contained in Z, and for all $C \in \mathbb{R}$.

The form of this conjecture is the same as the (known) theorem for holomorphic maps to Riemann surfaces. It has also been shown to hold, with a possibly weaker error term, for holomorphic maps $\mathbb{C}^d \to X$ if $d = \dim X$ and the jacobian determinant of the map is not identically zero; see [82] and [10]. The conjecture itself is attributed to Griffiths, although he seems not to have put it in print anywhere.

Conjecture 15.2 has been proved for curves (Theorem 23.2 and Corollary 29.7), but in higher dimensions very little is known. If $X = \mathbb{P}^n$ and D is a sum of hyperplanes, then the normal crossings condition is equivalent to the hyperplanes being in general position, and in that case the first part of the conjecture reduces to Cartan's theorem (Theorem 8.6). The second part is also known in this case [90].

A consequence of Conjecture 15.2 concerns holomorphic curves in varieties of general type, or of log general type.

Proposition 15.3. *Assume that either part of Conjecture 15.2 is true. If X is a smooth variety of general type, then a holomorphic curve $f \colon \mathbb{C} \to X$ cannot have Zariski-dense image. More generally, if X is a smooth variety, D is a normal crossings divisor on X, and $X \setminus D$ is a variety of log general type, then a holomorphic curve $f \colon \mathbb{C} \to X \setminus D$ cannot have Zariski-dense image.*

Proof. Assume that part (a) of the conjecture is true. The proof for (b) is similar and is left to the reader.

As was the case with (24) and (39), (62) can be rephrased as

$$N_f(D,r) \geq_{\text{exc}} T_{\mathcal{K}(D),f}(r) - O(\log^+ T_{\mathcal{A},f}(r)) - o(\log r). \tag{64}$$

In this case, since f misses D, the left-hand side is zero. By the definition of log general type, the line sheaf $\mathcal{K}(D) := \mathcal{K} \otimes \mathcal{O}(D)$ is big. Therefore, this inequality contradicts Proposition 12.11.   □

This consequence is also unknown in general. It is known, however, in the special case where X is a closed subvariety of a semiabelian variety and $D = 0$. Indeed, if X is a closed subvariety of a semiabelian variety and is not a translate of a semiabelian subvariety, then a holomorphic curve $f: \mathbb{C} \to X$ cannot have Zariski-dense image. See [41, 32, 78] for the case of abelian varieties, and [61] for the more general case of semiabelian varieties. All of these references build on work of Bloch [5].

Conjecture 15.2b is also known if X is an abelian variety and D is ample [79]. The theorem has been extended to semiabelian varieties again by Noguchi [63], but only applies to holomorphic curves whose image does not meet the divisor at infinity. Again, these proofs build on work of Bloch [5].

Conjecture 15.2 will be discussed further once its counterpart in number theory has been introduced. This, in turn, requires some definitions.

Definition 15.4. Let X be a nonsingular variety. A divisor D on X is said to have **strict normal crossings** if it is reduced, if each irreducible component of its support is nonsingular, and if those irreducible components meet transversally (i.e., their defining equations are linearly independent in the Zariski cotangent space at each point). We say that D has **normal crossings** if it has strict normal crossings locally in the étale topology. This means that for each $P \in X$ there is an étale morphism $\phi: X' \to X$ with image containing P such that $\phi^* D$ has strict normal crossings.

This definition is discussed more in [95, Sect. 7].

Definition 15.5. [83] Let X be a variety. A subset of $X(\bar{k})$ is **generic** if all infinite subsets are Zariski-dense in X.

The number-theoretic counterpart to Conjecture 15.2 is then the following.

Conjecture 15.6. Let k be a number field, let $S \supseteq S_\infty$ be a finite set of places of k, let X be a smooth projective variety over k, let D be a normal crossings divisor on X, let $\mathcal{K}$ be the canonical line sheaf on X, and let $\mathcal{A}$ be an ample line sheaf on X. Then:

(a) Let Σ be a generic subset of $X(k) \setminus \text{Supp} D$. Then the inequality

$$m_S(D,x) + h_{\mathcal{K},k}(x) \leq O(\log^+ h_{\mathcal{A},k}(x)) \tag{65}$$

holds for all $x \in \Sigma$.

(b) For any $\varepsilon > 0$ there is a proper Zariski-closed subset Z of X, depending only on X, D, $\mathscr{A}$, and ε, such that for all $C \in \mathbb{R}$ the inequality

$$m_S(D,x) + h_{\mathscr{K},k}(x) \leq \varepsilon h_{\mathscr{A},k}(x) + C \qquad (66)$$

holds for almost all $x \in (X \setminus Z)(k)$.

By Remark 10.12, one can replace $h_{\mathscr{A},k}$ in this conjecture with a big height (after possibly adjusting Z and ε in part (b)). One can then relax the condition on X to be just a smooth complete variety. The resulting conjecture actually would follow from the original Conjecture 15.6 by Chow's lemma, resolution of singularities, and Proposition 25.2 (without reference to $d_S(x)$ in the latter, since lifting a rational point to the cover does not involve passing to a larger number field in this case). This can also be done for Conjecture 15.2.

Except for error terms, the cases in which Conjecture 15.6 is known correspond closely to those cases for which Conjecture 15.2 is known. Indeed, Conjecture 15.6b is known for curves by Roth's theorem, by a theorem of Lang [44], Thm. 2, and by Faltings's theorem on the Mordell conjecture [21, 22], for genus 0, 1, and > 1, respectively. For curves, part (a) of the conjecture is identical to part (b) except for the error term. Also, Schmidt's Subspace Theorem (Theorem 8.10) proves Conjecture 15.6 except for the error term in part (a), and the assertion on the dependence of the set Z in part (b). As noted earlier, however, the latter assertion is also known (without the dependence on $\mathscr{A}$ and ε).

Remark 15.7. Conjecture 15.6 (and also Conjecture 15.2) are compatible with taking products. Indeed, let $X = X_1 \times_k X_2$ be the product of two smooth projective varieties, with projection morphisms $p_i \colon X \to X_i$ ($i = 1, 2$). Let D_1 and D_2 be normal crossings divisors on X_1 and X_2, respectively, and let $\mathscr{K}$, $\mathscr{K}_1$ and $\mathscr{K}_2$ be the canonical line sheaves on X, X_1, and X_2, respectively. We have $\mathscr{K} \cong p_1^* \mathscr{K}_1 \otimes p_2^* \mathscr{K}_2$. Then the conjecture for D_1 on X_1 and for D_2 on X_2 imply the conjecture for $p_1^* D_1 + p_2^* D_2$ on X.

Remark 15.8. One may ask whether one can make the same change to this conjecture as was done in going from Theorem 8.9 to 8.10 (and likewise in the Nevanlinna case). One can, but it would not make the conjecture any stronger. Indeed, suppose that $D_1, \ldots, D_q$ are normal crossings divisors on X. There exists a smooth projective variety X' over k and a proper birational morphism $\pi \colon X' \to X$ such that the support of the divisor $\sum \pi^* D_i$ has normal crossings. Let D' be the reduced divisor on X' for which $\operatorname{Supp} D' = \operatorname{Supp} \sum \pi^* D_i$, and let $\mathscr{K}'$ and $\mathscr{K}$ be the canonical line sheaves of X' and X, respectively. By Proposition 25.2, we have

$$\sum_{v \in S} \max\{\lambda_{D_i,v}(x) : i = 1, \ldots, q\} + h_{\mathscr{K},k}(x) \leq m_S(D',x') + h_{\mathscr{K}',k}(x') + O(1) \quad (67)$$

for all $x' \in X'(k)$, where $x = \pi(x')$. Therefore, if the left-hand side of (65) or (66) were replaced by the left-hand side of (67), then the resulting conjecture would be a consequence of Conjecture 15.6 applied to D' on X'.

Theorems 8.10 and 8.11 are still needed, though, because Conjectures 15.6 and 15.2 have not been proved for blowings-up of $\mathbb{P}^n$.

Corresponding to Proposition 15.3, we also have

Proposition 15.9. *Assume that either part of Conjecture 15.6 is true. Let k and S be as in Conjecture 15.6, let X be a smooth projective variety over k, and let D be a normal crossings divisor on X. Assume that $X \setminus D$ is of log general type. Then no set of S-integral k-rational points on $X \setminus D$ is Zariski dense.*

Proof. As in the earlier proof, (65) is equivalent to

$$N_S(D,x) \geq h_{\mathscr{K}(D),k}(x) - O(\log^+ h_{\mathscr{A},k}(x)),$$

and (66) can be rephrased similarly. For points x in a Zariski-dense set of k-rational S-integral points, $N_S(D,x)$ would be bounded, contradicting Proposition 10.11 since $\mathscr{K}(D)$ is big. $\square$

This proof shows how Conjecture 15.6 is tied to the Mordell conjecture.

As was the case in Nevanlinna theory, the conclusion of Proposition 15.9 has been shown to hold for closed subvarieties of semiabelian varieties, by Faltings [24] in the abelian case and Vojta [89] in the semiabelian case.

In addition, Conjecture 15.6b has been proved when X is an abelian variety and D is ample [23]. This has been extended to semiabelian varieties [92], but in that case (66) was shown only to hold for sets of integral points on the semiabelian variety.

In parts (b) of Conjectures 15.2 and 15.6, the exceptional set Z must depend on ε; this is because of the following theorem.

Theorem 15.10. [52] *There are examples of smooth projective surfaces X containing infinitely many rational curves Z_i for which the restrictions of (62) and (65) fail to hold.*

These examples do not contradict parts (b) of the conjectures of this section, since the degrees of the curves increase to infinity. Nor do they preclude the sharper error terms in parts (a) of the conjectures. However, they do prevent one from combining the two halves of each conjecture.

Lang [48, Chap. I, Sect. 3] has an extensive conjectural framework concerning how the exceptional set in part (b) of Conjectures 15.2 and 15.6 may behave, especially for varieties of general type (which would not include the examples of Theorem 15.10). See also Sect. 17. Note, however, that the exceptional sets of that section refer only to integral points (or holomorphic curves missing D), so the exceptional sets referenced here are more general.

As a converse of sorts, there are numerous examples of theorems in analysis that apply only to "very generic" situations; i.e., they exclude a countable union of proper analytic subsets. One could pose Conjecture 15.2a in such a setting as well. Such a change would not be meaningful for Conjecture 15.6a, however, since the set of rational (and even algebraic) points on a variety is at most countable.

The formulation of Conjecture 15.6 suggests that, in a higher-dimensional setting, the correct counterpart in number theory for a holomorphic curve with Zariski-dense image is not just an infinite set of rational (or algebraic) points, but an infinite *generic* set. Corresponding to holomorphic curves whose images need not be Zariski-dense, we also make the following definition.

Definition 15.11. Let X be a variety over a number field k. If Z is a closed sub-variety of X, then a Z-**generic subset** of $X(\bar{k})$ is a generic subset of $Z(\bar{k})$. Also, a **semi-generic** subset of $X(\bar{k})$ is a Z-generic subset of $X(\bar{k})$ for some closed subvariety Z of X.

A version of Conjecture 15.6 has also been posed for algebraic points. See Conjecture 25.1.

16 Function Fields

Although function fields are not emphasized in these lectures, they provide useful insights, especially when discussing Arakelov theory or use of models. They are briefly introduced in this section. Most results are stated without proof.

Mahler [54] and Osgood [64] showed that Roth's theorem is false for function fields of positive characteristic. Therefore these notes will discuss only function fields of characteristic zero.

For the purposes of these notes, a function field is a finitely-generated field extension of a "ground field" F, of transcendence degree 1. Such a field is called a "function field over F."

If k is a function field over F, it is the function field $K(B)$ for a unique (up to isomorphism) nonsingular projective curve B over F. For each closed point b on B, the local ring $\mathscr{O}_{B,b}$ is a discrete valuation ring whose valuation v determines a non-archimedean place of k with a corresponding norm given by $\|x\|_v = 0$ if $x = 0$ and by the formula

$$\|x\|_v = e^{-[\kappa(b):F]v(x)}$$

if $x \neq 0$. Here v is assumed to be normalized so that its image is $\mathbb{Z}$. We set $N_v = 0$, so that axioms (3) hold.

Let L be a finite extension of k. Then it, too, is a function field over F, and (4) and (5) hold in this context. If B' is the nonsingular projective curve over F corresponding to L, then the inclusion $k \subseteq L$ uniquely determines a finite morphism $B' \to B$ over F.

The field F is not assumed to be algebraically closed. In this context, note that the degree of a divisor on B is defined to be

$$\deg \sum n_b \cdot b = \sum n_b [\kappa(b) : F]. \tag{68}$$

This degree depends on F, since if $F \subseteq F' \subseteq k$ and k is also of transcendence degree 1 over F', then F' is necessarily finite over F,[1] and the degree is divided by $[F' : F]$ if it is taken relative to F' instead of to F.

With this definition of degree, principal divisors have degree 0, which implies that the Product Formula (6) holds, where the set M_k is the set of closed points on the corresponding nonsingular curve B. The Product Formula is the primary condition for k to be a **global field** (for the full set of conditions, see [2, Chap. 12]). There are many other commonalities between function fields and number fields; for example, the affine ring of any nonempty open affine subset of B is a Dedekind ring.

A function field is always implicitly assumed to be given with the subfield F, since (for example) $\mathbb{C}(x, y)$ can be viewed as a function field with either $F = \mathbb{C}(x)$ or $F = \mathbb{C}(y)$, with very different results.

> For the remainder of this section, k is a function field of characteristic 0 over a field F, and B is a nonsingular projective curve over F with $k = K(B)$.

A key benefit of working over function fields is the ability to explore diophantine questions using standard tools of algebraic geometry, using the notion of a model.

Definition 16.1. A **model** for a variety X over k is an integral scheme $\mathscr{X}$, given with a flat morphism $\pi \colon \mathscr{X} \to B$ of finite type and an isomorphism $\mathscr{X} \times_B \operatorname{Spec} k \cong X$ of schemes over k. The model is said to be **projective** (resp. **proper**) if the morphism π is projective (resp. proper).

If X is a projective variety over k, then a projective model can be constructed for it by taking the closure in $\mathbb{P}_B^N$. Likewise, a proper model for a complete variety exists, by Nagata's embedding theorem. In either case the model may be constructed so that any given finite collection of Cartier divisors and line sheaves extends to the same sorts of objects on the model [96].

If $\mathscr{X}$ is a proper model over B for a complete variety X over k, then rational points in $X(k)$ correspond naturally and bijectively to sections $i \colon B \to \mathscr{X}$ of $\pi \colon \mathscr{X} \to B$. Indeed, if i is such a section then it takes the generic point of B to a point on the generic fiber of π, which is X. Conversely, given a point in $X(k)$, one can take its closure to get a curve in $\mathscr{X}$; it is then possible to show that the restriction of π to this curve is an isomorphism.

More generally, if L is a finite extension of k, and B' is the nonsingular projective curve over F corresponding to L, then points in $X(L)$ correspond naturally and bijectively to morphisms $B' \to \mathscr{X}$ over B. This follows by applying the above argument to $\mathscr{X} \times_B B'$, which is a proper model for $X \times_k L$ over B'.

With this notation, we can define Weil functions in the function field case as follows.

---

[1] Let $t \in k$ be transcendental over F. Then $F(t)$ and F' are linearly disjoint over F, and therefore $[F' : F] = [F'(t) : F(t)] \leq [k : F(t)] < \infty$.

Definition 16.2. Let X be a complete variety over k, let D be a Cartier divisor on X, and let $\pi\colon \mathscr{X} \to B$ be a proper model for X. Assume that D extends to a Cartier divisor on $\mathscr{X}$, also denoted by D. Let L be a finite extension of k, let $x \in X(L)$ be a point not lying on $\operatorname{Supp} D$, let $i\colon B' \to \mathscr{X}$ be the corresponding morphism, as above, and let w be a place of L, corresponding to a closed point b' of B'. Then the image of i is not contained in $\operatorname{Supp} D$ on $\mathscr{X}$, so i^*D is a Cartier divisor on B'. Let n_w be the multiplicity of b' in i^*D. We then define

$$\lambda_{D,w}(x) = n_w[\kappa(b') : F].$$

(One may be tempted to require the notation to indicate the choice of F, but this is not necessary since the choice of F is encapsulated in the place w, which comes with a norm $\|\cdot\|_w$ that depends on F.)

It is possible to show that this definition satisfies the conditions of Definition 9.6 (where now k is a function field). Consequently, Theorems 9.8 and 9.10 hold in this context. Moreover, this definition is compatible with (50) (corresponding to changing B').

In the case of Theorem 9.8, though, a bit more is true: the M_k-constants are not necessary when one works with Cartier divisors on the model. Indeed, if D is an effective Cartier divisor on a model $\mathscr{X}$ of a k-variety X, then (in the notation of Definition 16.2) i^*D is an effective divisor on B', so $n_b \geq 0$ for all b, hence $\lambda_{D,w}(x) \geq 0$ for all w and all $x \notin \operatorname{Supp} D$. Similarly, suppose that D is a Cartier divisor on X, and that λ_D and λ'_D are two Weil functions obtained from extensions D and D' of D to models $\mathscr{X}$ and $\mathscr{X}'$, respectively, using Definition 16.2. We may reduce to the situation where the two models are the same: let $\mathscr{X}''$ be the closure of the graph of the birational map between $\mathscr{X}$ and $\mathscr{X}'$, and pull back D and D' to $\mathscr{X}''$. But now the difference $D - D'$ is a divisor on $\mathscr{X}$ which does not meet the generic fiber, so it is supported only on a finite sum of closed fibers of $\pi\colon \mathscr{X} \to B$. Therefore the Weil function associated to $D - D'$, and hence the difference between λ_D and λ'_D, is bounded by an M_k-constant.

Following Sect. 10, one can then define the height of points $x \notin \operatorname{Supp} D$, starting from a model $\mathscr{X}$ for X over B and a Cartier divisor D on $\mathscr{X}$: with notation as in Definition 16.2, we have

$$\begin{aligned}
h_{D,k}(x) &= \frac{1}{[L:k]} \sum_{w \in M_L} \lambda_{D,w}(x) \\
&= \frac{1}{[L:k]} \sum_{w \in M_L} n_w[\kappa(b') : F] \\
&= \frac{\deg i^*D}{[L:k]}
\end{aligned} \tag{69}$$

by (68).

Therefore, heights can be expressed using intersection numbers. It is this observation that led to the development of Arakelov theory, which defines models over $\mathscr{O}_k$

of varieties over number fields k, with additional information at archimedean places which again allows heights to be described using suitable intersection numbers.

Returning to function fields, heights defined as in (69) are defined exactly (given a model of the variety and an extension of the Cartier divisor to that model). Except for Theorem 10.8 (Northcott's theorem), all of the results of Sect. 10 extend to the case of heights defined as in (69). In particular, if $\mathscr{L}$ is a line sheaf on a model $\mathscr{X}$ for X, then

$$h_{\mathscr{L},k}(x) = \frac{\deg i^* \mathscr{L}}{[L:k]}. \tag{70}$$

Northcott's theorem is false over function fields (unless F is finite). Instead, however, it is true that the set (55) is parametrized by a scheme of finite type over F.

Models also provide a very geometric way of looking at integral points. For example, consider the situation with rational points. Let S be a finite set of places of k; this corresponds to a proper Zariski-closed subset of B, which we also denote by S. Let X, $\pi\colon \mathscr{X} \to B$, $x \in X(k)$, and $i\colon B \to \mathscr{X}$ be as in Definition 16.2 (with $L = k$), let D be an effective Cartier divisor on $\mathscr{X}$, and let λ_D be the corresponding Weil function as in Definition 16.2. Then, for any place $v \in M_k$, we have $\lambda_{D,v}(x) > 0$ if and only if $i(b)$ lies in the support of D, where $b \in B$ is the closed point associated to v. Thus, a rational point satisfies the condition of Definition 13.1 with λ_D as above and $c_v = 0$ if and only if it corresponds to a section of the map $\pi^{-1}(B \setminus S) \to B \setminus S$. A similar situation holds with algebraic points, which then correspond to multisections of $\pi^{-1}(B \setminus S) \to B \setminus S$.

Conversely, given a set of integral points as in Definition 13.1, by performing some blowings-up one can construct a model and an effective Cartier divisor on that model for which each of the given integral points corresponds to a section (or multisection) as above.

This formalism works also over number fields, without the need for Arakelov theory.

17 The Exceptional Set

The exceptional set mentioned in Conjectures 15.2b and 15.6b leads to interesting questions of its own, even when working only with rational points (or integral points) in the contexts of Propositions 15.3 or 15.9. This question has been explored in more detail by Lang; this is the main topic of this section. For references, see [47], [48, Chap. I, Sect. 3], and [51].

Definition 17.1. Let X be a complete variety over a field k.

(a) The **exceptional set** $\mathrm{Exc}(X)$ is the Zariski closure of the union of the images of all nonconstant rational maps $G \dashrightarrow X$, where G is a group variety over an extension field of k.

(b) If k is a finitely generated extension field of $\mathbb{Q}$, then the **diophantine exceptional set** $\mathrm{Exc}_{\mathrm{dio}}(X)$ is the smallest Zariski-closed subset Z of X such that $(X \setminus Z)(L)$ is finite for all fields L finitely generated over k.

(c) If $k = \mathbb{C}$ then the **holomorphic exceptional set** $\mathrm{Exc}_{\mathrm{hol}}(X)$ is the Zariski closure of the union of the images of all nonconstant holomorphic curves $\mathbb{C} \to X$.

Each of these sets (when defined) may be empty, all of X, or something in between. For each of these types of exceptional set, Lang has conjectured that the exceptional set is a proper subset if the variety X is of general type (but not conversely – see below). He also has conjectured that $\mathrm{Exc}_{\mathrm{dio}}(X) = \mathrm{Exc}(X)$ if k is a finitely-generated extension of $\mathbb{Q}$, and that $\mathrm{Exc}_{\mathrm{hol}}(X) = \mathrm{Exc}(X)$ if $k = \mathbb{C}$. And, finally, if k is finitely generated over $\mathbb{Q}$, then he conjectured that

$$\mathrm{Exc}_{\mathrm{dio}}(X) \times_k \mathbb{C} = \mathrm{Exc}_{\mathrm{hol}}(X \times_k \mathbb{C})$$

for all embeddings $k \hookrightarrow \mathbb{C}$.

The main example in which this conjecture is known is in the context of closed subvarieties of abelian varieties [41]:

Theorem 17.2. (Kawamata Structure Theorem) *Let X be a closed subvariety of an abelian variety A over $\mathbb{C}$. The **Kawamata locus** of X is the union $Z(X)$ of all translated abelian subvarieties of A contained in X. It is a Zariski-closed subset of X, and is a proper subset if and only if X is not fibered by (nontrivial) abelian subvarieties of A.*

This theorem is also true for semiabelian varieties, and by induction on dimension it follows from [61] that the image of a nonconstant holomorphic curve $\mathbb{C} \to X$ must be contained in $Z(X)$. Similarly, if X is a closed subvariety of a semiabelian variety A over a number field k, then any set of integral points on X can contain only finitely many points outside of $Z(X)$. It is also known that a closed subvariety X of a semiabelian variety A is of log general type if and only if it is not fibered by nontrivial semiabelian subvarieties of A. Thus, (restricting to A abelian) Lang's conjectures have been verified for closed subvarieties of abelian varieties.

In a similar vein, the finite collections of proper linear subspaces of positive dimension in Remark 8.12 are the same in Theorems 8.10 and 8.11.

In the context of integral points or holomorphic curves missing divisors, one can also define the same three types of exceptional sets. The changes are obvious, except possibly for $\mathrm{Exc}(X \setminus D)$: In this case it should be the Zariski closure of the union of the images of all non-constant strictly rational maps $G \dashrightarrow X \setminus D$, where G is a group variety. A **strictly rational map** [39, Sect. 2.12] is a rational map $X \dashrightarrow Y$ such that the closure of the graph is proper over X. This variation has not been studied much, though.

More conjectures relating the geometry of a variety and its diophantine properties are described by Campana [9]. He further classifies varieties in terms of fibrations. For example, let $X = C \times \mathbb{P}^1$ where C is a smooth projective curve of genus ≥ 2. This is an example of a variety which is not of general type, but for which all of

Lang's exceptional sets are the entire variety. Yet, for any given number field, $X(k)$ is not Zariski dense, and there are no Zariski-dense holomorphic curves in X. Campana's framework singles out the projection $X \to C$. This projection has general type base, and fibers have Zariski-dense sets of rational points over a large enough field (depending on the fiber).

For varieties of negative Kodaira dimension, the diophantine properties are studied in conjectures of Manin concerning the rate of growth of sets of rational points of height $\leq B$, as B varies. This is a very active area of number theory, but is beyond the scope of these notes.

18 Comparison of Problem Types

Before the analogy with Nevanlinna theory came on the scene, things were quite simple: You tried to prove something over number fields, and if you got stuck you tried function fields. If you succeeded over function fields, then you tried to translate the proof over to the number field case. For example, the Mordell conjecture was first proved for function fields by Manin, then Grauert modified his proof. But, those proofs used the absolute tangent bundle, which has no known counterpart over number fields. Ultimately, though, Faltings' proof of the Mordell conjecture did draw upon work over function fields, of Tate, Szpiro, and others.

The analogy with Nevanlinna theory gives a second way of working by analogy, although it is more distant. Also, more recent work has placed more emphasis on higher dimensional varieties, lending more importance to the exceptional set. Thus, a particular diophantine problem leads one to a number of related problems which may be easier and whose solutions may provide some insight into the original problem. These can be (approximately) linearly ordered, as follows. In each case, one looks at a class of pairs (X, D) consisting of a smooth complete variety X over the appropriate field and a normal crossings divisor D on X. Each class has been split into a qualitative part (A) and a quantitative part (B).

1A: Find the exceptional set $\mathrm{Exc}(X \setminus D)$.
  B: For each $\varepsilon > 0$, find the exceptional subset Z in Conjectures 15.2 and 15.6. This should be the Zariski closure of the union of all closed subvarieties $Y \subseteq X$ such that, after resolving singularities of Y and of $D\big|_Y$, the main inequality (63) or (66) on Y is weaker than that obtained by restricting the same inequality on X.
2A: Prove that, given any smooth projective curve Y over a field of characteristic zero, and any finite subset $S \subseteq Y$, the set of maps $Y \setminus S \to X \setminus D$ whose image is not contained in the exceptional set, is parametrized by a finite union of varieties.
  B: Prove Conjecture 15.6 in the split function field case of characteristic zero.
3A: Prove that all holomorphic curves $\mathbb{C} \to X \setminus D$ must lie in the exceptional set.
  B: Prove Conjecture 15.2 for holomorphic curves $\mathbb{C} \to X$.

4A: In the (general) function field case of characteristic zero, prove that the set of integral points on $X \setminus D$ outside of the exceptional set is finite.

B: Prove Conjecture 15.6 over function fields of characteristic zero.

5A: Prove over number fields that the set of integral points on $X \setminus D$ outside of the exceptional set is finite.

B: Prove Conjecture 15.6 (in the number field case).

In the case of the Mordell conjecture, for example, X would lie in the class of smooth projective curves of genus > 1 over the appropriate field, and D would be zero.

As another example, see Corollary 29.9, in going from 2A or 2B to 3A or 3B.

In each of the above items except 1A and 1B, one might also consider algebraic points of bounded degree (or holomorphic functions from a finite ramified covering of $\mathbb{C}$). See Sects. 25 and 27.

19 Embeddings

A major goal of these lectures is to describe recent work on partial proofs of Conjecture 15.6 (as well as Conjecture 15.2). One general approach is to use embeddings into larger varieties to sharpen the inequalities. This can only work if the conjecture is known on the larger variety, and if the exceptional set is also known. At the present time, all work on this has used Schmidt's Subspace Theorem (and Cartan's theorem).

This section will discuss some of the issues involved, before delving into some of the specific methods in following sections.

We begin by considering the example where $X = \mathbb{P}^2_k$ and D is a normal crossings divisor of degree ≥ 4. Then $X \setminus D$ is of log general type, and therefore (if k is a number field) integral sets of points on $X \setminus D$ cannot be Zariski dense, or (if $k = \mathbb{C}$) holomorphic curves $\mathbb{C} \to X \setminus D$ cannot have Zariski-dense image. From now on we will refer only to the number-theoretic case; the version in Nevanlinna theory is similar.

If D is a smooth divisor, then there is no clue on how to proceed. In the other extreme, if D is a sum of at least four lines (in general position), then Schmidt's Theorem gives the answer; see Sect. 14.

If D is a sum of three lines and a conic, some results are known. For example, if L_1, L_2, and L_3 are linear forms defining the lines and Q is a homogeneous quadratic polynomial defining the conic, then all L_i^2/Q must be units (or nearly so) at integral points. Since they are algebraically dependent, we can apply the unit theorem; see [31] and [87, Cor. 2.4.3].

More recently, nontrivial approximation results have been obtained for conics and higher-degree divisors in projective space by using r-uple embeddings.

For example, under the 2-uple embedding $\mathbb{P}^2 \hookrightarrow \mathbb{P}^5$, the image of a conic is contained in a hyperplane. Therefore Schmidt's Subspace Theorem can be applied to $\mathbb{P}^5$ to give an approximation result (provided there are sufficiently many other components in the divisor). Under the 3-uple embedding $\mathbb{P}^2 \hookrightarrow \mathbb{P}^9$, things are better: the image of a conic spans a linear subspace of codimension 3, so there are three linearly independent hyperplanes containing it.

More generally, suppose D is a divisor of degree d in $\mathbb{P}^n$. We consider its image under the r-uple embedding $\mathbb{P}^n \hookrightarrow \mathbb{P}^{\binom{r+n}{n}-1}$. This image spans a linear subspace of codimension $\binom{r+n-d}{n}$, because there are that many monomials of degree $r-d$ in $n+1$ variables (they then get multiplied by the form defining D to get homogeneous polynomials of degree r in the homogeneous coordinate variables in $\mathbb{P}^n$, hence hyperplanes in the image space).

Applying Schmidt's Subspace Theorem to $\mathbb{P}^{\binom{r+n}{n}-1}$ would would then give an inequality of the form

$$\binom{r+n-d}{n} m(D,x) + \cdots \le \left(\binom{r+n}{n} \cdot r + \varepsilon \right) h_k(x) + O(1)$$

for $x \in \mathbb{P}^n(k)$ outside of a finite union of proper subvarieties of degree $\le r$. The idea is to take r large. As $r \to \infty$, the ratio of the coefficients in the above inequality tends to 0, because

$$\frac{\binom{r+n-d}{n}}{\binom{r+n}{n} \cdot r} = \frac{(r-d+n)\cdots(r-d+1)}{(r+n)\cdots(r+1)r} = \frac{r^n + O(r^{n-1})}{r(r^n + O(r^{n-1}))} \to \frac{1}{r} \to 0. \qquad (71)$$

This is not useful, but we can try harder. Some hyperplanes in the image space can be made to contain D twice, or three times, etc. After taking this into account, the inequality improves to

$$\left(\binom{r+n-d}{n} + \binom{r+n-2d}{n} + \ldots \right) m(D,x) + \ldots$$
$$\le \left(\binom{r+n}{n} \cdot r + \varepsilon \right) h_k(x) + O(1).$$

To estimate the factor in front of $m(D,x)$, we have

$$\binom{r-kd+n}{n} = \frac{(r-kd+n)\cdots(r-kd+1)}{n!} = \frac{(r-kd)^n + O_n((r-kd)^{n-1})}{n!}$$

and therefore the coefficient in front of the proximity term is

$$\sum_{k=1}^{[r/d]} \binom{r-kd+n}{n} = \frac{(r-d)^n + \cdots + (r-[r/d]d)^n + O_{n,d}(r^n)}{n!}.$$

As $r \to \infty$, the ratio of this coefficient to the one in front of the height term now converges to

$$\frac{\sum_{k=1}^{[r/d]} \binom{r-kd+n}{n}}{r\binom{r+n}{n}} \approx \frac{\frac{r^{n+1}}{(n+1)d\cdot n!}}{\frac{r^{n+1}}{n!}} \rightarrow \frac{1}{(n+1)d}. \tag{72}$$

This indeed gives a nontrivial inequality

$$\frac{1}{d} m_S(D,x) + \cdots \le (n+1+\varepsilon)h_k(x) + O(1). \tag{73}$$

If $d = 1$ then this is best possible (but of course is not new, since then D is already a hyperplane in $\mathbb{P}^n$). If $d > 1$ then it is less than ideal, but is still new and noteworthy.

If D is the only component in the divisor, then (73) will never lead to a useful inequality, however, since the left-hand side is always bounded by $h_k(x)$. This approach only works if the divisor in question has more than one irreducible component. Having more than one component in the divisor, however, introduces some additional complications.

Suppose, for example, that there are two divisor components D_1 and D_2, and their images under the r-uple embedding span linear subspaces L_1 and L_2 of codimensions ρ_1 and ρ_2, respectively. We have

$$\operatorname{codim}(L_1 \cap L_2) \le \rho_1 + \rho_2 \tag{74}$$

(assuming $L_1 \cap L_2 \ne \emptyset$). If this inequality is strict then this causes problems. Indeed, let y denote the image of x under the d-uple embedding. If y is close to L_1 at some place $v \in S$, and also close to L_2 at that same place, then it is necessarily close to $L_1 \cap L_2$. If $L_1 \cap L_2$ is too large, though, then we will not be able to choose enough hyperplanes containing it to fully utilize both $m(D_1,x)$ and $m(D_2,x)$.

Indeed, choose ρ_1 generic hyperplanes containing L_1 and ρ_2 generic hyperplanes containing L_2. If these $\rho_1 + \rho_2$ hyperplanes are in general position then this implies

$$\operatorname{codim}(L_1 \cap L_2) \ge \rho_1 + \rho_2.$$

So if this inequality does not hold, then the $\rho_1 + \rho_2$ hyperplanes cannot (collectively) be in general position, and the max on the left-hand side of (44) will not be as large as one would hope.

So, in order to apply the reasoning leading up to (73) independently for each irreducible component of the divisor, equality must hold in (74) for each pair of components. (Similar considerations also apply to triples of components, etc.)

However, the standard computation of Hilbert functions using short exact sequences gives (for sufficiently large n) that the codimension of the linear span of $D_1 \cap D_2$ is

$$\binom{n-d_1+r}{n} + \binom{n-d_2+r}{n} - \binom{n-d_1-d_2+r}{n}.$$

This is too small by $\binom{n-d_1-d_2+r}{n}$. So, any use of this approach would have to take this into account, and would also have to incorporate the changes made in going from (71) to (72).

As noted earlier, the purpose of this section is not to actually prove anything, but merely to highlight the general idea, together with some of the stumbling blocks.

20 Schmidt's Subspace Theorem Implies Siegel's Theorem

One way to avoid the difficulties mentioned in the last section is to restrict to curves, since in that case irreducible divisors are just points, so they do not intersect. Of course, using Schmidt's theorem to imply Roth's theorem would not be interesting, since the latter is already a special case. However, if one restricts to a curve contained in projective space, then one can get a nontrivial result by applying the methods of Sect. 19. This was done by Corvaja and Zannier [13], and gave a new proof of Siegel's theorem.

Theorem 20.1. (Siegel) *Let C be a smooth affine curve over a number field k. Assume that C has at least 3 points at infinity (i.e., at least three points need to be added to obtain a nonsingular projective curve). Then all sets of integral points on C are finite.*

Proof. By expanding k, if necessary, we may assume that the points at infinity are k-rational. Let $\overline{C}$ be the nonsingular projective closure of C, let g be its genus, let $Q_1, \ldots, Q_r$ be the points at infinity, and let D be the divisor $Q_1 + \cdots + Q_r$. Pick N large, and embed $\overline{C}$ into $\mathbb{P}^{M-1}$ by the complete linear system $|ND|$; we have $M = Nr + 1 - g$. Assume that $\{P_1, P_2, \ldots\}$ is an infinite S-integral set of points on C, for some finite set $S \supseteq S_\infty$ of places of k. After passing to an infinite subsequence, we may assume that for each $v \in S$ there is an index $j(v) \in \{1, \ldots, r\}$ such that each P_i is at least as close to $Q_{j(v)}$ as to any other Q_j in the v-adic topology.

For all $\ell \in \mathbb{N}$, we have $h^0(\overline{C}, \mathscr{O}(ND - \ell Q_j)) \geq Nr - \ell + 1 - g$, so we can choose $Nr - \ell + 1 - g$ linearly independent hyperplanes in $\mathbb{P}^{M-1}$ vanishing to order $\geq \ell$ at Q_j. For each $v \in S$, do this with $j = j(v)$, obtaining one hyperplane vanishing to order $Nr - g$, a second vanishing to order $Nr - g - 1$, etc. Obtaining M hyperplanes in this way for each v, and applying Schmidt's Subspace Theorem, we obtain

$$\sum_{v \in S} \sum_{\ell=0}^{Nr-g} \ell \lambda_{Q_{j(v)}, v}(P_i) \leq (M + \varepsilon) h_k(P_i) + O(1) \tag{75}$$

outside of a finite union of proper linear subspaces of $\mathbb{P}^{M-1}$. Here the height $h_k(P_i)$ is taken in $\mathbb{P}^{M-1}$. The finitely many linear subspaces correspond to only finitely many points on C, and they can be removed from the set of integral points.

By the assumption on the distance from P_i to $Q_{j(v)}$ (and the fact that the points Q_j are separated by a distance independent of i), we have $\lambda_{Q_j, v}(P_i) = O(1)$ for all $j \neq j(v)$ and all $v \in S$. Therefore

$$\lambda_{Q_{j(v)}, v}(P_i) = \lambda_{D, v}(P_i) + O(1)$$

for all $v \in S$, with the constant independent of i. Also, since the embedding in $\mathbb{P}^{M-1}$ is obtained from the complete linear system $|ND|$, we have $h_k(P_i)=h_{ND,k}(P_i)+O(1)$. Therefore (75) becomes

$$\frac{(Nr-g)(Nr-g+1)}{2}m_S(D,P_i) \leq N(Nr-g+1+\varepsilon)h_{D,k}(P_i)+O(1).$$

Since the P_i are integral points, though, we have $m_S(D,P_i) = h_{D,k}(P_i)+O(1)$, and the inequality becomes

$$\left(\frac{(Nr-g)(Nr-g+1)}{2} - N(Nr-g+1) - N\varepsilon\right)h_{D,k}(P_i) \leq O(1).$$

If N is large and ε is small, then the quantity in parentheses is negative (since $r \geq 2$), leading to a contradiction since D is ample. $\qquad\square$

Of course, if $g \geq 1$ then Siegel only required $r > 0$. This can be proved by reducing to the above case. Indeed, embed $\overline{C}$ in its Jacobian and pull back by multiplication by 2. This gives an étale cover of $\overline{C}$ of degree at least 4, so the pull-back of D will have at least three points. Integral points on C will pull back to integral points on the pull-back of C in the étale cover, by the Chevalley-Weil theorem for integral points (see the end of Sect. 13).

21 The Corvaja-Zannier Method in Higher Dimensions

Corvaja and Zannier further developed their method to higher dimensions; see for example [14]. It did not provide the full strength of Conjecture 15.6, even when $X = \mathbb{P}^n$, but it did provide noteworthy new answers. The key to their method can be summarized in the following definition, which is due to Levin [51].

Definition 21.1. Let X be a nonsingular complete variety over a field. A divisor D on X is **very large** if D is effective and, for all $P \in X$, there is a basis B of $L(D)$ such that

$$\sum_{f\in B} \mathrm{ord}_E f > 0 \tag{76}$$

for all irreducible components E of D passing through P. A divisor D is **large** if it is effective and has the same support as some very large divisor.

In the following discussion, it will be useful to have the following functoriality property of large divisors.

Proposition 21.2. *Let X' and X be nonsingular complete varieties over fields L and k, respectively, with $L \supseteq k$, and let $\phi: X' \to X$ be a morphism of schemes such that the diagram*

$$X' \xrightarrow{\phi} X$$

$$\downarrow \qquad\qquad \downarrow$$

$$\operatorname{Spec} L \longrightarrow \operatorname{Spec} k$$

commutes. Let D be a very large divisor on X, and let $D' = \phi^* D$ be the corresponding divisor on X'. Assume that the natural map

$$\alpha : H^0(X, \mathscr{O}(D)) \otimes_k L \to H^0(X', \mathscr{O}(D')) \tag{77}$$

is an isomorphism. Then D' is very large.

Proof. Let P' be a point on X', and let B be a basis for $L(D)$ that satisfies (76) for D at $\phi(P') \in X$. We have a commutative diagram

$$L(D) \otimes_k L \xrightarrow{(\cdot 1_D) \otimes_k L} H^0(X, \mathscr{O}(D)) \otimes_k L$$

$$\downarrow \beta \qquad\qquad\qquad \downarrow \alpha$$

$$L(D') \xrightarrow{\cdot 1_{D'}} H^0(X', \mathscr{O}(D'))$$

in which β is an isomorphism because the other three arrows are isomorphisms. Therefore we let $B' = \{\beta(f \otimes 1) : f \in B\}$; it is a basis of $L(D')$.

Now let E' be an irreducible component of D' passing through P'. For each irreducible component E of D passing through $\phi(P')$, let n_E be the multiplicity of E' in $\phi^* E$. For all nonzero $f \in L(D)$, we have

$$\operatorname{ord}_{E'} \beta(f \otimes 1) \geq \sum_E n_E \operatorname{ord}_E f, \tag{78}$$

where the sum is over all irreducible components E of D passing through $\phi(P')$ (and in particular it includes all irreducible components of D containing $\phi(E')$). Indeed, to verify (78), note that if s is a nonzero element of $H^0(X, \mathscr{O}(D))$, then $\operatorname{ord}_{E'} \alpha(s \otimes 1) \geq \sum_E n_E \operatorname{ord}_E s$, with equality if $(s) = D$. (Strictness may arise if E' is exceptional for ϕ and s vanishes along prime divisors containing $\phi(E')$ that do not occur in D.)

By (78), we then have

$$\sum_{f' \in B'} \operatorname{ord}_{E'} f' \geq \sum_E n_E \sum_{f \in B} \operatorname{ord}_E f > 0$$

(since at least one n_E is strictly positive). Thus B' satisfies (76) for E'. $\qquad\square$

Corollary 21.3. *Let X be a nonsingular complete variety over a field k, let D be a divisor on X, let L be a field containing k, let $X_L = X \times_k L$ with projection $\phi : X_L \to X$, and let $D_L = \phi^* D$. If D is large (resp. very large), then so is D_L.*

Proof. Indeed, we may assume that D is very large, and note that (77) is an isomorphism because L is flat over k [36, III Prop. 9.3]. □

Corollary 21.4. *Let* $\phi: X' \to X$ *be a proper birational morphism*[2] *of nonsingular complete varieties over a field, and let D be a divisor on X. If D is large (resp. very large), then $\phi^* D$ is a large (resp. very large) divisor on X'.*

Proof. Again, we may assume that D is very large. In this case (77) is an isomorphism because $\phi_* \mathcal{O}_{X'} = \mathcal{O}_X$; see [36, proof of III Cor. 11.4 and III Remark 8.8.1]. □

Remark 21.5. More generally, for a linear subspace V of $L(D)$, one may define V-very large. For this definition, Proposition 21.2 holds with the weaker assumption that (77) is injective.

We also note that the definition of largeness is (vacuously) true for $D = 0$.

Having discussed the definition of large divisor, the theorem that makes the definition useful is the following.

Theorem 21.6. (Corvaja-Zannier) *Let k be a number field, let $S \supseteq S_\infty$ be a finite set of places of k, let X be a nonsingular complete variety over k, and let D be a nonzero large divisor on X. Then any set of (D, S)-integral points on X is not Zariski-dense.*

Proof. By Proposition 13.2, we may assume that D is very large.

Since $D \neq 0$, there is a component E as in the definition of very large, so $\ell(D) > 1$. Therefore there is a nontrivial rational map $\Phi: X \dashrightarrow \mathbb{P}_k^{\ell(D)-1}$.

We may assume that Φ is a morphism. Indeed, let X' be a desingularization of the closure of the graph of Φ. Replace X with X' and D with its pull-back. By Corollary 21.4 the pull-back remains very large. Moreover, the notion of integral set of points remains unchanged, by functoriality of Weil functions (or by the fact that the desingularization may be chosen such that the map $X' \to X$ is an isomorphism away from the support of D).

Now suppose that there is a Zariski-dense set $\{P_i\}$ of integral points. After passing to a Zariski-dense subset, we may assume that for each $v \in S$ there is a point $P_v \in X(k_v)$ such that the P_i converge to P_v in the v-topology. For each such v, let $B_v = \{f_{v,1}, \ldots, f_{v,\ell(D)}\}$ be a basis for $L(D)$ that satisfies (76) at the point P_v. Let $H_{v,1}, \ldots, H_{v,\ell(D)}$ be the corresponding hyperplanes in $\mathbb{P}_k^{\ell(D)-1}$. (Corollary 21.3 is not really needed here, but can be used if the reader prefers. When finding P_v, one should think of P_v and the P_i as points on the (complex, real, or p-adic) manifold $X(k_v)$, and when defining B_v one should realize that P_v is a morphism from $\operatorname{Spec} k_v$ to X, whose image is a point in X (which may be the generic point).)

To obtain a contradiction, it will suffice to find an $\varepsilon > 0$ such that

$$\sum_{v \in S} \sum_{j=1}^{\ell(D)} \lambda_{H_{v,j},v}(\Phi(P_i)) \geq (\ell(D) + \varepsilon) h_k(\Phi(P_i)) + O(1) \tag{79}$$

---

[2] This requires ϕ to be a morphism, not just a rational map.

for all i. Indeed, Schmidt's Subspace Theorem would then imply that the $\Phi(P_i)$ are contained in a finite union of proper linear subspaces of $\mathbb{P}_k^{\ell(D)-1}$, contradicting the fact that the P_i are Zariski-dense in X and that $\Phi(X)$ is not contained in any proper linear subspace of $\mathbb{P}_k^{\ell(D)-1}$.

Let μ be the largest multiplicity of a component of D, and let $\varepsilon = 1/\mu$, so that $\varepsilon \operatorname{ord}_E D \leq 1$ for all irreducible components E of D.

Let $v \in S$, and let E be an irreducible component of D. First suppose that E contains P_v. Then $\sum \operatorname{ord}_E \Phi^* H_{v,j} > \ell(D) \operatorname{ord}_E D$ (by (76)), so

$$\sum_{j=1}^{\ell(D)} \operatorname{ord}_E \Phi^* H_{v,j} \geq \ell(D) \operatorname{ord}_E D + 1 \geq (\ell(D) + \varepsilon) \operatorname{ord}_E D$$

and therefore

$$\sum_{j=1}^{\ell(D)} (\operatorname{ord}_E \Phi^* H_{v,j}) \lambda_{E,v}(P_i) \geq (\ell(D) + \varepsilon)(\operatorname{ord}_E D) \lambda_{E,v}(P_i) + O(1)$$

since $\lambda_{E,v}(P_i) \geq O(1)$. This latter inequality also holds if E does not contain P_v, since in that case $\lambda_{E,v}(P_i) = O(1)$. Therefore, we have

$$\sum_{j=1}^{\ell(D)} \lambda_{H_{v,j},v}(\Phi(P_i)) \geq \sum_{j=1}^{\ell(D)} \sum_E (\operatorname{ord}_E \Phi^* H_{v,j}) \lambda_{E,v}(P_i) + O(1)$$

$$\geq (\ell(D) + \varepsilon) \sum_E (\operatorname{ord}_E D) \lambda_{E,v}(P_i) + O(1)$$

$$= (\ell(D) + \varepsilon) \lambda_{D,v}(P_i) + O(1),$$

where the sums over E are sums over all irreducible components E of D.

Summing over $v \in S$ then gives

$$\sum_{v \in S} \sum_{j=1}^{\ell(D)} \lambda_{H_{v,j},v}(\Phi(P_i)) \geq (\ell(D) + \varepsilon) m_S(D, P_i) + O(1) = (\ell(D) + \varepsilon) h_{D,k}(P_i) + O(1)$$

since the P_i are (D, S)-integral. This is equivalent to (79) by functoriality of heights. $\qquad\square$

This proof does not carry over directly to Nevanlinna theory, because it relies on the finiteness of S; moreover, the whole idea of passing to a subsequence is unsuited to Nevanlinna theory. In fact, the proof does not even work for function fields over infinite fields, since in such cases the local fields are not locally compact.

The following lemma, essentially due to Levin [51], works around this problem. The key idea is that it suffices to use only finitely many bases in Definition 21.1.

Lemma 21.7. *Let X be a nonsingular complete variety over $\mathbb{C}$, and let D be an effective divisor on X. Let σ_0 be the set of prime divisors occurring in D, and let Σ be the set of subsets σ of σ_0 for which $\bigcap_{E \in \sigma} E$ is nonempty. For each $\sigma \in \Sigma$ let D_σ be the sum of those components of D not lying in σ, with the same multiplicities as they have in D. Pick a Weil function for each such D_σ. Then there is a constant C, depending only on X and D, such that*

$$\min_{\sigma \in \Sigma} \lambda_{D_\sigma}(P) \le C \tag{80}$$

for all $P \in X(\mathbb{C})$.

Proof. The conditions imply that

$$\bigcap_{\sigma \in \Sigma} \operatorname{Supp} D_\sigma = \emptyset,$$

since for all $P \in X$ the set $\sigma := \{E \in \sigma_0 : E \ni P\}$ is an element of Σ, and then $P \notin \operatorname{Supp} D_\sigma$. The lemma then follows from Lemma 9.9, since Σ is a finite set. $\square$

Lemma 21.8. *Let X be a nonsingular complete variety over $\mathbb{C}$, let D be a very large divisor on X whose complete linear system is base-point-free, and let $\Phi: X \to \mathbb{P}^{\ell(D)-1}$ be a corresponding morphism to projective space. Then there is a finite collection $H_1, \ldots, H_q$ of hyperplanes and $\varepsilon > 0$ such that, given choices $\lambda_{H_1}, \ldots, \lambda_{H_q}$ and λ_D of Weil functions on $\mathbb{P}^{\ell(D)-1}$ and X, respectively, we have*

$$\max_J \sum_{j \in J} \lambda_{H_j}(\Phi(P)) \ge (\ell(D) + \varepsilon)\lambda_D(P) + O(1), \tag{81}$$

where the implicit constant in $O(1)$ is independent of $P \in X(\mathbb{C})$. Here, as in Theorem 8.11, J varies over all subsets of $\{1, \ldots, q\}$ corresponding to subsets of $\{H_1, \ldots, H_q\}$ that lie in general position.

Proof. Let Σ be the (finite) set of Lemma 21.7, and for each $\sigma \in \Sigma$ let B_σ be a basis for $L(D)$ that satisfies (76) at some (and hence all) points $P \in \bigcap_{E \in \sigma} E$. Let $H_1, \ldots, H_q$ be the distinct hyperplanes in $\mathbb{P}^{\ell(D)-1}$ corresponding to elements of the union $\bigcup_{\sigma \in \Sigma} B_\sigma$, and choose Weil functions λ_{H_j} for them. For each $\sigma \in \Sigma$ let D_σ be as in Lemma 21.7, and choose a Weil function λ_{D_σ} for it. Let C be a constant that satisfies (80). Finally, choose Weil functions λ_E for each prime divisor E occurring in D.

Let μ be the largest multiplicity of a component of D, and let $\varepsilon = 1/\mu$.

Now let $P \in X(\mathbb{C})$. Pick $\sigma \in \Sigma$ for which

$$\lambda_{D_\sigma} \le C, \tag{82}$$

and let $J \subseteq \{1, \ldots, q\}$ be the subset for which $\{H_j : j \in J\}$ are the hyperplanes corresponding to the elements of B_σ. As before, (76) applied to B_σ implies that $\sum_{j \in J} \operatorname{ord}_E \Phi^* H_j > \ell(D) \operatorname{ord}_E D$ for all $E \in \sigma$; hence

$$\sum_{j \in J} \operatorname{ord}_E \Phi^* H_j \geq (\ell(D) + \varepsilon) \operatorname{ord}_E D,$$

and therefore

$$\sum_{j \in J} (\operatorname{ord}_E \Phi^* H_j) \lambda_E(P) \geq (\ell(D) + \varepsilon)(\operatorname{ord}_E D) \lambda_E(P) + O(1) \tag{83}$$

since $\lambda_E(P) \geq O(1)$. Also

$$D = D_\sigma + \sum_{E \in \sigma} (\operatorname{ord}_E D) \cdot E. \tag{84}$$

By (83), (82), and (84), we then have

$$\sum_{j \in J} \lambda_{H_j}(\Phi(P)) \geq \sum_{j \in J} \sum_{E \in \sigma} (\operatorname{ord}_E \Phi^* H_j) \lambda_E(P) + O(1)$$

$$\geq (\ell(D) + \varepsilon)(\lambda_{D_\sigma}(P) - C) + (\ell(D) + \varepsilon) \sum_{E \in \sigma} (\operatorname{ord}_E D) \lambda_E(P) + O(1)$$

$$= (\ell(D) + \varepsilon) \lambda_D(P) + O(1).$$

In the above, the constants in $O(1)$ depend only on the choices of B_σ and the choices of Weil functions, and on σ (which has only finitely many choices). Since J is one of the sets in (81), the lemma then follows. □

This then leads to the theorem in Nevanlinna theory corresponding to Theorem 21.6:

Theorem 21.9. [51] *Let X be a nonsingular complete variety over $\mathbb{C}$, let D be a nonzero large divisor on X, and let $f \colon \mathbb{C} \to X$ be a holomorphic curve whose image is disjoint from D. Then the image of f is not Zariski dense.*

Proof. As in the proof of Theorem 21.6, we may assume that D is very large and base point free. Let $\Phi \colon X \to \mathbb{P}_{\mathbb{C}}^{\ell(D)-1}$ be a morphism corresponding to a complete linear system of D. Let $f \colon \mathbb{C} \to X$ be a holomorphic curve whose image does not meet D. Let $H_1, \ldots, H_q$ and ε be as in Lemma 21.8. By that lemma, we then have

$$\int_0^{2\pi} \max_J \sum_{j \in J} \lambda_{H_j}(\Phi(f(re^{i\theta}))) \frac{d\theta}{2\pi} \geq (\ell(D) + \varepsilon) m_f(D, r) + O(1)$$

$$= (\ell(D) + \varepsilon) T_{\Phi \circ f}(r) + O(1)$$

for all $r > 0$. This contradicts Theorem 8.11 unless the image of $\Phi \circ f$ is contained in a proper linear subspace of $\mathbb{P}_{\mathbb{C}}^{\ell(D)-1}$ (since $\ell(D) > 1$). This in turn implies that the image of f cannot be Zariski dense. □

This proof can also be adapted back to the number field case, and it also works over function fields.

Some concrete examples of large divisors follow. First, in order to show that Theorem 20.1 is a consequence of Theorem 21.6, we have the following.

Proposition 21.10. *Let C be a smooth projective curve over a field k, and let D be an effective divisor supported on (distinct) rational points $Q_1, \dots, Q_r$. If $r = 0$ or $r \geq 3$ then D is large.*

Proof. If $r = 0$ then $D = 0$, which is already known to be large.

Assume $r \geq 3$. It will suffice to show that if $D = N(Q_1 + \dots + Q_r)$ then D is very large for sufficiently large integers N. As in the proof of Theorem 20.1, we have $h^0(C, \mathscr{O}(D - \ell Q_j)) \geq Nr - \ell + 1 - g$ for all $\ell \in \mathbb{N}$ and all $j = 1, \dots, r$, where g is the genus of C. For each such j there is a basis $(s_1, \dots, s_{Nr+1-g})$ of $H^0(C, \mathscr{O}(D))$ such that s_ℓ vanishes to order $\geq \ell - 1$ at Q_j. Dividing each such s_ℓ by the canonical section 1_D then gives a basis $(f_1, \dots, f_{Nr+1-g})$ of $L(D)$ such that $\operatorname{ord}_{Q_j} f_\ell \geq \ell - 1 - N$ for all ℓ. Thus

$$\sum_{\ell=1}^{Nr+1-g} \operatorname{ord}_{Q_j} f_\ell \geq (Nr + 1 - g)\left(\frac{Nr - g}{2} - N\right)$$

if $N > (g-1)/r$, and is strictly positive if also $N > g/(r-2)$. $\square$

Proposition 21.11. *Let X_1 and X_2 be smooth complete varieties over a field k, and let D_1 and D_2 be divisors on X_1 and X_2, respectively. If D_1 and D_2 are large (resp. very large), then $p_1^* D_1 + p_2^* D_2$ is a large (resp. very large) divisor on $X_1 \times_k X_2$, where $p_i \colon X_1 \times_k X_2 \to X_i$ is the projection ($i = 1, 2$).*

Proof. It will suffice to show that if D_1 and D_2 are very large, then so is $p_1^* D_1 + p_2^* D_2$. Write $D = p_1^* D_1 + p_2^* D_2$.

We first claim that the natural map

$$H^0(X_1, \mathscr{O}(D_1)) \otimes_k H^0(X_2, \mathscr{O}(D_2)) \longrightarrow H^0(X_1 \times_k X_2, \mathscr{O}(D)) \qquad (85)$$

is an isomorphism. Indeed, the projection formula [36, II Ex. 5.1d] gives an isomorphism

$$\mathscr{O}(D_1) \otimes_k H^0(X_2, \mathscr{O}(D_2)) \xrightarrow{\;\sim\;} (p_1)_* \mathscr{O}(p_1^* D_1 + p_2^* D_2),$$

of sheaves on X_1, and taking global sections gives (85).

To show that D is very large, let $P \in X_1 \times_k X_2$. Let B_1 and B_2 be bases of $L(D_1)$ and $L(D_2)$ satisfying (76) with respect to the points $p_1(P)$ and $p_2(P)$, respectively. By (85), $\{p_1^* f_1 \cdot p_2^* f_2 : f_1 \in B_1, f_2 \in B_2\}$ is a basis for $L(D)$; call it B. Let E be an irreducible component of D passing through P. If $E = p_1^* E_1$ for a component E_1 of D_1, then

$$\sum_{h \in B} \operatorname{ord}_E h = \ell(D_2) \sum_{f \in B_1} \operatorname{ord}_{E_1} f > 0,$$

so (76) is satisfied for E. Otherwise, we must have $E = p_2^* E_2$ for an irreducible component E_2 of D_2, and (76) is satisfied for a symmetrical reason. Thus D is very large. $\square$

The following gives a slightly more complicated example of large divisors. For this example, recall that a Cartier divisor D on a complete variety X over a field k is **nef** ("numerically effective") if $\deg j^* \mathcal{O}(D) \geq 0$ for all maps $j \colon C \to X$ from a curve C over k to X.

Theorem 21.12. [51, Thm. 9.2] *Let X be a nonsingular projective variety of dimension q, and let $D = \sum D_i$ be a divisor on X for which all D_i are effective and nef. Assume also that all irreducible components of D are nonsingular, and that $D^q > 2q D^{q-1} D_P$ for all $P \in \operatorname{Supp} D$, where $D_P = \sum_{\{i : D_i \ni P\}} D_i$. Then D is large.*

For the proof, see [51]. Note that this generalizes Proposition 21.10.

As another example of this method, we note another theorem of Levin.

Definition 21.13. A variety V over a number field k is **Mordellic** if for all number fields $L \supseteq k$ and all finite sets $S \supseteq S_\infty$ of places of L, there are no infinite sets of S-integral L-rational points on $V_L := V \times_k L$. A variety V over k is **quasi-Mordellic** if there is a proper Zariski-closed subset Z of V such that, for all L and S as above, and for all S-integral sets of L-rational points on V_L, almost all points in the set are contained in Z_L.

Theorem 21.14. [51, Thm. 9.11A] *Let X be a projective variety over a number field k. Let $D = \sum_{i=1}^r D_i$ be a divisor on X such that each D_i is an effective Cartier divisor, and the intersection of any $m + 1$ of the supports of the D_i is empty. Then:*

(a) If D_i is big for all i and $r > 2m \dim X$ then $X \setminus D$ is quasi-Mordellic.
(b) If D_i is ample for all i and $r > 2m \dim X$ then $X \setminus D$ is Mordellic.

The proof of this theorem, as well as its counterpart in Nevanlinna theory, appear in [51].

Again, we note that if X is a nonsingular curve, then this reduces to the combination of Theorem 21.6 and Proposition 21.10.

22 Work of Evertse and Ferretti

Evertse and Ferretti also found a way of using Schmidt's Subspace Theorem in combination with d-uple embeddings to get partial results on more general varieties, with respect to more general divisors. Their method is based on using Mumford's degree of contact, which was originally developed to study moduli spaces, but which is also well suited for this application. It uses a bit more machinery than the method of Corvaja and Zannier, and this machinery makes direct comparisons more difficult.

Because of the machinery, we offer here only a sketch of the methods, without proofs.

The idea originated from a paper of Ferretti [25], and was further developed jointly with Evertse; see for example [20]. This work was translated into Nevanlinna theory by Ru [70], solving a conjecture of Shiffman.

Throughout this section, k is a field of characteristic 0 and $X \subseteq \mathbb{P}_k^N$ is a projective variety over k of dimension n and degree Δ.

Definition 22.1. The **Chow form** of X is the unique (up to scalar multiple) polynomial

$$F_X \in k[\mathbf{u}_0, \dots, \mathbf{u}_n] = k[u_{00}, \dots, u_{0N}, u_{10}, \dots, u_{nN}],$$

homogeneous of degree Δ in each block $\mathbf{u}_i$, characterized by the condition

$$F_X(\mathbf{u}_0, \dots, \mathbf{u}_n) = 0 \iff X \cap H_{\mathbf{u}_0} \cap \dots \cap H_{\mathbf{u}_n} \neq \emptyset,$$

where $H_{\mathbf{u}_i}$ is the hyperplane in $\mathbb{P}_k^N$ corresponding to $\mathbf{u}_i \in (\mathbb{P}_k^N)^*$.

For more details on Chow forms, see Hodge and Pedoe [38, Vol. II, Chap. X, Sect. 6–8].

Definition 22.2. Let $\mathbf{c} = (c_0, \dots, c_N) \in \mathbb{R}^{N+1}$ and let F_X be as above. For an indeterminate t, write

$$F_X(t^{c_0} u_{00}, \dots, t^{c_N} u_{0N}, \dots, t^{c_N} u_{nN}) = t^{e_0} G_0(\mathbf{u}_0, \dots, \mathbf{u}_n) + \dots + t^{e_r} G_r(\mathbf{u}_0, \dots, \mathbf{u}_n),$$

where $G_0, \dots, G_r$ are nonzero polynomials in $k[u_{00}, \dots, u_{nN}]$ and $e_0 > \dots > e_r$. Then the **Chow weight** of X with respect to $\mathbf{c}$ is $e_X(\mathbf{c}) = e_0$.

If I is the (prime) homogeneous ideal in $k[x_0, \dots, x_N]$ corresponding to $X \subseteq \mathbb{P}_k^N$, then recall that the **Hilbert function** $H_X(m)$ for $m \in \mathbb{N}$ is defined by

$$H_X(m) = \dim_k k[x_0, \dots, x_N]_m / I_m,$$

where the subscript m denotes the homogeneous part of degree m.

Recall also that the Hilbert polynomial of X (which agrees with the Hilbert function for $m \gg 0$) has leading term $\Delta m^n / n!$.

Definition 22.3. Let I be as above, and let $\mathbf{c} \in \mathbb{R}^{N+1}$. The the **Hilbert weight** of X with respect to $\mathbf{c}$ is

$$S_X(m, \mathbf{c}) = \max \left(\sum_{\ell=1}^{H_X(m)} \mathbf{a}_\ell \cdot \mathbf{c} \right),$$

where the max is taken over all collections $(\mathbf{a}_1, \dots, \mathbf{a}_{H_X(m)})$ with $\mathbf{a}_\ell \in \mathbb{N}^{N+1}$ for all ℓ, whose corresponding monomials $\mathbf{x}^{\mathbf{a}_1}, \dots, \mathbf{x}^{\mathbf{a}_{H_X(m)}}$ give a basis (over k) when mapped to $k[x_0, \dots, x_N]_m / I_m$. (Here $\mathbf{x}^{\mathbf{a}_\ell}$ denotes $x_0^{a_{\ell 0}} \cdots x_N^{a_{\ell N}}$, and the conditions necessarily imply that $a_{\ell 0} + \dots + a_{\ell N} = m$ for all ℓ.)

Mumford showed that

$$S_X(m, \mathbf{c}) = e_X(\mathbf{c}) \cdot \frac{m^{n+1}}{(n+1)!} + O(m^n),$$

and Evertse and Ferretti showed further that if $m > \Delta$ then

$$\frac{S_X(m, \mathbf{c})}{H_X(m)} \geq \frac{m}{(n+1)\Delta} e_X(\mathbf{c}) - (2n+1)\Delta \max_{0 \leq j \leq N} c_j \tag{86}$$

[20, Prop. 3.2]. (To compare these two inequalities, note that $H_X(m) \sim \Delta m^n / n!$.)

In diophantine applications, k is a number field and $S \supseteq S_\infty$ is a finite set of places of k. The following is a slight simplification of the main theorem of Evertse and Ferretti [20].

Theorem 22.4. *Assume that $n = \dim X > 0$. For each $v \in S$ let $D_0^{(v)}, \ldots, D_n^{(v)}$ be a system of effective divisors on $\mathbb{P}_k^N$ satisfying*

$$X \cap \bigcap_{j=0}^{n} \operatorname{Supp} D_j^{(v)} = \emptyset.$$

Then for all ε with $0 < \varepsilon \leq 1$, there are hypersurfaces $G_1, \ldots, G_u$ in $\mathbb{P}^N$, not containing X and of degree

$$\deg G_i \leq 2(n+1)(2n+1)(n+2)\Delta d^{n+1}\varepsilon^{-1}, \tag{87}$$

where d is the least common multiple of the degrees of the $D_j^{(v)}$, such that all solutions $x \in X(k)$ of the inequality

$$\sum_{v \in S} \sum_{j=0}^{n} \frac{\lambda_{D_j^{(v)}, v}(x)}{\deg D_j^{(v)}} \geq (n + 1 + \varepsilon)h_k(x) + O(1) \tag{88}$$

lie in the union of the G_i. In particular, these solutions are not Zariski-dense.

(Evertse and Ferretti also prove a more quantitative version of this theorem, which gives explicit bounds on u, if one ignores solutions of (88) of height below a given explicit bound. They obtain a weaker bound than (87), because of this added strength.)

Note that this result is weaker than Conjecture 15.6, since the latter conjecture does not divide the Weil functions by the degrees of the divisors. It is also stronger, though, in the sense that the sum of the divisors does not have to have normal crossings.

Here we will restate this theorem in a way that translates more readily into Nevanlinna theory. The proof will roughly follow Evertse and Ferretti [20], with substantial simplifications since we are not bounding u. In particular, the "twisted heights" (which are related to the first successive minima in Schmidt's original proof) are not needed here. Because of these simplifications, it may be easier to follow Ru [70], even though he is working in Nevanlinna theory.

Theorem 22.5. *Assume that* $n = \dim X > 0$. *Let* $D_0, \ldots, D_q$ *be effective divisors on* $\mathbb{P}^N_k$, *whose supports do not contain* X. *Let* $\mathscr{J}$ *be the set of all* $(n+1)$-*element subsets* J *of* $\{0, \ldots, q\}$ *for which*

$$X \cap \bigcap_{j \in J} \operatorname{Supp} D_j = \emptyset, \tag{89}$$

and assume that $\mathscr{J}$ *is not empty. Then for all* ε *with* $0 < \varepsilon \leq 1$, *all constants* $C \in \mathbb{R}$, *and all choices of Weil functions* $\lambda_{D_j,v}$, *there are hypersurfaces* $G_1, \ldots, G_u$, *as before, such that all solutions* $x \in X(k)$ *of the inequality*

$$\sum_{v \in S} \max_{J \in \mathscr{J}} \sum_{j \in J} \frac{\lambda_{D_j,v}(x)}{\deg D_j} \geq (n+1+\varepsilon) h_k(x) + C \tag{90}$$

lie in the union of the G_i.

Proof (sketch). First, by replacing each D_j with a suitable positive integer multiple, we may assume that all of the D_j have the same degree d.

Next, we reduce to $d = 1$, as follows. Let $\phi : \mathbb{P}^N_k \to \mathbb{P}^M_k$ be the d-uple embedding, where $M = \binom{N+d}{N} - 1$, and let $Y = \phi(X)$. Then Y has dimension n, degree Δd^n, and ϕ multiplies the projective height by d. Moreover, there are hyperplanes $E_0, \ldots, E_q$ on $\mathbb{P}^M_k$ such that $\phi^* E_j = D_j$ for all j. Thus if $y = \phi(x)$ then (90) is equivalent to

$$\sum_{v \in S} \max_J \sum_{j \in J} \lambda_{E_j,v}(y) \geq (n+1+\varepsilon) h_k(y) + C' \tag{91}$$

for a suitable constant C' independent of x. Applying Theorem 22.5 with $d = 1$ to Y and $E_0, \ldots, E_q$ then gives hypersurfaces $G'_1, \ldots, G'_u$ in $\mathbb{P}^M_k$, of degrees bounded by $(8n+6)(n+2)^2 \Delta d^n \varepsilon^{-1}$, not containing Y, but containing all solutions $y \in Y(k)$ of the inequality (91). Pulling these hypersurfaces back to $\mathbb{P}^N_k$ multiplies their degrees by d, so these pull-backs satisfy (91).

By a further linear embedding of $\mathbb{P}^N_k$, we may assume that $D_0, \ldots, D_q$ are the coordinate hyperplanes $x_0 = 0, \ldots, x_q = 0$, respectively. We may also assume that all of the Weil functions occurring in (90) are nonnegative.

Now assume, by way of contradiction, that the set of solutions of (90) is not contained in a finite union of hypersurfaces G_i satisfying (87). By a partitioning argument [20, Lemma 5.3] there is a subset Σ of $X(k)$, not contained in a finite union of hypersurfaces G_i as before, $(n+1)$-element subsets J_v of $\{0, \ldots, q\}$ satisfying (89) for each $v \in S$, and nonnegative real constants $c_{j,v}$ for all $v \in S$ and all $j \in J_v$, such that

$$\sum_{v \in S} \sum_{j \in J_v} c_{j,v} = 1 \tag{92}$$

and such that the inequality

$$\lambda_{D_j,v}(x) \geq c_{j,v}\left(n+1+\frac{\varepsilon}{2}\right)h_k(x) \tag{93}$$

holds for all $v \in S$, all $j \in J_v$, and all $x \in \Sigma$. Also let $c_{j,v} = 0$ if $v \in S$ and $j \notin J_v$, so that (93) holds for all $v \in S$ and all $j = 0,\dots,q$.

Now for some $m > \Delta$ (its exact value is given by (103)), let

$$\phi_m : \mathbb{P}_k^N \to \mathbb{P}_k^{R_m}$$

be the m-uple embedding; here $R_m = \binom{N+m}{m} - 1$. Let X_m be the linear subspace of $\mathbb{P}_k^{R_m}$ spanned by $\phi_m(X)$. We have

$$\dim X_m = H_X(m) - 1.$$

For each $v \in S$ let

$$\mathbf{c}_v = (c_{0,v},\dots,c_{q,v},0,\dots,0) \in \mathbb{R}_{\geq 0}^{N+1},$$

and let $\mathbf{a}_{1,v},\dots,\mathbf{a}_{H_X(m),v}$ be the elements of $\mathbb{N}^{N+1}$ for which the monomials $\mathbf{x}^{\mathbf{a}_{\ell,v}}$, $\ell = 1,\dots,H_X(m)$, give a basis for $k[x_0,\dots,x_N]_m/I_m$ satisfying

$$S_X(m,\mathbf{c}_v) = \sum_{\ell=1}^{H_X(m)} \mathbf{a}_{\ell,v} \cdot \mathbf{c}_v. \tag{94}$$

For each v the monomials $\mathbf{x}^{\mathbf{a}_{\ell,v}}$, $\ell = 1,\dots,H_X(m)$, define linear forms $L_{\ell,v}$ in the homogeneous coordinates on $\mathbb{P}_k^{R_m}$ which are linearly independent on $\phi_m(X)$, and therefore on X_m. For each v and ℓ choose Weil functions $\lambda_{L_{\ell,v}}$ on $\mathbb{P}_k^{R_m}$. We have

$$\lambda_{L_{\ell,v},v}(\phi_m(x)) \geq \sum_{j=0}^{q} a_{\ell,v,j}\lambda_{D_j,v}(x) + O(1)$$

for all v and ℓ. After adjusting the Weil function, we may assume that the $O(1)$ term is not necessary.

By (93) and (94), we then have

$$\sum_{\ell=1}^{H_X(m)} \lambda_{L_{\ell,v},v}(\phi_m(x)) \geq S_X(m,\mathbf{c}_v)\left(n+1+\frac{\varepsilon}{2}\right)h_k(x) \tag{95}$$

for all $x \in \Sigma$ and all $v \in S$. Assume for now that there is an integer $m \geq \Delta$ such that

$$m \leq \frac{2(n+1)(2n+1)(n+2)\Delta}{\varepsilon} \tag{96}$$

and such that

$$\left(n+1+\frac{\varepsilon}{2}\right)\sum_{v\in S}S_X(m,\mathbf{c}_v) > mH_X(m).\tag{97}$$

Then, for sufficiently small $\varepsilon' > 0$, (95) will imply

$$\sum_{v\in S}\sum_{\ell=1}^{H_X(m)}\lambda_{L_{\ell,v},v}(\phi_m(x)) \geq (H_m(x)+\varepsilon')mh_k(x)+O(1)\tag{98}$$

for all $x \in \Sigma$. Note that $h_k(\phi_m(x)) = mh_k(x)+O(1)$. Applying Schmidt's Subspace Theorem to X_m (via some chosen isomorphism $X_m \cong \mathbb{P}_k^{H_X(m)-1}$) it follows that there is a finite union of hyperplanes in X_m, and hence in $\mathbb{P}_k^{R_m}$, containing $\phi_m(\Sigma)$. These pull back to give homogeneous polynomials G_i of degree m on $\mathbb{P}_k^N$; they satisfy (87) by (96).

We now show that (97) holds for some m satisfying (96).

By (86), we have

$$\sum_{v\in S}S_X(m,\mathbf{c}_v) \geq H_X(m)\sum_{v\in S}\left(\frac{m}{(n+1)\Delta}e_X(\mathbf{c}_v) - (2n+1)\Delta\max_{j\in J_v}c_{j,v}\right).\tag{99}$$

By [20, Lemma 5.1], we have

$$e_X(\mathbf{c}_v) \geq \Delta\sum_{j\in J_v}c_{j,v},\tag{100}$$

and therefore

$$\sum_{v\in S}e_X(\mathbf{c}_v) \geq \Delta\sum_{v\in S}\sum_{j\in J_v}c_{j,v} = \Delta\tag{101}$$

by (92). Thus (99) becomes

$$\begin{aligned}\sum_{v\in S}S_X(m,\mathbf{c}_v) &\geq H_X(m)\left(\frac{m}{n+1} - (2n+1)\Delta\sum_{v\in S}\max_{j\in J_v}c_{j,v}\right)\\ &\geq H_X(m)\left(\frac{m}{n+1} - (2n+1)\Delta\right).\end{aligned}\tag{102}$$

Now let

$$m = \left\lfloor\frac{2(n+1)(2n+1)(n+2)\Delta}{\varepsilon}\right\rfloor.\tag{103}$$

This clearly satisfies (96); in addition, we have

$$m < \left(n+1+\frac{\varepsilon}{2}\right)\left(\frac{m}{n+1} - (2n+1)\Delta\right).\tag{104}$$

Thus (102) becomes

$$\left(n+1+\frac{\varepsilon}{2}\right)\sum_{v\in S}S_X(m,\mathbf{c}_v)>mH_X(m),$$

which is (97).   □

In Nevanlinna theory, the counterpart to Theorem 22.5 was proved by Ru [70]. Here we give a slightly stronger version of his theorem: X is not required to be nonsingular, and we incorporate the set $\mathscr{J}$. This stronger version still follows from his proof without essential changes, though.

Theorem 22.6. *Let $k=\mathbb{C}$ and assume that $n=\dim X>0$. Let $D_0,\dots,D_q$ be effective divisors on $\mathbb{P}^N_{\mathbb{C}}$, whose supports do not contain X. Let $\mathscr{J}$ be the set of all $(n+1)$-element subsets J of $\{0,\dots,q\}$ for which*

$$X\cap\bigcap_{j\in J}\operatorname{Supp}D_j=\emptyset,$$

and assume that $\mathscr{J}$ is not empty. Fix $\varepsilon\in\mathbb{R}$ with $0<\varepsilon\le 1$, fix $C\in\mathbb{R}$, and choose Weil functions λ_{D_j} for all j. Let $f\colon\mathbb{C}\to X(\mathbb{C})$ be a holomorphic curve whose image is not contained in any hypersurface in $\mathbb{P}^N$ not containing X of degree $\le 2(n+1)(2n+1)(n+2)\Delta d^{n+1}\varepsilon^{-1}$. Then

$$\int_0^{2\pi}\max_{J\in\mathscr{J}}\sum_{j\in J}\frac{\lambda_{D_j}(f(re^{i\theta}))}{\deg D_j}\frac{d\theta}{2\pi}\le_{\mathrm{exc}}(n+1+\varepsilon)T_f(r)+C. \tag{105}$$

Proof (sketch). This proof uses the same general outline as the proof of Theorem 22.5, but there is an essential difference in that one cannot take a subsequence in order to define constants $c_{j,v}$, since the interval $[0,2\pi]$ is not a finite set. Instead, however, it is possible to drop the condition (92); then (101) is no longer valid. However, (100) still holds, and is homogeneous in the components of $\mathbf{c}$. Therefore, we may omit the step of dividing by the height, and just let the components of $\mathbf{c}$ be the Weil functions themselves (assumed nonnegative).

In detail, as before we assume that $D_0,\dots,D_q$ are restrictions of the coordinate hyperplanes $x_0=0,\dots,x_q=0$, and that the Weil functions λ_{D_j} are nonnegative.

For each $r>0$ and $\theta\in[0,2\pi]$ let $J_{r,\theta}$ be an element of $\mathscr{J}$ for which

$$\sum_{j\in J_{r,\theta}}\lambda_{D_j}(f(re^{i\theta}))$$

is maximal, for each $j\in J_{r,\theta}$ let

$$c_{j,r,\theta}=\lambda_{D_j}(f(re^{i\theta})),$$

and for each $j\in\{0,\dots,N\}\setminus J_{r,\theta}$ let $c_{j,r,\theta}=0$. Let

$$\mathbf{c}_{r,\theta}=(c_{0,r,\theta},\dots,c_{N,r,\theta})\in\mathbb{R}^{N+1}_{\ge 0}.$$

Then, as before, [20, Lemma 5.1] gives

$$e_X(\mathbf{c}_{r,\theta}) \geq \Delta \sum_{j \in J_{r,\theta}} c_{j,r,\theta}. \tag{106}$$

Let m be as in (103). By (86), (106), and nonnegativity of $c_{j,r,\theta}$, we have

$$\frac{1}{H_X(m)} \int_0^{2\pi} S_X(m, \mathbf{c}_{r,\theta}) \frac{d\theta}{2\pi} \geq \int_0^{2\pi} \left(\frac{m}{(n+1)\Delta} e_X(\mathbf{c}_{r,\theta}) - (2n+1)\Delta \max_{j \in J_{r,\theta}} c_{j,r,\theta} \right) \frac{d\theta}{2\pi}$$

$$\geq \int_0^{2\pi} \left(\frac{m}{n+1} \sum_{j \in J_{r,\theta}} c_{j,r,\theta} - (2n+1)\Delta \max_{j \in J_{r,\theta}} c_{j,r,\theta} \right) \frac{d\theta}{2\pi}$$

$$\geq \left(\frac{m}{n+1} - (2n+1)\Delta \right) \int_0^{2\pi} \left(\sum_{j \in J_{r,\theta}} c_{j,r,\theta} \right) \frac{d\theta}{2\pi}. \tag{107}$$

By (104) there is an $\varepsilon' > 0$ such that

$$\frac{m}{n+1} - (2n+1)\Delta \geq \frac{m}{n+1+\varepsilon/2} \cdot \frac{H_X(m) + \varepsilon'}{H_X(m)};$$

hence (107) becomes

$$\left(n+1+\frac{\varepsilon}{2} \right) \int_0^{2\pi} S_X(m, \mathbf{c}_{r,\theta}) \frac{d\theta}{2\pi} \geq m(H_X(m) + \varepsilon') \int_0^{2\pi} \left(\sum_{j \in J_{r,\theta}} c_{j,r,\theta} \right) \frac{d\theta}{2\pi}. \tag{108}$$

This corresponds to (97) in the earlier proof.

Now let $\phi_m \colon \mathbb{P}_{\mathbb{C}}^N \to \mathbb{P}_{\mathbb{C}}^{R_m}$ be the m-uple embedding. Then the image of $\phi_m \circ f$ is not contained in any hyperplane not also containing $\phi_m(X)$. As before, let X_m be the linear subspace of $\mathbb{P}_{\mathbb{C}}^{R_m}$ spanned by $\phi_m(X)$.

As before, for each r and θ there are $\mathbf{a}_{1,r,\theta}, \ldots, \mathbf{a}_{H_X(m),r,\theta} \in \mathbb{N}^{N+1}$, corresponding to a basis of $\mathbb{C}[x_0, \ldots, x_N]_m / I_m$, such that

$$S_X(m, \mathbf{c}_{r,\theta}) = \sum_{\ell=1}^{H_X(m)} \mathbf{a}_{\ell,r,\theta} \cdot \mathbf{c}_{r,\theta}. \tag{109}$$

These correspond to linear forms $L_{\ell,r,\theta}$ on $\mathbb{P}_{\mathbb{C}}^{R_m}$, $\ell = 1, \ldots, H_X(m)$, which are linearly independent on X_m for each r and θ, and which satisfy

$$\lambda_{L_{\ell,r,\theta}}(\phi_m(f(re^{i\theta}))) \geq \sum_{j=0}^{q} a_{\ell,r,\theta,j} \lambda_{D_j}(f(re^{i\theta})) \tag{110}$$

for suitable choices of Weil functions $\lambda_{L_{\ell,r,\theta}}$.

By the definitions of $J_{r,\theta}$ and of $c_{j,r,\theta}$, by (108), by (109), by (110), and by applying Cartan's Theorem 8.11 to X_m, we then have

$$
\int_0^{2\pi} \max_{J \in \mathscr{J}} \sum_{j \in J} \lambda_{D_j}(f(re^{i\theta})) \frac{d\theta}{2\pi}
$$

$$
= \int_0^{2\pi} \sum_{j \in J_{r,\theta}} \lambda_{D_j}(f(re^{i\theta})) \frac{d\theta}{2\pi}
$$

$$
= \int_0^{2\pi} \left(\sum_{j \in J_{r,\theta}} c_{j,r,\theta} \right) \frac{d\theta}{2\pi}
$$

$$
\leq \frac{n+1+\varepsilon/2}{m(H_X(m)+\varepsilon')} \int_0^{2\pi} S_X(m, \mathbf{c}_{r,\theta}) \frac{d\theta}{2\pi}
$$

$$
= \frac{n+1+\varepsilon/2}{m(H_X(m)+\varepsilon')} \int_0^{2\pi} \left(\sum_{\ell=1}^{H_X(m)} \mathbf{a}_{\ell,r,\theta} \cdot \mathbf{c}_{r,\theta} \right) \frac{d\theta}{2\pi}
$$

$$
\leq \frac{n+1+\varepsilon/2}{m(H_X(m)+\varepsilon')} \int_0^{2\pi} \left(\sum_{\ell=1}^{H_X(m)} \lambda_{L_{\ell,r,\theta}}(\phi_m(f(re^{i\theta}))) \right) \frac{d\theta}{2\pi}
$$

$$
\leq_{\mathrm{exc}} \frac{n+1+\varepsilon/2}{m} T_{\phi_m \circ f}(r) + C'
$$

$$
\leq \left(n+1+\frac{\varepsilon}{2} \right) T_f(r) + C,
$$

where C' is chosen so that the last inequality holds. $\qquad\square$

(This is better than (105) by $\varepsilon/2$, since we have removed the partitioning argument. Thus, the bound (87) can be improved by a factor of 2. This can be done in the number field case too, also by eliminating the partitioning argument there. We decided to keep the partitioning argument in that case, though, since such arguments are common in number theory and it is useful to know how to translate them into Nevanlinna theory.)

23 Truncated Counting Functions and the abc Conjecture

Many results in Nevanlinna theory, when expressed in terms of counting functions instead of proximity functions, hold also in strengthened form using what are called *truncated counting functions*. As usual, one can define truncated counting functions in the number field case as well, and this leads to deep conjectures of high current interest. Perhaps the best-known such conjecture is the abc conjecture of Masser and Oesterlé.

Definition 23.1. Let X be a complete complex variety, let D be an effective Cartier divisor on X, let $f: \mathbb{C} \to X$ be a holomorphic curve whose image is not contained in the support of D, and let $n \in \mathbb{Z}_{>0}$. Then the n-**truncated counting function** with respect to D is

$$N_f^{(n)}(D,r) = \sum_{0<|z|<r} \min\{\mathrm{ord}_z f^* D, n\} \log \frac{r}{|z|} + \min\{\mathrm{ord}_0 f^* D, n\} \log r.$$

As with the earlier counting function, the n-truncated counting function is functorial and nonnegative. It is not additive in D, though, due to the truncation. We only have an inequality: If D_1 and D_2 are effective, then

$$N_f^{(n)}(D_1 + D_2, r) \leq N_f^{(n)}(D_1, r) + N_f^{(n)}(D_2, r).$$

In Nevanlinna theory, the Second Main Theorem for curves has been extended to truncated counting functions:

Theorem 23.2. *Let X be a smooth complex projective curve, let D be a reduced effective divisor on X, let $\mathcal{K}$ be the canonical line sheaf on X, and let $\mathcal{A}$ be an ample line sheaf on X. Then the inequality*

$$N_f^{(1)}(D,r) \geq_{\mathrm{exc}} T_{\mathcal{K}(D),f}(r) - O(\log^+ T_{\mathcal{A},f}(r)) - o(\log r)$$

holds for all nonconstant holomorphic curves $f \colon \mathbb{C} \to X$.

Of course, X needs to be a curve of genus ≤ 1 for this to be meaningful, since otherwise there is no function f. However, if the domain is a finite ramified covering of $\mathbb{C}$, then X can have large genus; see Conjecture 27.5 and the discussion following it.

Also, in Theorem 8.6, the counting functions in (39) can be replaced by n-truncated counting functions:

$$\sum_{j=1}^{q} N_f^{(n)}(H_j, r) \geq_{\mathrm{exc}} (q - n - 1) T_f(r) - O(\log^+ T_f(r)) - o(\log r) \tag{111}$$

Theorem 8.11 is not suitable for using truncated counting functions, though, due to its emphasis on the proximity function.

Conjecture 15.2, though, should also be true with counting functions. The question arises, however: truncation to what? Note that (111) is false if the terms $N_f^{(n)}(H_j, r)$ are replaced by $N_f^{(1)}(H_j, r)$, unless one allows the exceptional set to contain hypersurfaces of degree greater than 1 (see Example 23.6). I do believe that Conjecture 15.2 should be true with 1-truncated counting functions, though, even though it would involve substantial complications.

The translation of the above into number theory is straightforward (except that the counterparts to Theorem 23.2 and 111 are only conjectural).

Definition 23.3. Let k be a number field, let $S \supseteq S_\infty$ be a finite set of places of k, let X be a complete variety over k, let D be an effective Cartier divisor on X, and let $n \in \mathbb{Z}_{>0}$. For every place $v \notin S$ (necessarily non-archimedean), let $\mathfrak{p}_v$ denote the corresponding prime ideal in $\mathscr{O}_k$. Then the n-**truncated counting function** with respect to D is

$$N_S^{(n)}(D,x) = \sum_{v \notin S} \min\{\lambda_{D,v}(x), n\log(\mathscr{O}_k : \mathfrak{p}_v)\}$$

for all $x \in X(k)$ not lying in the support of D. If $x \in X(\bar{k})$ lies outside the support of D, then we let $L = \kappa(x)$, let T be the set of places of L lying over S, and define

$$N_S^{(n)}(D,x) = \frac{1}{[L:k]} \sum_{w \notin T} \min\{\lambda_{D,v}(x), n\log(\mathscr{O}_L : \mathfrak{p}_w)\}. \tag{112}$$

Truncation does not respect (48) at ramified places, so (112) is not independent of the choice of L. It is independent of the choice of Weil function, up to $O(1)$. As in the case of Nevanlinna theory, this truncated counting function is functorial and nonnegative, and obeys an inequality

$$N_S^{(n)}(D_1 + D_2, x) \leq N_S^{(n)}(D_1, x) + N_S^{(n)}(D_2, x)$$

if D_1 and D_2 are effective.

We conjecture that a counterpart to Conjecture 15.6 holds with truncated counting functions:

Conjecture 23.4. Let k be a number field, let $S \supseteq S_\infty$ be a finite set of places of k, let X be a smooth projective variety over k, let D be a normal crossings divisor on X, let $\mathscr{K}$ be the canonical line sheaf on X, and let $\mathscr{A}$ be an ample line sheaf on X. Then:

(a) Let Σ be a generic subset of $X(k) \setminus \mathrm{Supp}\, D$. Then the inequality

$$N_S^{(1)}(D,x) \geq h_{\mathscr{K}(D),k}(x) - O\left(\sqrt{h_{\mathscr{A},k}(x)}\right) \tag{113}$$

holds for all $x \in \Sigma$.

(b) For any $\varepsilon > 0$ there is a proper Zariski-closed subset Z of X, depending only on $X, D, \mathscr{A}$, and ε, such that for all $C \in \mathbb{R}$ the inequality

$$N_S^{(1)}(D,x) \geq h_{\mathscr{K}(D),k}(x) - \varepsilon h_{\mathscr{A},k}(x) - C \tag{114}$$

holds for almost all $x \in (X \setminus Z)(k)$.

Note that the error term in (113) is weaker than in (65); see Stewart and Tijdeman [81] and van Frankenhuijsen [86].

Unlike the situation in Nevanlinna theory, this conjecture is not known in *any* case over number fields (other than those for which $X(k)$ is not Zariski dense).

The simplest (nontrivial) case of this conjecture is when $X = \mathbb{P}^1_k$ and D is a divisor consisting of three points, say $D = [0] + [1] + [\infty]$. In that case, it is equivalent to the "abc conjecture" of Masser and Oesterlé [62, (9.5)]. This conjecture can be stated (over $\mathbb{Q}$ for simplicity, and with a weaker error term) as follows.

Conjecture 23.5. Fix $\varepsilon > 0$. Then there is a constant C_ε such that there are only finitely many triples $(a, b, c) \in \mathbb{Z}^3$ satisfying $a + b + c = 0$, $\gcd(a, b, c) = 1$, and

$$\log \max\{|a|, |b|, |c|\} \le (1 + \varepsilon) \sum_{p \mid abc} \log p + C_\varepsilon. \tag{115}$$

To see the equivalence with the above-mentioned special case of Conjecture 23.4, let (a, b, c) be a triple of relatively prime rational integers satisfying $a + b + c = 0$, and let $x \in \mathbb{P}^2_{\mathbb{Q}}$ be the corresponding point with homogeneous coordinates $[a : b : c]$. Then the left-hand side of (115) is just the height $h_{\mathbb{Q}}(x)$.

Now let D be the divisor consisting of the coordinate hyperplanes H_0, H_1, and H_2 in $\mathbb{P}^2_{\mathbb{Q}}$ (defined respectively by $x_0 = 0$, $x_1 = 0$, and $x_2 = 0$). Since $\gcd(a, b, c) = 1$, we have

$$\lambda_{H_0, p}(x) = -\log \frac{\|a\|_p}{\max\{\|a\|_p, \|b\|_p, \|c\|_p\}} = \operatorname{ord}_p(a) \log p,$$

for all (finite) rational primes p, where $\operatorname{ord}_p(a)$ denotes the largest integer m for which $p^m \mid a$. Thus

$$N^{(1)}_{\{\infty\}}(H_0, x) = \sum_{p \mid a} \log p.$$

Similarly

$$N^{(1)}_{\{\infty\}}(H_1, x) = \sum_{p \mid b} \log p \qquad \text{and} \qquad N^{(1)}_{\{\infty\}}(H_2, x) = \sum_{p \mid c} \log p.$$

Therefore, by relative primeness,

$$N^{(1)}_{\{\infty\}}(D, x) = \sum_{p \mid abc} \log p,$$

so (115) is equivalent to

$$h_{\mathbb{Q}}(x) \le (1 + \varepsilon) N^{(1)}_{\{\infty\}}(D, x) + C_\varepsilon.$$

Since $a + b + c = 0$, the points $[a : b : c]$ all lie on the line $x_0 + x_1 + x_2 = 0$ in $\mathbb{P}^2_{\mathbb{Q}}$. Choosing an isomorphism of this line with $\mathbb{P}^1_{\mathbb{Q}}$ such that the restriction of D corresponds to the divisor $[0] + [1] + [\infty]$ on $\mathbb{P}^1_{\mathbb{Q}}$, it follows by functoriality of $h_{\mathbb{Q}}$ and $N^{(1)}_{\{\infty\}}(D, x)$ that Conjecture 23.5 is equivalent to the special case of Conjecture 23.4 with $k = \mathbb{Q}$, $S = \{\infty\}$, $X = \mathbb{P}^1_{\mathbb{Q}}$, $D = [0] + [1] + [\infty]$, and with a weaker error term.

Some effort has been expended on finding a higher-dimensional counterpart to the abc conjecture (in the spirit of Cartan's and Schmidt's theorems). One major decision, for example, is how to extend the condition on relative primeness. For the equation $a + b + c = 0$, the condition $\gcd(a, b, c) = 1$ is equivalent to pairwise relative primeness, since $p \mid a$ and $p \mid b$ easily implies $p \mid c$. With more terms, such as $a + b + c + d = 0$, though, the two variants are no longer equivalent. For the sake of the present discussion, we use weaker condition of overall relative primeness. This is all that is needed for the largest absolute value to be equivalent to the (multiplicative) height.

So let $a_0, \ldots, a_{n+1}$ be integers with $\gcd(a_0, \ldots, a_{n+1}) = 1$ and $a_0 + \cdots + a_{n+1} = 0$. Such an $(n+2)$-tuple gives a point $x := [a_0 : \ldots : a_n] \in \mathbb{P}^n_{\mathbb{Q}}$ with

$$h(x) = \log\max\{|a_0|, \ldots, |a_{n+1}|\} + O(1).$$

Let D be the divisor on $\mathbb{P}^n$ consisting of the sum of the coordinate hyperplanes and the hyperplane $x_0 + \cdots + x_n = 0$. Then we have

$$N^{(1)}_{\{\infty\}}(D, x) = \sum_{p \mid a_0 \cdots a_{n+1}} p + O(1).$$

Since the canonical line sheaf $\mathscr{K}$ on $\mathbb{P}^n$ is $\mathscr{O}(-n-1)$ and D has degree $n+2$, we have $\mathscr{K}(D) \cong \mathscr{O}(1)$ and therefore (114) would (if true) give

$$\sum_{p \mid a_0 \cdots a_{n+1}} \log p \geq (1 - \varepsilon) \log\max\{|a_0|, \ldots, |a_{n+1}|\} - C \tag{116}$$

for all $\varepsilon > 0$ and all C, for almost all rational points $x = [a_0 : \ldots : a_n]$ outside of some proper Zariski-closed subset of $\mathbb{P}^n_{\mathbb{Q}}$ depending on ε.

The following example, due to Brownawell and Masser [8, p. 430], then shows that an obvious extension of the abc conjecture to hyperplanes in $\mathbb{P}^n$ is false with 1-truncated counting functions, unless one allows exceptional hypersurfaces of degree > 1.

Example 23.6. Let $n \in \mathbb{Z}_{>0}$, and consider the map $\phi : \mathbb{P}^1_{\mathbb{Q}} \to \mathbb{P}^n_{\mathbb{Q}}$ given by

$$\phi([x_0 : x_1]) = \left[x_0^n : \binom{n}{1} x_0^{n-1} x_1 : \binom{n}{2} x_0^{n-2} x_1^2 : \ldots : x_1^n \right].$$

Let D be the divisor $(x_0 x_1 (x_0 + x_1)) = [0] + [-1] + [\infty]$ on $\mathbb{P}^1_{\mathbb{Q}}$ and let D' be the (similar) divisor $(y_0 \cdots y_n (y_0 + \cdots + y_n))$ on $\mathbb{P}^n_{\mathbb{Q}}$. Note that $\operatorname{Supp} \phi^* D' = \operatorname{Supp} D$, so that

$$N^{(1)}_{\{\infty\}}(D', \phi([x_0 : x_1])) = N^{(1)}_{\{\infty\}}(D, [x_0 : x_1])$$

if $x_0 \neq 0$ and $x_1 \neq 0$, and that $h_{\mathbb{Q}}(\phi([x_0 : x_1])) = n h_{\mathbb{Q}}([x_0 : x_1]) + O(1)$. It is known that there are infinitely many pairs (a, b) of relatively prime integers for which

$$N_{\{\infty\}}^{(1)}(D, [a:b]) \leq h_{\mathbb{Q}}([a:b]) ;$$

therefore we have

$$N_{\{\infty\}}^{(1)}(D', \phi([a:b])) \leq \frac{1}{n} h_{\mathbb{Q}}(\phi([a:b])) + O(1),$$

contrary to (116). The points $\phi([a:b])$ are not contained in any hyperplane in $\mathbb{P}_{\mathbb{Q}}^n$. They are, of course, contained in the image of ϕ, hence are not Zariski dense.

The abc conjecture is still unsolved, and is regarded to be quite deep. This is so even though its counterpart in Nevanlinna theory is already known (and has been known for decades). The remainder of these notes discuss extensions of Conjecture 15.6 that all have the property of implying the abc conjecture.

24 On Discriminants

This section discusses some facts about discriminants of number fields. These will be used to formulate a diophantine conjecture for algebraic points in Sect. 25.

Definition 24.1. Let $L \supseteq k$ be number fields, and let D_L denote the absolute discriminant of L. Then the **logarithmic discriminant** of L (relative to k) is

$$d_k(L) = \frac{1}{[L:k]} \log |D_L| - \log |D_k|.$$

Also, if X is a variety over k and $x \in X(\bar{k})$, then let

$$d_k(x) = d_k(\kappa(x)).$$

The discriminant of a number field k is related to the different $\mathscr{D}_{k/\mathbb{Q}}$ of k over $\mathbb{Q}$ by the formulas

$$|D_k| = (\mathbb{Z} : N_{\mathbb{Q}}^k \mathscr{D}_{k/\mathbb{Q}}) = (\mathscr{O}_k : \mathscr{D}_{k/\mathbb{Q}}).$$

By multiplicativity of the different in towers, we therefore have

$$\begin{aligned} d_k(L) &= \frac{1}{[L:k]} \log(\mathscr{O}_L : \mathscr{D}_{L/k}) \\ &= \frac{1}{[L:k]} \sum_{\substack{\mathfrak{q} \in \operatorname{Spec} \mathscr{O}_L \\ \mathfrak{q} \neq (0)}} \operatorname{ord}_{\mathfrak{q}} \mathscr{D}_{L/k} \cdot \log(\mathscr{O}_L : \mathfrak{q}) \end{aligned} \tag{117}$$

for number fields $L \supseteq k$.

This expression can be used to define a "localized" log discriminant term:

Definition 24.2. Let $L \supseteq k$ be number fields, and let $S \supseteq S_\infty$ be a finite set of places of k. Let $\mathcal{O}_{L,S}$ denote the localization of $\mathcal{O}_L$ away from (finite) places of L lying over places in S; note that $\mathcal{O}_{L,S} = \mathcal{O}_L \otimes_{\mathcal{O}_k} \mathcal{O}_{k,S}$ (cf. (58)). Then we define

$$d_S(L) = \frac{1}{[L:k]} \sum_{\substack{\mathfrak{q} \in \mathrm{Spec}\, \mathcal{O}_{L,S} \\ \mathfrak{q} \neq (0)}} \mathrm{ord}_{\mathfrak{q}}\, \mathscr{D}_{L/k} \cdot \log(\mathcal{O}_{L,S} : \mathfrak{q}). \tag{118}$$

Also, if X is a variety over k and $x \in X(\bar{k})$, then let

$$d_S(x) = d_S(\kappa(x)).$$

By (117), if $S = S_\infty \subseteq M_k$ then $d_S(L) = d_k(L)$. Other than in this section, these notes will be concerned only with $d_k(\cdot)$. In fact, for any rational prime p the portion of $(\log |D_k|)/[k:\mathbb{Q}]$ coming from primes over p is bounded by $(1 + \log_p[k:\mathbb{Q}]) \log p$ [73, Chap. III, Remark 1 following Prop. 13]. These notes will be primarily concerned with number fields of bounded degree over $\mathbb{Q}$, so the difference between d_S and d_k is bounded and can be ignored. However, for this section (only) the general situation will be considered, since the results may be useful elsewhere and may not be available elsewhere.

For the remainder of this section, k is a number field and $S \supseteq S_\infty$ is a finite set of places of k.

The following lemma shows that the discriminant is not increased by taking the compositum with a given field.

Lemma 24.3. *Let*

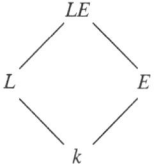

be a diagram of number fields. Then

$$d_S(LE) - d_S(E) \leq d_S(L).$$

Proof. We first show that

$$\mathscr{D}_{LE/E} \mid \mathscr{D}_{L/k} \cdot \mathcal{O}_{LE}, \tag{119}$$

which really amounts to showing that $\mathscr{D}_{L/k} \subseteq \mathscr{D}_{LE/E}$. Recall that $\mathscr{D}_{L/k}$ is the ideal in $\mathcal{O}_L$ generated by all elements $f'(\alpha)$, as α varies over the set $\{\alpha \in \mathcal{O}_L : L = k(\alpha)\}$

and f is the (monic) irreducible polynomial for α over k. Such α also lie in $\mathscr{O}_{LE}$, and generate LE over E. Let g be the irreducible polynomial for α over E; we note that $g \mid f$ and therefore $f = gh$ for a monic polynomial $h \in \mathscr{O}_E[t]$. We also have $f'(\alpha) = g'(\alpha)h(\alpha)$. Since $h(\alpha) \in \mathscr{O}_{LE}$, it then follows that $f'(\alpha) \in \mathscr{D}_{LE/E}$, which then implies (119).

It then suffices to show that

$$\operatorname{ord}_{\mathfrak{q}} \mathscr{D}_{L/k} \cdot \log(\mathscr{O}_L : \mathfrak{q}) = \frac{1}{[LE:L]} \sum_{\substack{Q \in \operatorname{Spec} \mathscr{O}_{LE} \\ Q \mid \mathfrak{q}}} \operatorname{ord}_Q(\mathscr{D}_{L/k} \mathscr{O}_{LE}) \cdot \log(\mathscr{O}_{LE} : Q)$$

for all nonzero $\mathfrak{q} \in \operatorname{Spec} \mathscr{O}_L$. But this follows immediately from the classical fact that $[LE:L] = \sum e_{Q/\mathfrak{q}} f_{Q/\mathfrak{q}}$. $\qquad\square$

The Chevalley-Weil theorem may be generalized to a situation where ramification is allowed. This involves a proximity function for the ramification divisor, which is defined as follows.

Definition 24.4. Let $\phi \colon X \to Y$ be a generically finite, dominant morphism of non-singular complete varieties over a field k. Assume that the function field extension $K(X)/K(Y)$ is separable. Then the natural map $\phi^* \Omega_{Y/k} \to \Omega_{X/k}$ induces a natural map $\phi^* \mathscr{K}_Y \to \mathscr{K}_X$ of canonical sheaves, which in turn defines a natural map $\mathscr{O}_X \to \mathscr{K}_X \otimes \phi^* \mathscr{K}_Y^\vee$. This latter map defines a section of $\mathscr{K}_X \otimes \phi^* \mathscr{K}_Y^\vee$, whose divisor is the **ramification divisor** of X over Y. It is an effective divisor R, and we have $\mathscr{K}_X \cong \mathscr{K}_Y(R)$.

(The remainder of this section will be quite technical. Most readers will likely be interested only in the statement of Theorem 24.11, and should now skip to the statement of that theorem and to Theorem 24.13.)

The following definition will also be needed to generalize Chevalley-Weil.

Definition 24.5. Let M be a finitely generated module over a noetherian ring R, and let

$$R^m \xrightarrow{f} R^n \longrightarrow M \longrightarrow 0 \tag{120}$$

be a presentation of M. Then the 0th **Fitting ideal** of M is the ideal $F_0(M)$ in R generated by the determinants of all $n \times n$ submatrices of the $n \times m$ matrix representing f. It is independent of the presentation [17, 20.4]. This globalizes: if $\mathscr{F}$ is a coherent sheaf on a noetherian scheme X, then locally one can form presentations (120) and glue them to give a sheaf of ideals $F_0(\mathscr{F})$.

Lemma 24.6. Let $\mathscr{F}$ be a coherent sheaf on a noetherian scheme X.

(a) Forming the Fitting ideal commutes with pull-back: Let $\phi \colon X' \to X$ be a morphism of noetherian schemes. Then

$$F_0(\phi^* \mathscr{F}) = \phi^* F_0(\mathscr{F}) \cdot \mathscr{O}_{X'}. \tag{121}$$

(b) If $\mathscr{F} \twoheadrightarrow \mathscr{F}'$ is a surjection, then

$$F_0(\mathscr{F}') \supseteq F_0(\mathscr{F}).$$

Proof. Let $\phi \colon X' \to X$ be as in (a). Since tensoring is right exact, a presentation of $\mathscr{F}$ pulls back to give a presentation of $\phi^* \mathscr{F}$ on X', and (121) follows directly.

If $\mathscr{F} \twoheadrightarrow \mathscr{F}'$ is a surjection, then one can use the same middle terms in the local presentations of $\mathscr{F}$ and of $\mathscr{F}'$, and the first term in the presentation of $\mathscr{F}$ can be a direct summand of the first term in the presentation of $\mathscr{F}'$. In this case, the resulting generators of $F_0(\mathscr{F})$ are a subset of the generators of $F_0(\mathscr{F}')$. This gives (b). □

The ramification divisor can be described using a Fitting ideal.

Lemma 24.7. *Let $\phi \colon X \to Y$ be as in Definition 24.4. Then $F_0(\Omega_{X/Y})$ is a sheaf of ideals, locally principal and generated by functions f which locally generate the ramification divisor as a Cartier divisor.*

Proof. Indeed, the first exact sequence of differentials

$$\phi^* \Omega_{Y/k} \longrightarrow \Omega_{X/k} \longrightarrow \Omega_{X/Y} \longrightarrow 0$$

gives a locally free presentation of $\Omega_{X/Y}$, and the two initial terms have the same rank. □

The proof of Theorem 24.11 will also need the notion of a *model* of a variety over a number field (corresponding to Definition 16.1 in the function field case).

Definition 24.8. Let X be a variety over a number field k. A **model** for X over $\mathscr{O}_k$ is an integral scheme $\mathscr{X}$, flat over $\mathscr{O}_k$, together with an isomorphism $X \cong \mathscr{X} \times_{\mathscr{O}_k} k$.

If X is a complete variety, then a model $\mathscr{X}$ can be constructed using Nagata's embedding theorem. Moreover, $\mathscr{X}$ can be chosen to be proper over Spec $\mathscr{O}_k$. On such a model, rational points correspond naturally and bijectively to sections $\mathscr{O}_k \to \mathscr{X}$ by the valuative criterion of properness, and algebraic points Spec $L \to X$ correspond naturally and bijectively to morphisms Spec $\mathscr{O}_L \to \mathscr{X}$ over $\mathscr{O}_k$.

As is the case over function fields, a key benefit of working with models is the fact that Weil functions, and therefore the proximity, counting, and height functions, can be defined exactly once one has chosen an extension of the given Cartier divisor D to the model. (Such an extension may not always exist, but the model can be chosen so that it does exist.) At archimedean places, these definitions rely on the additional data specified in Arakelov theory. These definitions, however, will not be described in these notes.

For the purposes of this section, however, we do need to define the proximity function relative to a sheaf of ideals.

Definition 24.9. Let X be a complete variety over a number field k, let $\mathscr{X}$ be a proper model for X over $\mathscr{O}_k$, let $\mathscr{I}$ be a sheaf of ideals on $\mathscr{X}$, and let $S \supseteq S_\infty$ be a finite set of places of k. Let $x \in X(\bar{k})$, and assume that x does not lie in the closed subscheme of $\mathscr{X}$ defined by $\mathscr{I}$. Then the **counting function** $N_S(\mathscr{I},x)$ is defined as follows. Let L be some number field containing $\kappa(x)$, let $i \colon \operatorname{Spec} \mathscr{O}_L \to \mathscr{X}$ be the morphism over $\operatorname{Spec} \mathscr{O}_k$ corresponding to x, and let I be the ideal in $\mathscr{O}_L$ corresponding to the ideal sheaf $i^* \mathscr{I} \cdot \mathscr{O}_{\operatorname{Spec} \mathscr{O}_L}$ on $\operatorname{Spec} \mathscr{O}_L$. Then we define

$$N_S(\mathscr{I},x) = \frac{1}{[L:k]} \log(\mathscr{O}_{L,S} : I\mathscr{O}_{L,S})$$

$$= \frac{1}{[L:k]} \sum_{\substack{\mathfrak{q} \in \operatorname{Spec} \mathscr{O}_{L,S} \\ \mathfrak{q} \neq (0)}} \operatorname{ord}_{\mathfrak{q}} I \cdot \log(\mathscr{O}_{L,S} : \mathfrak{q}).$$

It can be shown (although we will not do so here) that if $\mathscr{X}$ is a proper model over $\operatorname{Spec} \mathscr{O}_k$ for a complete variety X over k, if $\mathscr{I}$ is an ideal sheaf on $\mathscr{X}$, and if the restriction of $\mathscr{I}$ to X corresponds to a Cartier divisor D, then

$$N_S(\mathscr{I},x) = N_S(D,x) + O(1) \tag{122}$$

for all $x \in X(\bar{k})$ not lying in $\operatorname{Supp} D$.

Counting (and proximity and height) functions relative to sheaves of ideals were first introduced by Yamanoi [102], in the context of Nevanlinna theory.

The different can be described via differentials as well. Indeed, $\mathscr{D}_{L/k}$ is the annihilator of the sheaf of relative differentials:

$$\mathscr{D}_{L/k} = \operatorname{Ann} \Omega_{\mathscr{O}_L/\mathscr{O}_k}. \tag{123}$$

In this case $\Omega_{\mathscr{O}_L/\mathscr{O}_k}$ is a torsion sheaf locally generated by one element; hence (118) can be rewritten as

$$d_S(L) = \frac{1}{[L:k]} \sum_{\substack{\mathfrak{q} \in \operatorname{Spec} \mathscr{O}_{L,S} \\ \mathfrak{q} \neq (0)}} \operatorname{length}_{\mathfrak{q}} \Omega_{\mathscr{O}_L/\mathscr{O}_k} \cdot \log(\mathscr{O}_{L,S} : \mathfrak{q})$$

$$= \frac{1}{[L:k]} \log \# H^0(\mathscr{O}_{L,S}, \Omega_{\mathscr{O}_{L,S}/\mathscr{O}_{k,S}}). \tag{124}$$

The following lemma does most of the work in generalizing the Chevalley-Weil theorem. It has been stated as a separate lemma because it will also be used in the Nevanlinna case.

Lemma 24.10. *Let $A \to B$ be a local homomorphism of discrete valuation rings, with B finite over A, let $\phi \colon X \to Y$ be a generically finite morphism of schemes, and let*

$$\mathrm{Spec}\, B \xrightarrow{\; j \;} X$$

$$\downarrow \qquad\qquad \downarrow{\scriptstyle \phi} \qquad\qquad (125)$$

$$\mathrm{Spec}\, A \xrightarrow{\;\;\;\;\;} Y$$

be a commutative diagram. Assume also that the image of j is not contained in the support of $\Omega_{X/Y}$, that $j^* \mathscr{O}_X$ generates the fraction field of B over the fraction field of A, and that the fraction field of B is separable over the fraction field of A. Then

$$\mathscr{D}_{B/A} \supseteq j^* F_0(\Omega_{X/Y}) \cdot B. \qquad (126)$$

Proof. The map j factors through the product $\mathrm{Spec}\, A \times_Y X$:

$$\mathrm{Spec}\, B \xrightarrow{\; j'' \;} \mathrm{Spec}\, A \times_Y X \xrightarrow{\; q \;} X$$

$$\downarrow \qquad\qquad\qquad \downarrow$$

$$\mathrm{Spec}\, A \xrightarrow{\;\;\;\;\;\;\;\;\;} Y \;;$$

here $j = q \circ j''$. We may replace $\mathrm{Spec}\, A \times_Y X$ in this diagram with an open affine neighborhood $\mathrm{Spec}\, B''$ of $j''(b)$, where b denotes the closed point of $\mathrm{Spec}\, B$. By Lemma 24.6a, we have

$$j^* F_0(\Omega_{X/Y}) \cdot B = (j'')^* F_0(\Omega_{B''/A}) \cdot B. \qquad (127)$$

The map $\mathrm{Spec}\, B \to \mathrm{Spec}\, B''$ corresponds to a ring homomorphism $B'' \to B$; let B' denote its image. Then j'' factors through $j' : \mathrm{Spec}\, B \to \mathrm{Spec}\, B'$ and a closed immersion $\mathrm{Spec}\, B' \to \mathrm{Spec}\, B''$. By the second exact sequence for differentials, the map $\Omega_{B''/A}\big|_{B'} \to \Omega_{B'/A}$ is surjective, and by Lemma 24.6b

$$(j'')^* F_0(\Omega_{B''/A}) \cdot B \subseteq (j')^* F_0(\Omega_{B'/A}) \cdot B. \qquad (128)$$

By [43, Def. 10.1 and p. 166], $F_0(\Omega_{B'/A})$ is the Kähler different $\mathfrak{d}_K(B'/A)$. Since B' is finite over A and the fraction field of B' is separable over the fraction field of A, [43, Prop. 10.22] gives

$$F_0(\Omega_{B'/A}) \subseteq \mathscr{D}_{B'/A} \qquad (129)$$

(note that $\mathscr{D}_{B'/A}$ is $\mathfrak{d}_D(B'/A)$ in Kunz's notation). Finally, since $B' \subseteq B$, it follows directly from the definition (see for example [43, G.9a]) that

$$\mathscr{D}_{B'/A} \subseteq \mathscr{D}_{B/A}. \qquad (130)$$

Combining (127)–(130) then gives (126).   □

The generalized Chevalley-Weil theorem can now be stated as follows. This was originally proved in [87, Thm. 5.1.6], but the proof there is only valid if X and Y

have good reduction everywhere, or if the points $x \in X(\bar{k})$ have bounded degree over k. Therefore we will give a general proof here.

Theorem 24.11. *Let* $\phi: X \to Y$ *be a generically finite, dominant morphism of non-singular complete varieties over k, with ramification divisor R. Then for all $x \in X(\bar{k})$ not lying on* $\operatorname{Supp} R$, *we have*

$$d_S(x) - d_S(\phi(x)) \leq N_S(R,x) + O(1). \tag{131}$$

Proof. Let $\mathscr{X}$ and $\mathscr{Y}$ be models for X and Y over $\mathcal{O}_k$, respectively. By replacing $\mathscr{X}$ with the closure of the graph of the rational map $\phi: \mathscr{X} \to \mathscr{Y}$ if necessary, we may assume that ϕ extends to a morphism $\mathscr{X} \to \mathscr{Y}$ over $\operatorname{Spec} \mathcal{O}_k$. By Lemma 24.7 and (122), it will then suffice to show that

$$d_S(x) - d_S(\phi(x)) \leq N_S(F_0(\Omega_{\mathscr{X}/\mathscr{Y}}),x). \tag{132}$$

Fix $x \in X(\bar{k})$ as above, and let $E = \kappa(x)$ and $L = \kappa(\phi(x))$. Then $d_S(x) = d_S(E)$ and $d_S(\phi(x)) = d_S(L)$. Let w be a place of E with $w \nmid S$, and let v be a place of L lying under w. Let $\mathcal{O}_w$ and $\mathcal{O}_v$ denote the localizations of $\mathcal{O}_E$ and $\mathcal{O}_L$ at the primes corresponding to w and v, respectively. Let $j: \operatorname{Spec} \mathcal{O}_w \to \mathscr{X}$ be the restriction of the map $\operatorname{Spec} \mathcal{O}_E \to \mathscr{X}$ over $\mathcal{O}_k$ corresponding to x. By (118), multiplicativity of the different in towers, Definition 24.9, and compatibility of various things with localization, it suffices to show that

$$\mathscr{D}_{\mathcal{O}_w/\mathcal{O}_v} \supseteq j^* F_0(\Omega_{\mathscr{X}/\mathscr{Y}}) \cdot \mathcal{O}_w. \tag{133}$$

The point $\phi(x) \in Y(L)$ determines a map $\operatorname{Spec} \mathcal{O}_v \to \mathscr{Y}$, so there is a diagram

$$
\begin{array}{ccc}
\operatorname{Spec} \mathcal{O}_w & \xrightarrow{\ j\ } & \mathscr{X} \\
\downarrow & & \downarrow \\
\operatorname{Spec} \mathcal{O}_v & \longrightarrow & \mathscr{Y}.
\end{array}
$$

This satisfies the conditions of Lemma 24.10, which implies (133). $\qquad\square$

Remark 24.12. More generally, the above proof shows that if one replaces (131) with (132), then Theorem 24.11 still holds without the assumptions that X and Y are nonsingular.

The counterpart to this theorem in Nevanlinna theory is the following. (This is much easier in the special case $\dim X = \dim Y = 1$. The general case is more complicated because then the ramification divisor may not be easy to describe.)

Theorem 24.13. *Let B be a connected (nonempty) Riemann surface, let $\phi: X \to Y$ be a generically finite, dominant morphism of smooth complete complex varieties,* *with ramification divisor R, and let $f\colon B \to X$ be a holomorphic function whose image is not contained in $\phi(\operatorname{Supp} R)$. Then there is a connected Riemann surface B', a proper surjective holomorphic map $\pi\colon B' \to B$ of degree bounded by $[K(X) : K(Y)]$, and a holomorphic function $g\colon B' \to X$ such that the diagram*

$$
\begin{array}{ccc}
B' & \xrightarrow{\ g\ } & X \\
{\scriptstyle \pi}\big\downarrow & & \big\downarrow{\scriptstyle \phi} \\
B & \xrightarrow{\ f\ } & Y
\end{array}
\tag{134}
$$

commutes. Moreover, if e is the ramification index of B' over B at at any given point of B', then $e - 1$ is bounded by the multiplicity of the analytic divisor g^*R at that point. (In other words, the ramification divisor of π is bounded by g^*R, relative to the cone of effective analytic divisors.)

Proof. Let $d = [K(X) : K(Y)]$. Let $B_0 = \{(b,x) \in B \times X : f(b) = \phi(x)\}$; it is an analytic variety, of degree d over B (i.e., fibers of the projection $B_0 \to B$ have at most d points, and some fibers have exactly d points). Let B' be the normalization of B_0 [35, R13]; again B' is of degree d over B. After replacing B' with one of its connected components, we may assume that B' is connected (of degree $\leq d$ over B). We then have holomorphic functions $\pi\colon B' \to B$ and $g\colon B' \to X$ as in (134). Also, B' is a Riemann surface [35, Q13].

Now fix $b' \in B'$, and let $b = \pi(b') \in B$. Fix holomorphic local coordinates z' at b' and z at b, vanishing at the respective points. Via the local coordinate z, we identify an open neighborhood of b in B with an open neighborhood of 0 in $\mathbb{C}$, and identify the ring $\mathcal{O}$ of germs of holomorphic functions on B at b with the ring of germs of holomorphic functions on $\mathbb{C}$ at 0. Also let $\mathcal{O}'$ be the ring of germs of holomorphic functions on B' at b', and let e be the ramification index of π at b'. Then the germ of the analytic variety B' at b' is a finite branched covering of the germ of B at b, of covering order e, and we identify $\mathcal{O}$ with a subring of $\mathcal{O}'$ via π. By [35, C5 and C8], there is a canonically defined monic polynomial $P \in \mathcal{O}[t]$ of degree e such that $P(z') = 0$, and

$$
\mathcal{O}' \cong \mathcal{O}[t]/P(t).
$$

Since B' is regular at b', the germ of the variety B' at b' is irreducible, so $\mathcal{O}'$ is an entire ring [35, B6]. Therefore $P(t)$ is irreducible, and by the Weierstrass Preparation Theorem [35, A4] it is a Weierstrass polynomial. This means that all non-leading coefficients vanish at b. A straightforward computation then gives

$$
\Omega_{\mathcal{O}'/\mathcal{O}} = \mathcal{O}'/(z')^{e-1}.
$$

By [35, A8 and G20], $\mathcal{O}$ and $\mathcal{O}'$ are discrete valuation rings, hence Dedekind, and then (123) gives

$$
\mathscr{D}_{\mathcal{O}'/\mathcal{O}} = (z')^{e-1}.
$$

Moreover, we have a commutative diagram

$$\begin{array}{ccc} \operatorname{Spec} \mathscr{O}' & \xrightarrow{\ j\ } & X \\ \downarrow & & \downarrow \phi \\ \operatorname{Spec} \mathscr{O} & \longrightarrow & Y \end{array}$$

which satisfies the conditions of Lemma 24.10. Therefore,

$$\mathscr{D}_{\mathscr{O}'/\mathscr{O}} \supseteq j^* F_0(\Omega_{X/Y}) \cdot \mathscr{O}'.$$

This, together with Lemma 24.7, implies the theorem.   □

Remark 24.14. As was the case in Remark 24.12, Theorem 24.13 remains true when X and Y are allowed to be singular, provided that the conclusion is replaced by the assertion that the ramification divisor of π is bounded by the analytic divisor associated to $g^* F_0(\Omega_{X/Y})$.

25   A Diophantine Conjecture for Algebraic Points

This section describes an extension of Conjecture 15.6 to allow algebraic points instead of rational points. This comes at a cost of adding a discriminant term to the inequality.

   This conjecture is subject to some doubt: see Remark 27.6.

Conjecture 25.1. Let k be a number field, let $S \supseteq S_\infty$ be a finite set of places of k, let X be a smooth projective variety over k, let D be a normal crossings divisor on X, let $\mathscr{K}$ be the canonical line sheaf on X, let $\mathscr{A}$ be an ample line sheaf on X, and let r be a positive integer. Then:

(a) Let Σ be a generic subset of $X(\bar{k}) \setminus \operatorname{Supp} D$ such that $[\kappa(x) : k] \leq r$ for all $x \in \Sigma$. Then the inequality

$$m_S(D,x) + h_{\mathscr{K},k}(x) \leq d_k(x) + O(\log^+ h_{\mathscr{A},k}(x)) \tag{135}$$

holds for all $x \in \Sigma$.

(b) For any $\varepsilon > 0$ there is a proper Zariski-closed subset Z of X, depending only on $X, D, \mathscr{A}$, and ε, such that for all $C \in \mathbb{R}$ the inequality

$$m_S(D,x) + h_{\mathscr{K},k}(x) \leq d_k(x) + \varepsilon h_{\mathscr{A},k}(x) + C \tag{136}$$

holds for almost all $x \in (X \setminus Z)(\bar{k})$ with $[\kappa(x) : k] \leq r$.

   When $r = 1$, this just reduces to Conjecture 15.6, since then $\kappa(x) = k$ for all x. Other than with $r = 1$, no case of this conjecture is known (for number fields). Over function fields, some parts are known.

One may also ask if the conjecture is true without the bound on r. This would require changing the quantization of C in part (b): for example, there are infinitely many roots of unity, which all have height zero. It would also require changing the $d_k(x)$ terms to $d_S(x)$. Other than that, though, it seems to be a reasonable conjecture.

As with any mathematical statement, it is often useful to be aware of how its strength varies with the parameters. For this conjecture, replacing k with a larger number field (and S with the corresponding set), adding places to S, increasing D, increasing r, or (in the case of part (b)) decreasing ε results in a stronger statement.

As was the case with Conjecture 15.6, this Conjecture 25.1 can also be posed for smooth complete varieties X, provided that $h_{\mathscr{A},k}$ is replaced by a big height.

Remark 15.7 does not extend trivially to Conjecture 25.1, though, since the discriminant terms may not add up.

By [87, Prop. 5.4.1],[3] Conjecture 25.1b with $D = 0$ implies the full Conjecture 25.1b. This uses a covering construction.

Next, we show how Conjecture 25.1 relates to generically finite ramified covers.

Proposition 25.2. *Let k and S be as in Conjecture 25.1, let $\pi\colon X' \to X$ be a surjective generically finite morphism of complete nonsingular varieties over k, and let D be a normal crossings divisor on X. Let $D' = (\pi^*D)_{\mathrm{red}}$ (this means the reduced divisor with the same support as π^*D), and assume that it too has normal crossings. Let $\mathscr{K}$ and $\mathscr{K}'$ denote the canonical line sheaves on X and X', respectively. Then, for all $x \in X'(\bar{k})$ not lying on $\mathrm{Supp}\, D'$ or on the support of the ramification divisor,*

$$m_S(D, \pi(x)) + h_{\mathscr{K},k}(\pi(x)) - d_S(\pi(x)) \le m_S(D', x) + h_{\mathscr{K}',k}(x) - d_S(x) + O(1).$$

$$(137)$$

In particular, since the pull-back of any big line sheaf on X to X' remains big, either part of Conjecture 25.1 for D' on X' implies that same part for D on X.

Proof. Let R be the ramification divisor for X' over X; since $\mathscr{K}' \cong \pi^*\mathscr{K} \otimes \mathscr{O}(R)$, we then have

$$h_{\mathscr{K},k}(\pi(x)) - h_{\mathscr{K}',k}(x) = -m_S(R, x) - N_S(R, x) + O(1).$$

Also, Theorem 24.11 gives

$$d_S(x) - d_S(\pi(x)) \le N_S(R, x) + O(1).$$

Finally, by [87, Lemma 5.2.2], $\pi^*D - (\pi^*D)_{\mathrm{red}} \le R$ (relative to the cone of effective divisors). Therefore

$$m_S(D, \pi(x)) - m_S(D', x) \le m_S(R, x) + O(1).$$

---

[3] The proposition is actually valid in more generality than its statement indicates. However, the proof has an error. The functions $f_1, \ldots, f_n$ must be chosen such that each point of $\mathrm{Supp}\, D$ has an open neighborhood U such that $D = (f_i)$ on U for some i.

Adding this equation and the two inequalities then gives (137).   □

It was this proposition that motivated the original version of Conjecture 25.1 [87, p. 63].

Finally, we note that Conjecture 25.1 can also be posed with truncated counting functions.

Conjecture 25.3. Let k, S, X, D, $\mathscr{K}$, $\mathscr{A}$, and r be as in Conjecture 25.1. Then:

(a) Let Σ be a generic subset of $X(\bar{k}) \setminus \operatorname{Supp} D$ such that $[\kappa(x) : k] \leq r$ for all $x \in \Sigma$. Then the inequality

$$N_S^{(1)}(D,x) + d_k(x) \geq h_{\mathscr{K}(D),k}(x) - O(\log^+ h_{\mathscr{A},k}(x)) \qquad (138)$$

holds for all $x \in \Sigma$.

(b) For any $\varepsilon > 0$ there is a proper Zariski-closed subset Z of X, depending only on X, D, $\mathscr{A}$, and ε, such that for all $C \in \mathbb{R}$ the inequality

$$N_S^{(1)}(D,x) + d_k(x) \geq h_{\mathscr{K}(D),k}(x) - \varepsilon h_{\mathscr{A},k}(x) - C \qquad (139)$$

holds for almost all $x \in (X \setminus Z)(\bar{k})$ with $[\kappa(x) : k] \leq r$.

Remark 25.4. Using a covering construction, it has been shown that Conjecture 25.3b would follow from Conjecture 25.1b [91]. As noted earlier in this section, the latter would then follow from Conjecture 25.1b with $D = 0$, again using a covering construction. In both of these cases, the coverings involved are generically finite, so the implication holds for varieties of any given dimension. Thus, as is noted in the next section, Conjecture 25.3b has been fully proved for curves over function fields of characteristic 0.

Proposition 25.2 does not extend to the situation of truncated counting functions, though.

26   The $1 + \varepsilon$ Conjecture and the abc Conjecture

The special case of Conjecture 25.1 in which $\dim X = 1$ and $D = 0$ is perhaps the most approachable unsolved special case, and has drawn some attention. It is called the "$1 + \varepsilon$ conjecture."

Conjecture 26.1. Let k be a number field, let X be a smooth projective curve over k, let $\mathscr{K}$ denote the canonical line sheaf on X, let r be a positive integer, let $\varepsilon > 0$, and let $C \in \mathbb{R}$. Then

$$h_{\mathscr{K},k}(x) \leq (1+\varepsilon)d_k(x) + C$$

for almost all $x \in X(\bar{k})$ with $[\kappa(x) : k] \leq r$.

This conjecture was recently proved over function fields of characteristic 0 by McQuillan [57] and (independently) by Yamanoi [103]. See also McQuillan [58] and Gasbarri [27]. Thus, Conjectures 25.1 and 25.3 hold for curves over function fields of characteristic 0, by Remark 25.4; see also [103, Thm. 5]

Conjecture 26.1 is known to imply the abc conjecture [87, pp. 71–72].

Proposition 26.2. *If Conjecture 26.1 holds, then so does the abc conjecture.*

Proof. Let $\varepsilon > 0$, and let a, b, c be relatively prime integers with $a + b + c = 0$. For large integers n, there is an associated point

$$P_n = \left[\sqrt[n]{a} : \sqrt[n]{b} : \sqrt[n]{c} \right] \in X_n(\mathbb{Q}),$$

where X_n is the nonsingular curve $x_0^n + x_1^n + x_2^n = 0$ in $\mathbb{P}^2_{\mathbb{Q}}$. This point has height

$$h_{\mathscr{K},\mathbb{Q}}(P_n) = \frac{n-3}{n} \log\max\{|a|, |b|, |c|\} + O(1),$$

since the canonical line sheaf $\mathscr{K}$ on X_n is the restriction of $\mathscr{O}(n-3)$. Here the implicit constant depends only on n. We also have

$$d_{\mathbb{Q}}(P_n) \leq \frac{n-1}{n} \sum_{p|abc} \log p + O(1),$$

where the implicit constant depends only on n. Therefore, applying Conjecture 26.1 to points P_n on X_n gives, for all n and all $\varepsilon' > 0$ a constant $C_{n,\varepsilon'} \in \mathbb{R}$ such that

$$\frac{n-3}{n} \log\max\{|a|, |b|, |c|\} \leq \left(\frac{n-1}{n} + \varepsilon' \right) \sum_{p|abc} \log p + C_{n,\varepsilon'}.$$

The proof concludes by taking n sufficiently large and ε' sufficiently small so that

$$\left(\frac{n-1}{n} + \varepsilon' \right) \Big/ \frac{n-3}{n} < 1 + \varepsilon,$$

and noting that the constants in the above discussion are independent of the triple (a, b, c). $\square$

27 Nevanlinna Theory of Finite Ramified Coverings

In Nevanlinna theory, changing the domain of the holomorphic function from $\mathbb{C}$ to a *finite ramified covering* is the counterpart to working with algebraic points of bounded degree.

References on finite ramified coverings include Lang and Cherry [49], Chap. III and Yamanoi [103].

> Throughout this section, B is a connected Riemann surface, $\pi: B \to \mathbb{C}$ is a proper surjective holomorphic map, X is a smooth complete complex variety, and $f: B \to X$ is a holomorphic function.

Note that π has a well-defined, finite degree, denoted $\deg \pi$.

We again refer to f as a holomorphic curve.

Definition 27.1. Define

$$B[r] = \{b \in B : |\pi(b)| \le r\},$$
$$B(r) = \{b \in B : |\pi(b)| < r\}, \quad \text{and}$$
$$B\langle r \rangle = \{b \in B : |\pi(b)| = r\}.$$

On $B\langle r \rangle$, let σ be the measure

$$\sigma = \frac{1}{\deg \pi} \pi^* \left(\frac{d\theta}{2\pi} \right).$$

Definition 27.2. Let D be an effective divisor on X whose support does not contain the image of f, and let λ_D be a Weil function for D. Then the **proximity function** of f with respect to D is

$$m_f(D,r) = \int_{B\langle r \rangle} \lambda_D \circ f \cdot \sigma.$$

Definition 27.3.

(a) The **counting function** for an analytic divisor $\Delta = \sum_{b \in B} n_b \cdot b$ on B is

$$N_\Delta(r) = \frac{1}{\deg \pi} \left(\sum_{b \in B(r) \setminus \pi^{-1}(0)} n_b \log \frac{r}{|\pi(b)|} + \sum_{b \in \pi^{-1}(0)} n_b \log r \right).$$

(b) If D is a divisor on X whose support does not contain the image of f, then the **counting function** for D is the function

$$N_f(D,r) = N_{f^*D}(r).$$

(c) The **ramification counting function** for π is the counting function for the ramification divisor of π. It is denoted $N_{\text{Ram}(\pi)}(r)$.

If $B = \mathbb{C}$ and π is the identity mapping, then the proximity function of Definition 27.2 and the counting function of Definition 27.3b extend those of Definitions 12.1

and 12.2, respectively. They also satisfy additivity, functoriality, and boundedness properties, as in Proposition 12.3.

If B' is another connected Riemann surface and $\pi' : B' \to B$ is another proper surjective holomorphic map, then

$$m_{f \circ \pi'}(D, r) = m_f(D, r) \qquad \text{and} \qquad N_{f \circ \pi'}(D, r) = N_f(D, r). \tag{140}$$

This holds in particular if $B = \mathbb{C}$ and π is the identity map. It is the counterpart to the fact that the proximity and counting functions in number theory are independent of the choice of number field used in Definition 11.1.

In this situation (but without the assumption $B = \mathbb{C}$), we also have

$$N_{\mathrm{Ram}(\pi \circ \pi')}(r) = N_{\mathrm{Ram}(\pi)}(r) + N_{\mathrm{Ram}(\pi')}(r). \tag{141}$$

(Note that the first term on the right-hand side is a counting function on B, while the others are on B'.) This corresponds to multiplicativity of the different in towers. Note also that, although in general $\mathrm{Ram}(\pi)$ has infinite support, its support in any given set $B(r)$ is finite, in parallel with the fact that any given extension of number fields has only finitely many ramified primes. However, given a sequence of algebraic points of bounded degree, the corresponding sequence of number fields will in general have no bound on the number of ramified primes, corresponding to the fact that $\mathrm{Ram}(\pi)$ may have infinite support.

The height is defined similarly to Definitions 12.4 and 12.7:

Definition 27.4. If D is an effective divisor on X whose support does not contain the image of f, then the **height** of f relative to D is defined up to $O(1)$ by

$$T_{D,f}(r) = m_f(D, r) + N_f(D, r).$$

If $\mathscr{L}$ is a line sheaf on X, then the height $T_{\mathscr{L},f}(r)$ is defined to be $T_{D,f}(r)$ for any divisor D on X for which $\mathscr{O}(D) \cong \mathscr{L}$ and whose support does not contain the image of f.

A First Main Theorem holds for the height as defined here, so the height relative to a line sheaf is well defined [49, III Thm. 2.1]. Theorem 12.8 also holds for heights in this context, as do Propositions 12.10 and 12.11. Corollary 12.9 is not meaningful in this context, since B need not be algebraic.

If $\pi' : B' \to B$ is as in (140) and D and $\mathscr{L}$ are as in Definition 27.4, then

$$T_{D,f \circ \pi'}(r) = T_{D,f}(r) + O(1) \qquad \text{and} \qquad T_{\mathscr{L},f \circ \pi'}(r) = T_{\mathscr{L},f}(r) + O(1).$$

Griffiths' conjecture (Conjecture 15.2) can be posed in this context, without any changes other than the domain of the holomorphic curve f, and adding terms $N_{\mathrm{Ram}(\pi)}(r)$ to the right-hand sides of (62) and (63). It will not be repeated here.

The variant with truncated counting functions reads as follows.

Conjecture 27.5. Let X be a smooth complex projective variety, let D be a normal crossings divisor on X, let $\mathcal{K}$ be the canonical line sheaf on X, and let $\mathcal{A}$ be an ample line sheaf on X. Then:

(a) The inequality

$$N_f^{(1)}(D,r) + N_{\mathrm{Ram}(\pi)}(r) \geq_{\mathrm{exc}} T_{\mathcal{K}(D),f}(r) - O(\log^+ T_{\mathcal{A},f}(r)) - o(\log r) \quad (142)$$

holds for all holomorphic curves $f \colon B \to X$ with Zariski-dense image.

(b) For any $\varepsilon > 0$ there is a proper Zariski-closed subset Z of X, depending only on $X, D, \mathcal{A}$, and ε, such that the inequality

$$N_f^{(1)}(D,r) + N_{\mathrm{Ram}(\pi)}(r) \geq_{\mathrm{exc}} T_{\mathcal{K}(D),f}(r) - \varepsilon T_{\mathcal{A},f}(r) - C \quad (143)$$

holds for all nonconstant holomorphic curves $f \colon B \to X$ whose image is not contained in Z, and for all $C \in \mathbb{R}$.

This is proved in many of the same situations where Conjecture 15.2 is proved (except possibly for the level of truncation of the counting functions). See for example [49], and also Corollary 29.7 for the case when $\dim X = 1$.

Remark 27.6. This conjecture, and therefore also Conjecture 25.1, is doubted by some. For example, McQuillan [56, Example V.1.5] notes that if X is a quotient of the unit ball in $\mathbb{C}^2$, if $f \colon B \to X$ is a one-dimensional geodesic, and if $\pi \colon B \to \mathbb{C}$ is a proper surjective holomorphic map, then

$$T_{\mathcal{K}_X,f}(r) = N_{\mathrm{Ram}(\pi)}(r) + o(T_{\mathcal{K}_X,f}(r)).$$

However, *loc. cit.* does not address how to show that a suitable map π exists, and in subsequent communications McQuillan has referred only to proper ramified coverings of the unit disk. Therefore, this is not strictly speaking a counterexample, but McQuillan finds it persuasive.

28 The $1 + \varepsilon$ Conjecture in the Split Function Field Case

This section describes how the $1 + \varepsilon$ conjecture can be easily proved in what is called the "split function field case," following early work of de Franchis [46, p. 223].

Throughout this section, F is a field, B is a smooth projective curve over F, and $k = K(B)$ is the function field of B.

If L is a finite separable extension of k, corresponding to a smooth projective curve B' over F and a finite morphism $B' \to B$ over F, then the logarithmic discriminant term in the function field case is defined as

$$d_k(L) = \frac{\deg \mathscr{K}'}{[L:k]} - \deg \mathscr{K}, \qquad (144)$$

where $\mathscr{K}'$ and $\mathscr{K}$ are the canonical line sheaves of B' and B, respectively, and degrees are taken relative to F. As before, we then define $d_k(x) = d_k(\kappa(x))$ for $x \in X(\bar{k})$. The discriminant can also be written

$$d_k(L) = \frac{1}{[L:k]} \dim_F H^0(B', \Omega_{B'/B}) \qquad (145)$$

(cf. (124)).

This definition is valid for general function fields.

The remainder of this section will restrict to the **split function field case**. This refers to the situation in which X is of the form $X \cong X_0 \times_F k$ for a smooth projective curve X_0 over F, the model $\mathscr{X}$ is a product $X_0 \times_F B$ (so that the model *splits* into a product), and $\pi \colon \mathscr{X} \to B$ is the projection morphism to the second factor.

Following early work of de Franchis [46, p. 223], it is fairly easy to prove the $1 + \varepsilon$ conjecture in the split function field case of characteristic 0.

Theorem 28.1. *Let F be a field of characteristic 0, let X_0 be a smooth projective curve over F, let $X = X_0 \times_F k$, and let $\mathscr{X} = X_0 \times_F B$. Let $\mathscr{K}$ be the canonical line sheaf on X_0, and let $\mathscr{A}$ be an ample line sheaf on X_0. View both of these line sheaves as line sheaves on X or on $\mathscr{X}$ by pulling back via the projection morphisms. Then*

$$h_{\mathscr{K},k}(x) \leq d_k(x) + \deg \Omega_{B/F} \qquad (146)$$

for all $x \in X(\bar{k})$.

Proof. The proof is particularly easy if F is algebraically closed.

In that case, let $q \colon \mathscr{X} \to X_0$ denote the projection morphism. If $q \circ i$ is a constant morphism, then by (70) the left-hand side of (146) is zero. Since the right-hand side is nonnegative by (145), the inequality is true in this case.

If $q \circ i$ is nonconstant, then it is finite and surjective, and we have

$$h_{\mathscr{K},k}(x) = \frac{(2g(X_0) - 2)\deg(q \circ i)}{[K(B') : k]} \leq \frac{2g(B') - 2}{[K(B') : k]} = d_k(x) + 2g(B) - 2.$$

by (70), the Riemann-Hurwitz formula (twice), and by (144). (Here, as usual, $g(B)$, $g(B')$, and $g(X_0)$ denote the genera of these curves.)

The general case proceeds by reducing to the above special case. First, we may assume that F is algebraically closed in k (i.e., that k/F is a *regular* field extension). Indeed, replacing F with a finite extension divides both sides of (146) by the degree of that extension, due to the fact that all quantities are expressed in terms of degrees of line sheaves, which depend in that way on F.

Let x be an algebraic point on X, let B' be the smooth projective curve over F corresponding to $\kappa(x)$, and let $i\colon B' \to \mathscr{X}$ be the morphism over B corresponding to x. Again, $K(B')$ need not be a regular extension of F; let F' be the algebraic closure of F in $K(B')$. We may replace X_0 with $X_0 \times_F \overline{F}$, B with $B \times_F \overline{F}$, $\mathscr{X}$ with $\mathscr{X} \times_F \overline{F}$, and B' with $B' \times_{F'} \overline{F}$. This again does not affect the validity of (146), since both sides are divided by $[F' : F]$. Indeed, replacing B' with $B' \times_F \overline{F}$ would have left both sides of the inequality unchanged, but B' would now be a disjoint union of $[F' : F]$ smooth projective curves. Choosing one of those curves amounts to taking $B' \times_{F'} \overline{F}$ instead of $B' \times_F \overline{F}$ for some choice of embedding $F' \hookrightarrow \overline{F}$. Also, these changes do not affect the fact that $\mathscr{X} = X_0 \times_F B$.

This reduces to the case in which F is algebraically closed. $\square$

A way to look at this proof is to think of the derivative of the map $i\colon B' \to \mathscr{X}$. It takes values in the absolute tangent bundle $T_{\mathscr{X}/F}$. Since $\mathscr{X}$ is a product, though, the tangent bundle is also a product $p^*T_{X_0/F} \times q^*T_{B/F}$, where p and q are the projection morphisms. This allows us to project onto the second factor $T_{X_0/F}$, which gives a way to bound $T_{\mathscr{X},k}(x)$.

In the general (non-split) function field case, there is first of all no bundle $T_{X_0/F}$ to project to. Instead, we have only the relative tangent bundle $T_{\mathscr{X}/B}$. This is a subbundle of the absolute tangent bundle, not a quotient, and there is no canonical projection from $T_{\mathscr{X}/F}$ to $T_{\mathscr{X}/B}$. McQuillan's proof works mainly because, for points of large height, the tangent vectors giving the derivative of $i\colon B' \to \mathscr{X}$ are "more vertical" than for points of smaller height. Therefore, two arbitrarily chosen ways of projecting the absolute tangent bundle to the relative tangent bundle will differ by a smaller amount, measured relative to the size of the tangent vector. This is sufficient to make the argument carry over.

29 Derivatives in Nevanlinna Theory

Generally speaking, proofs of theorems in Nevanlinna theory rely upon either of two methods for their basic proofs. Historically, the first was Nevanlinna's "Lemma on the Logarithmic Derivative" (Theorem 29.1). Slightly more recently, methods using differential geometry, especially focusing on curvature, have also been used. Although the latter has obvious geometric appeal, the method of the lemma on the logarithmic derivative has also been phrased in geometric terms, and (at present) is the preferred method for comparisons with number theory.

Throughout this section, B is a connected Riemann surface and $\pi\colon B \to \mathbb{C}$ is a proper surjective holomorphic map.

Nevanlinna's original Lemma on the Logarithmic Derivative (LLD) is the following.

Theorem 29.1. (Lemma on the Logarithmic Derivative) *Let f be a meromorphic function on $\mathbb{C}$. Then*

$$\int_0^{2\pi} \log^+ \left| \frac{f'(re^{i\theta})}{f(re^{i\theta})} \right| \frac{d\theta}{2\pi} \leq_{\mathrm{exc}} O(\log^+ T_f(r)) + o(\log r). \tag{147}$$

More generally, if f is a meromorphic function on B, then

$$\int_{B\langle r\rangle} \log^+ \left| \frac{df/\pi^* dz}{f} \right| \sigma \leq_{\mathrm{exc}} O(\log T_f(r) + \log r). \tag{148}$$

Proof. For the first part, see [60, IX 3.3], or [76, Thm. 3.11] for the error term given here. The second part follows from [3, Thm. 2.2]. □

A geometrical adaptation of this lemma has recently been discovered by Kobayashi, McQuillan, Wong, and others. This first requires a definition.

Definition 29.2. Let X be a smooth complex projective variety, and let D be a normal crossings divisor on X. Then the sheaf $\Omega_X^1(\log D)$ is the subsheaf of the sheaf of meromorphic sections of Ω_X^1 generated by the holomorphic sections and the local sections of the form df/f, where f is a local holomorphic function that vanishes only on D [15, II 3.1]. This is locally free of rank $\dim X$. The log tangent sheaf $T_X(-\log D)$ is its dual. There are corresponding vector bundles, of the same names.

Theorem 29.3. (Geometric Lemma on the Logarithmic Derivative) *Let X be a smooth complex projective variety, let D be a normal crossings divisor on X, and let $f: B \to X$ be a holomorphic curve whose image is not contained in $\operatorname{Supp} D$. Let $\mathscr{A}$ be an ample line sheaf on X. Finally, let $|\cdot|$ be a hermitian metric on the log tangent bundle $T_X(-\log D)$, and let $d_D f: B \to T_X(-\log D)$ denote the canonical lifting of f (as a meromorphic function). Then*

$$\int_{B\langle r\rangle} \log^+ \left| d_D f(re^{i\theta}) \right| \sigma \leq_{\mathrm{exc}} O(\log T_{\mathscr{A},f}(r) + \log r). \tag{149}$$

Proof (Wong). The general idea of the proof is that one can work locally on finitely many open sets to reduce the question to finitely many applications of the classical LLD. This proof presents a geometric rendition of this idea.

We first note that the assertion is independent of the choice of metric, since by compactness any two metrics are equivalent up to nonzero constant factors.

Next, note that the special case $X = \mathbb{P}^1$, $D = [0] + [\infty]$, is equivalent to the classical Lemma on the Logarithmic Derivative. Indeed, in this case $T_X(-\log D) \cong X \times \mathbb{C}$ is just the trivial vector bundle of rank 1. Choose the metric on $T_X(-\log D)$ to be the one corresponding to the obvious metric on $X \times \mathbb{C}$; then (149) reduces to (147).

The assertion of the theorem is preserved by taking products. Indeed, if it holds for holomorphic curves $f_1 : B \to X_1$ and $f_2 : B \to X_2$ relative to normal crossings divisors D_1 and D_2 on smooth complex projective varieties X_1 and X_2, respectively, then it is true for the product $(f_1, f_2) : B \to X_1 \times X_2$ relative to the normal crossings divisor $D := p_1^* D_1 + p_2^* D_2$, where $p_j : X_1 \times X_2 \to X_j$ are the projection morphisms $(j = 1, 2)$. This follows by choosing the obvious metric on $T_{X_1 \times X_2}(-\log D)$ and applying (31); details are left to the reader.

Finally, let D' be a normal crossings divisor on a smooth complex projective variety X', let Z be a closed subvariety of X that contains the image of f, and let $\phi : Z \dashrightarrow X'$ be a rational map. Assume that there is a nonempty Zariski-open subset U of Z and a constant $C > 0$ such that $|\phi_*(v)| \geq C|v|$ for all $v \in T_Z(-\log D)$ lying over U, that the holomorphic curve f meets U, and that Theorem 29.3 holds for the holomorphic curve $\phi \circ f$ in X' relative to D'. We then claim that the theorem also holds for f. Indeed, the left-hand side of (149) does not decrease by more than $\log C$ if f is replaced by $\phi \circ f$, and the right-hand sides in the two cases are comparable by Proposition 12.11 and properties of big line sheaves.

Therefore, we may assume that the divisor D has *strict* normal crossings. Indeed, there is a smooth complex projective variety X' and a birational morphism $\pi : X' \to X$, isomorphic over $X \setminus \operatorname{Supp} D$, such that $D' := (\pi^* D)_{\mathrm{red}}$ is a strict normal crossings divisor. This is true because one can resolve the singularities of each component of D. Since π_* induces a holomorphic map $T_{X'}(-\log D') \to T_X(-\log D)$, the inverse rational map π^{-1} satisfies the conditions of the claim.

Thus, to prove the theorem, it suffices to let Z be the Zariski closure of the image of f, and find nonzero elements $f_1, \ldots, f_n \in K(Z)$ for which the corresponding rational map $\phi : Z \dashrightarrow (\mathbb{P}^1)^n$ satisfies the conditions of the claim (for suitable $U \subseteq Z$ and $C > 0$) relative to the divisor $D' = \sum_{j=1}^n p_j^*([0] + [\infty])$, where $p_j : (\mathbb{P}^1)^n \to \mathbb{P}^1$ is the projection morphism to the jth factor, $1 \leq j \leq n$.

To satisfy the conditions of the claim, it suffices to find a finite set $\mathscr{G}$ of nonzero functions in $K(X)$ such that at each closed point $z \in Z$ there is a subset $\mathscr{G}_z \subseteq \mathscr{G}$ such that for each $g \in \mathscr{G}_z$ the differential dg/g determines a regular section of $\Omega_X^1(\log D)$ in a neighborhood of z in X, and such that as g varies over $\mathscr{G}_z$ the differentials dg/g generate $\Omega_X^1(\log D)$ at z.

To construct $\mathscr{G}$, let z be a closed point of Z. For some open neighborhood V of z in X there are regular functions $g_1, \ldots, g_r$ on V whose vanishing determines the components of D passing through z in V. Letting $\mathfrak{m}$ denote the maximal ideal of z in X, the strict normal crossings condition implies that the g_j are linearly independent in the complex vector space $\mathfrak{m}/\mathfrak{m}^2$ (the Zariski cotangent space). After shrinking V if necessary, we may choose regular functions $g_{r+1}, \ldots, g_d$ on V, such that the functions $g_1, \ldots, g_r, g_{r+1} - 1, \ldots, g_d - 1$ all vanish at z, and such that their images in $\mathfrak{m}/\mathfrak{m}^2$ form a basis. Then $dg_1/g_1, \ldots, dg_d/g_d$ determine regular sections of $\Omega_X^1(\log D)$ in a neighborhood of z in X, and generate the sheaf on that neighborhood. By a compactness argument, one then obtains a finite collection $\mathscr{G}$ satisfying the condition everywhere on Z. $\square$

Remark 29.4. There is no $N_{\mathrm{Ram}(\pi)}(r)$ term in either of these theorems; it appears subsequently. The same is true of the exceptional set (it appears later still).

Remark 29.5. When $B = \mathbb{C}$ and π is the identity map, the error term in (149) can be sharpened to $O(\log^+ T_{\mathscr{A},f}(r)) + o(\log r)$. This will also be true in subsequent results, but will not be explicitly mentioned.

The Geometric LLD leads to an inequality, due originally to McQuillan [55, Thm. 0.2.5]. This inequality presently shows more promise for possible diophantine analogies, since it omits some of the information on the derivative, and since it may be related to parts of the proof of Schmidt's Subspace Theorem.

Before stating the theorem, we note that for the purposes of these notes, if $\mathscr{E}$ is a quasi-coherent sheaf on a scheme X, then

$$\mathbb{P}(\mathscr{E}) = \mathbf{Proj} \bigoplus_{d \geq 0} S^d \mathscr{E}$$

(as in EGA). In particular, if $\mathscr{E}$ is a vector sheaf, then points on $\mathbb{P}(\mathscr{E})$ correspond bijectively to *hyperplanes* (not lines) in the fiber over the corresponding point on X. This scheme comes with a **tautological line sheaf** $\mathscr{O}(1)$, which gives rise to the name of McQuillan's inequality.

If X and D are as in Theorem 29.3, then $\Omega^1_X(\log D)$ is defined as a locally free sheaf on X as an analytic space. This is a coherent sheaf, hence by GAGA [72], it comes from a coherent sheaf on X as a scheme. This latter sheaf is denoted $\Omega_{X/\mathbb{C}}(\log D)$. In fact it is locally free – see the introduction to Sect. 30.

Theorem 29.6. (McQuillan's "Tautological Inequality") *Let X, D, $f : B \to X$, and $\mathscr{A}$ be as in Theorem 29.3. Assume also that f is not constant. Let*

$$f' : B \to \mathbb{P}(\Omega_{X/\mathbb{C}}(\log D))$$

be the canonical lifting of f, associated to the nonzero map from $f^* \Omega^1_X(\log D)$ to the cotangent sheaf of B. Then

$$T_{\mathscr{O}(1),f'}(r) \leq_{\mathrm{exc}} N_f^{(1)}(D,r) + N_{\mathrm{Ram}(\pi)}(r) + O(\log T_{\mathscr{A},f}(r) + \log r). \tag{150}$$

Proof. Let

$$V = \mathbb{V}(\Omega_{X/\mathbb{C}}(\log D)) = \mathbf{Spec} \bigoplus_{d \geq 0} S^d \Omega_{X/\mathbb{C}}(\log D).$$

This is the total space of $T_X(-\log D)$. Also let

$$\overline{V} = \mathbb{P}(\Omega_{X/\mathbb{C}}(\log D) \oplus \mathscr{O}_X).$$

We have a natural embedding $V \hookrightarrow \overline{V}$ that realizes $\overline{V}$ as the projective closure on fibers of V.

Let $[\infty]$ denote the (reduced) divisor $\overline{V} \setminus V$. The integrand of (149) can be viewed as a proximity function for $[\infty]$, and the strategy of the proof is to use this to get a bound on $T_{\mathscr{O}(1),f'}(r)$, via the rational map $\overline{V} \dashrightarrow \mathbb{P}(\Omega_{X/\mathbb{C}}(\log D))$. To compare the geometries of these two objects, we use the closure of the graph of this rational map in $\overline{V} \times_X \mathbb{P}(\Omega_{X/\mathbb{C}}(\log D))$, which is the blowing-up of $\overline{V}$ along the image $[0]$ of the zero section. Let $p\colon P \to \overline{V}$ be this blowing-up, let E be its exceptional divisor. Let $q\colon P \to \mathbb{P}(\Omega_{X/\mathbb{C}}(\log D))$ be the projection to the second factor. We have a diagram

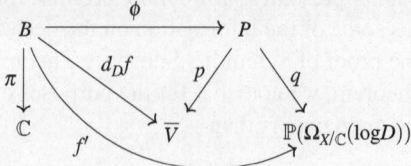

There is a unique lifting $\phi\colon B \to P$ that satisfies $d_D f = p \circ \phi$ and $f' = q \circ \phi$. We also have

$$p^* \mathscr{O}(1) \cong q^* \mathscr{O}(1) \otimes \mathscr{O}(E)$$

(where the first $\mathscr{O}(1)$ is on $\overline{V}$ and the second one is on $\mathbb{P}(\Omega_{X/\mathbb{C}}(\log D))$). This is because any given nonzero rational section s of $\Omega_{X/\mathbb{C}}(\log D)$ on X gives a rational section $(s,1)$ of $\mathscr{O}(1)$ on $\overline{V}$, and also a rational section of $\mathscr{O}(1)$ on $\mathbb{P}(\Omega_{X/\mathbb{C}}(\log D))$. Their pull-backs to P coincide except that the first one also vanishes to first order along E.

We also have $\mathscr{O}([\infty]) \cong \mathscr{O}(1)$ on $\overline{V}$, because the divisor $[\infty]$ is cut out by the section $(0,1)$ of $\Omega_{X/\mathbb{C}}(\log D) \oplus \mathscr{O}_X$.

Thus, we have

$$\begin{aligned}
T_{\mathscr{O}(1),f'}(r) &= T_{q^* \mathscr{O}(1),\phi}(r) + O(1) \\
&= T_{\mathscr{O}(1),d_D f}(r) - T_{\mathscr{O}(E),\phi}(r) + O(1) \\
&= N_{d_D f}([\infty],r) - T_{\mathscr{O}(E),\phi}(r) + m_{d_D f}([\infty],r) + O(1) \\
&\leq_{\mathrm{exc}} N_f^{(1)}(D,r) + N_{\mathrm{Ram}(\pi)}(r) + O(\log T_{\mathscr{A},f}(r) + \log r).
\end{aligned}$$

To explain the last step above, $m_{d_D f}([\infty],r)$ is bounded by Theorem 29.3. Since E is effective and does not contain the image of ϕ (since f is not constant), $T_{\mathscr{O}(E),\phi}(r)$ is bounded from below. (It can also be used to subtract a term $N_{\mathrm{Ram}(f)}(r)$ from the right-hand side of (150).)

Now consider $N_{d_D f}([\infty],r)$. Fix a point $b \in B$, let w be a local coordinate on B at b, let z be the coordinate on $\mathbb{C}$, and let $z_1,\dots,z_n$ be local coordinates on X at $f(b)$ such that D is locally given by $z_1 \cdots z_r = 0$ nearby. Then, near $f(b)$, $\overline{V}$ has homogeneous coordinate functions $dz_1/z_1,\dots,dz_r/z_r, dz_{r+1},\dots,dz_n,1$. Relative to these coordinates, the value of $d_D f$ in a punctured neighborhood of b is given by

$$\begin{aligned}
&\left[\frac{d(z_1 \circ f)/dz}{z_1 \circ f} : \dots : \frac{d(z_r \circ f)/dz}{z_r \circ f} : \frac{d(z_{r+1} \circ f)}{dz} : \dots : \frac{d(z_n \circ f)}{dz} : 1 \right] \\
&= \left[\frac{d(z_1 \circ f)/dw}{z_1 \circ f} : \dots : \frac{d(z_r \circ f)/dw}{z_r \circ f} : \frac{d(z_{r+1} \circ f)}{dw} : \dots : \frac{d(z_n \circ f)}{dw} : \frac{dz}{dw} \right].
\end{aligned}$$

Then $d_D f$ will meet $[\infty]$ to the extent that there are poles among the first n coordinates or a zero in the last coordinate. Poles among the first n coordinates can only occur in the first r coordinates (using the second representation above), and in that case they will at most be simple poles and will only occur if $f(b) \in \mathrm{Supp}\, D$. Thus the contribution to $N_{d_D f}([\infty], r)$ from poles in these coordinates is bounded by $N_f^{(1)}(D, r)$. The contribution coming from zeroes in the last coordinate is bounded by $N_{\mathrm{Ram}(\pi)}(r)$. □

As a sample application of this theorem, it implies the Second Main Theorem with truncated counting functions for maps to Riemann surfaces, including the case in which the domain is a finite ramified cover of $\mathbb{C}$. This is the (proved) case $\dim X = 1$ of Conjecture 27.5.

Corollary 29.7. *Let X be a smooth complex projective curve, let D be an effective reduced divisor on X, and let $f: B \to X$ be a non-constant holomorphic curve. Then*

$$N_f^{(1)}(D, r) + N_{\mathrm{Ram}(\pi)}(r) \geq_{\mathrm{exc}} T_{\mathscr{K}(D), f}(r) - O(\log T_{\mathscr{A}, f}(r) + \log r). \qquad (151)$$

Proof. Since X is a curve, the vector sheaf $\Omega_{X/\mathbb{C}}(\log D)$ is isomorphic to the line sheaf $\mathscr{K}(D)$. Therefore the canonical projection $p: \mathbb{P}(\Omega_{X/\mathbb{C}}(\log D)) \to X$ is an isomorphism, $\mathscr{O}(1) \cong p^* \mathscr{K}(D)$, and $f' = p^{-1} \circ f$. Thus

$$T_{\mathscr{O}(1), f'}(r) = T_{\mathscr{K}(D), f}(r) + O(1),$$

so (151) is equivalent to (150). □

Remark 29.8. In fact, when $\dim X = 1$, McQuillan's inequality is directly equivalent to Conjecture 27.5, as can be seen from the above proof. This is not true in higher dimension, though (McQuillan's inequality is proved, but Conjecture 27.5 is not).

Cartan's theorem (Theorem 8.6) can also be proved using McQuillan's inequality, but this requires more work than can be included here. See [97]. The modified version (Theorem 8.11) requires a modified form of McQuillan's inequality (involving the same type of change).

It is hoped that other key results in Nevanlinna theory can also be proved using Theorem 29.6.

We end the section with another corollary, which often has applications in Nevanlinna theory. It generalizes the Schwarz lemma, which has played an important role in Nevanlinna theory for a long time; see [80, Thm. 3], where it is proved for jet differentials. The introduction of *op. cit.* also describes some of the history of this result. See also [53, Sect. 4], [93, Cor. 5.2], and [94].

Corollary 29.9. *Let X be a smooth complex projective variety, let D be a normal crossings divisor on X, let $f: B \to X$ be a holomorphic map, let $\mathscr{A}$ be an ample line sheaf on X, let $\mathscr{L}$ be a line sheaf on X, let d be a positive integer, and let ω be a* *global section of* $S^d \Omega_{X/\mathbb{C}}(\log D)$. *If* $f^* \omega \neq 0$ *(i.e., it does not vanish everywhere on* B*), then*

$$\frac{1}{d} T_{\mathscr{L},f}(r) \leq_{\mathrm{exc}} N_f^{(1)}(D,r) + N_{\mathrm{Ram}(\pi)}(r) + O(\log T_{\mathscr{A},f}(r) + \log r).$$

Proof. Let $f' \colon B \to \mathbb{P}(\Omega_{X/\mathbb{C}}(\log D))$ be as in Theorem 29.6, and let

$$p \colon \mathbb{P}(\Omega_{X/\mathbb{C}}(\log D)) \to X$$

be the canonical projection. Then ω corresponds to a global section

$$\omega' \in \Gamma\left(\mathbb{P}(\Omega_{X/\mathbb{C}}(\log D)), \mathcal{O}(d) \otimes p^* \mathscr{L}^{\vee}\right),$$

and $(f')^* \omega' = f^* \omega$. Thus the image of f' is not contained in the base locus of $\mathcal{O}(d) \otimes \mathscr{L}^{\vee}$, so

$$T_{\mathcal{O}(d),f'}(r) \geq T_{\mathscr{L},f}(r) + O(1)$$

by Theorem 12.8c. The result then follows immediately from (150).   □

One can think of this result as a generalization of the fact that if $f \colon C \to X$ is a nonconstant map from a nonsingular projective curve of genus g to a smooth complete variety X, then $\deg f^* \mathscr{K}_X \leq 2g - 2$, where $\mathscr{K}_X$ is the canonical line sheaf of X. Thus, it is useful in carrying over results from the split function field case to Nevanlinna theory (see Sect. 18). It is used in this manner in the proof of [93, Thm. 5.3].

30   Derivatives in Number Theory

Whether one uses the Lemma on the Logarithmic Derivative or curvature, Nevanlinna theory depends in an essential way on the ability to take the derivative of a holomorphic function. In the number field case, on the other hand, there is currently no known counterpart to the derivative. Even in the function field case, the derivative lives in the absolute tangent bundle, but any counterpart to the derivative as in Nevanlinna theory should live in the *relative* tangent bundle. McQuillan gets around this in his proof of the $1 + \varepsilon$ conjecture, by noting that for points of large height the derivative has an approximate projection to the relative tangent bundle that is precise enough to be useful (see the end of Sect. 28). Although this method shows a great deal of promise, it will not be explored further here.

Instead, this section will describe a conjecture in number theory based on McQuillan's tautological inequality. Because of its origin, the name "tautological conjecture" is too good to pass up.

If X is a smooth complete variety over a field k, and if D is a normal crossings divisor on X, then an algebraic definition of $\Omega_{X/k}(\log D)$ is given in [40, 1.7]. Kato

[40, 1.8] also shows this to be locally free in the étale topology of rank $\dim X$. This then descends to a quasi-coherent sheaf on X in the Zariski topology by [33, VIII Thm. 1.1]. It is a vector sheaf (with non-obvious generators) by [34, IV 2.5.2].

Conjecture 30.1. (Tautological Conjecture) Let k be a number field, let $S \supseteq S_\infty$ be a finite set of places of k, let X be a nonsingular complete variety over k with $\dim X > 0$, let D be a normal crossings divisor on X, let r be a positive integer, let $\mathscr{A}$ be an ample line sheaf on X, and let $\varepsilon > 0$. Then, for all $x \in X(\bar{k}) \setminus \operatorname{Supp} D$ with $[\kappa(x) : k] \leq r$, there is a closed point $x' \in \mathbb{P}(\Omega_{X/k}(\log D))$ lying over x such that

$$h_{\mathscr{O}(1),k}(x') \leq N_S^{(1)}(D,x) + d_k(x) + \varepsilon h_{\mathscr{A},k}(x) + O(1). \tag{152}$$

Moreover, given a finite collection of rational maps $g_i : X \dashrightarrow W_i$ to varieties W_i, there are finite sets Σ_i of closed points on W_i for each i with the following property. For each x as above, x' may be chosen so that, for each i, if x lies in the domain of g_i and if $g_i(x) \notin \Sigma_i$, then x' lies in the domain of the induced rational map $\mathbb{P}(\Omega_{X/k}(\log D)) \dashrightarrow \mathbb{P}(\Omega_{W_i/k})$. Moreover, the constant implicit in the $O(1)$ term depends only on k, S, X, D, r, $\mathscr{A}$, ε, the rational maps g_i, and the choices of height and counting functions.

This extra condition (involving the rational maps g_i) should perhaps be explained a bit. This condition seems to be necessary in order to ensure that the points x' behave more like derivatives. For example, consider the special case in which $r = 1$, $D = 0$, and X is a product $X_1 \times X_2$. Then the points x must be rational points, and (152) for X is the sum of the same inequality for X_1 and X_2. But then, without the last condition in the conjecture, the conjecture would hold if it held for *either* factor, since one could take x' tangent to the copy of X_1 or X_2 sitting inside of X. This seems a bit unnatural. In addition, the last condition is useful for applications.

McQuillan's work is not the only support for this conjecture. For some time, it has been known that parts of Schmidt's proof of his Subspace Theorem correspond to a proof of Cartan's theorem due to H. and J. Weyl [100], further developed by Ahlfors [1]. Both of these proofs can be divided up into an "old" part (corresponding to an extension to higher dimensions of the proofs of Roth and Nevanlinna of the earlier case on $\mathbb{P}^1$), and a "new" part. In Ahlfors' case, the "new" part consists of working with the associated curves (Frenet formalism); in Schmidt's case, it consists of working with Minkowski's theory of successive minima. In either case, the proof involves geometric constructions on $\bigwedge^p \mathbb{C}^{n+1}$ or $\bigwedge^p k^{n+1}$, respectively. This strongly suggested that the theory of successive minima may be related to the use of derivatives in number theory [87, Chap. 6]. This has been recently refined [97] to more explicitly involve a variant of Conjecture 30.1, and also to use the geometry of flag varieties.

The tie-in between successive minima and the tautological conjecture proceeds as follows. Let X be a nonsingular complete variety over a number field k, let D be a normal crossings divisor on X, let $Y = \operatorname{Spec} \mathscr{O}_k$, and let $\mathscr{X} \to Y$ be a proper model for X. Then we have a relative tangent sheaf $T_{\mathscr{X}/Y}(-\log D)$ on $\mathscr{X}$. This is not necessarily a vector sheaf, since $\mathscr{X}$ need not be smooth over Y and D need not extend as a normal crossings divisor. However, we shall ignore that distinction for the sake of discussion. Via the mechanisms of Arakelov theory, one can assign a Hermitian metric to the sheaf at all archimedean places. If $i\colon \operatorname{Spec} Y \to \mathscr{X}$ is the section of $\mathscr{X} \to Y$ corresponding to a given rational point, then $i^* T_{\mathscr{X}/Y}(-\log D)$ is a vector sheaf on $Y = \operatorname{Spec}\mathscr{O}_k$, which can then be viewed as a lattice in $\mathbb{R}^n$ (if $k = \mathbb{Q}$), or as a lattice in $(k \otimes_{\mathbb{Q}} \mathbb{R})^n$ (generally). Bounds on the metrics at archimedean places correspond to giving a convex symmetric body in $(k \otimes_{\mathbb{Q}} \mathbb{R})^n$, and therefore Minkowski's theory of successive minima can be translated into Arakelov theory as a search for linearly independent nonzero sections of $i^* T_{\mathscr{X}/Y}(-\log D)$, obeying certain upper bounds on its metric at each infinite place. See, for example, [28]. Or, in the function field case, it is known that Minkowski's theory corresponds to a search for nonzero global sections with bounded poles at certain places (an application of Riemann-Roch).

Giving a nonzero section as the first successive minimum is basically equivalent to giving a line subsheaf of largest degree (unless the first two successive minima are close). This corresponds to giving a quotient subbundle $\mathscr{L}$ of $i^* \Omega_{\mathscr{X}/Y}(\log D)$ of smallest degree. This, in turn, corresponds to giving a point $x' \in \mathbb{P}(\Omega_{X/Y}(\log D))$ lying over the rational point in question [36, II 7.12]. The sheaf $\mathscr{L}$ is none other than the pull-back of the tautological bundle $\mathscr{O}(1)$ to Y via the section $i'\colon Y \to \mathbb{P}(\Omega_{\mathscr{X}/Y}(\log d))$ corrsponding to the point x'. Therefore, the degree of $\mathscr{L}$ is the height $h_{\mathscr{O}(1),k}(x')$ (see (70)). This again leads to Conjecture 30.1.

We emphasize that specific bounds on the $O(1)$ term in (152) are not given, so Conjecture 30.1 is meaningful only for an infinite set of points – or, better yet, for a generic or semi-generic set of points (Definitions 15.5 and 15.11).

Remark 30.2. The assertions of Remark 29.8 also apply in the arithmetic situation. When $\dim X = 1$, the extra conditions involving the rational functions g_i in the Tautological Conjecture are automatically satisfied. Therefore, in this case the Tautological Conjecture is equivalent to Conjecture 25.3b (for the same reasons as in Remark 29.8). Thus, by Remark 25.4, the Tautological Conjecture is proved for curves over function fields of characteristic 0. As in Remark 29.8, though, Conjectures 30.1 and 25.3b are not as closely related when $\dim X > 1$.

We also note that points of low height as in Conjecture 30.1 behave like derivatives in the following sense.

Proposition 30.3. (Arithmetic Chain Rule) *Let $f\colon X_1 \to X_2$ be a morphism of complete varieties over a number field k. Then, for all $x \in X_1(\bar{k})$ where f is étale, and for all closed points $x' \in \mathbb{P}(\Omega_{X_1/k})$ lying over x, the rational map $f_*\colon \mathbb{P}(\Omega_{X_1/k}) \dashrightarrow \mathbb{P}(\Omega_{X_2/k})$ takes x' to a point x_2' (lying over $f(x)$) for which*

$$h_{\mathscr{O}(1),k}(x_2') \le h_{\mathscr{O}(1),k}(x') + O(1).$$

Moreover, assume that X_1 and X_2 are projective, with ample line sheaves $\mathscr{A}_1$ and $\mathscr{A}_2$, respectively, and that $\varepsilon_1 > 0$ is a positive number for which (152) holds for all x in some set $\Sigma \subseteq X_1(\bar{k})$ (with respect to $\mathscr{A}_1$ and ε_1). Then there is an $\varepsilon_2 > 0$ such that (152) holds for $f(x) \in X_2(\bar{k})$ for all $x \in \Sigma$, with respect to $\mathscr{A}_2$ and ε_2.

Proof. Let $\mathscr{X}_1$ and $\mathscr{X}_2$ be models for X_1 and X_2 over $Y := \operatorname{Spec} \mathscr{O}_k$, respectively, chosen such that f extends as a morphism $f \colon \mathscr{X}_1 \to \mathscr{X}_2$, let $Y' = \operatorname{Spec} \mathscr{O}_{\kappa(x')}$, and let $i \colon Y' \to \mathscr{X}_1$ be the multisection corresponding to x (which factors through $\operatorname{Spec} \mathscr{O}_{\kappa(x)}$). Then x' corresponds to a surjection $i^* \Omega_{\mathscr{X}_1/Y} \twoheadrightarrow \mathscr{L}$ for a line sheaf $\mathscr{L}$ on Y', and $h_{\mathscr{O}(1),k}(x')$ is the Arakelov degree of $\mathscr{L}$ divided by $[\kappa(x') : k]$. We also have a morphism $f^* \Omega_{\mathscr{X}_2/Y} \to \Omega_{\mathscr{X}_1/Y}$, isomorphic at x. This gives a nonzero map $(f \circ i)^* \Omega_{\mathscr{X}_2/Y} \to \mathscr{L}$, so $h_{\mathscr{O}(1),k}(x_2') \leq h_{\mathscr{O}(1),k}(x')$ (with heights defined using these models).

The second assertion is immediate from the first assertion, by Proposition 10.13.

$\square$

A similar result holds for closed immersions (but without the assumption on étaleness).

The name "Arithmetic Chain Rule" comes from the fact that this result shows that the "derivatives" x' and x_2' are related in the expected way.

31 Another Conjecture Implies abc

Conjecture 23.4, involving truncated counting functions, is of course a vast generalization of the abc conjecture, and Conjecture 25.1, which involved algebraic points, also rather easily implies abc. Actually, though, the (seemingly) weaker Conjecture 15.6 has also been shown to imply the abc conjecture [93]. This implication, however, (necessarily) needs to use varieties of dimension > 1, whereas knowing either of the former two conjectures even for curves would suffice.

This section sketches the proof of the implication mentioned above.

Theorem 31.1. *For any $\varepsilon > 0$ there is a nonsingular projective variety X_ε over $\mathbb{Q}$, a normal crossings divisor D_ε on X_ε, and a real number $\varepsilon' > 0$, such that if Conjecture 15.6b holds for X_ε, D_ε, and ε', then the abc conjecture (Conjecture 23.5) holds for ε.*

Proof (sketch). Fix an integer $n > 3/\varepsilon + 3$. Let X_n be the closed subvariety in $\left(\mathbb{P}^2 \right)^n$ in coordinates

$$([x_1 : y_1 : z_1], \ldots, [x_n : y_n : z_n])$$

given by the equation

$$\prod_{i=1}^n x_i^i + \prod_{i=1}^n y_i^i + \prod_{i=1}^n z_i^i = 0.$$

There is a rational map $X_n \dashrightarrow \mathbb{P}^2$ given by

$$([x_1 : y_1 : z_1], \ldots, [x_n : y_n : z_n]) \mapsto \left[\prod x_i^i : \prod y_i^i : \prod z_i^i \right]. \tag{153}$$

Let Γ_n be the closure of the graph of this rational map in $X_n \times \mathbb{P}^2$, and let $\phi \colon \Gamma_n \to \mathbb{P}^2$ be the projection to the second factor. The image of ϕ is a line, which we identify with $\mathbb{P}^1$.

Given relatively prime integers a, b, c with $a + b + c = 0$, define a point in $X_n(\mathbb{Q})$ as follows. Let

$$x_n = \prod_p p^{[(\mathrm{ord}_p a)/n]} \quad \text{and} \quad x_i = \prod_{\mathrm{ord}_p a \equiv i \pmod{n}} p \quad (i < n).$$

(The brackets in the definition of x_n denote the greatest integer function.) With these definitions, we have $a = \prod x_i^i$, with x_n as large as possible subject to all x_i being integers. Similarly define $y_1, \ldots, y_n$ using b and $z_1, \ldots, z_n$ using c. This point lifts to a unique point in $\Gamma_n(\mathbb{Q})$, which we denote $P_{a,b,c}$.

Let D be the effective Cartier divisor on Γ_n obtained by pulling back the divisor $x_1 \cdots x_n y_1 \cdots y_n z_1 \cdots z_n = 0$ from X_n, and let E be the divisor on the image $\mathbb{P}^1$ of ϕ, obtained by restricting the coordinate hyperplanes on $\mathbb{P}^2$. The latter divisor is the sum of the points $[1 : -1 : 0]$, $[0 : 1 : -1]$, and $[-1 : 0 : 1]$. Let $S = \{\infty\} \subseteq M_{\mathbb{Q}}$. It is possible to show that if p is a rational prime and v is the corresponding place of $\mathbb{Q}$, then

$$\lambda_{E,v}([a : b : c]) = \mathrm{ord}_p(abc) \log p$$

and

$$\lambda_{D,v}(P_{a,b,c}) = \mathrm{ord}_p(x_1 \cdots x_n y_1 \cdots y_n z_1 \cdots z_n) \log p,$$

using Weil functions suitably defined using (40). It then follows that

$$N_S(D, P_{a,b,c}) \leq \sum_{p \mid abc} \log p + \frac{1}{n} N_S(E, \phi(P_{a,b,c})).$$

One would like to apply Conjecture 15.6 to the divisor D on Γ_n, but this is not possible since Γ_n is singular. However, there is a nonsingular projective variety Γ_n', a normal crossings divisor D' on Γ_n', and a proper birational morphism $\psi : \Gamma_n' \to \Gamma_n$, such that $\mathrm{Supp}\, D' = \psi^{-1}(\mathrm{Supp}\, D)$, ψ is an isomorphism over a suitably large set, and $\mathscr{K}_{\Gamma_n'}(D') \geq \psi^* \phi^* \mathscr{O}(1)$ relative to the cone of effective divisors. For details see [93, Lemma 3.9].

Let $P_{a,b,c}'$ be the point on Γ_n' lying over $P_{a,b,c}$. Then one can show that

$$\begin{aligned}
h_{\mathbb{Q}}([a : b : c]) &= h_{\psi^* \phi^* \mathscr{O}(1), \mathbb{Q}}(P_{a,b,c}') + O(1) \\
&\leq h_{\mathscr{K}_{\Gamma_n'}(D'), \mathbb{Q}}(P_{a,b,c}') + O(1) \\
&\leq N_S(D', P_{a,b,c}') + \varepsilon' h_{\mathscr{A}, \mathbb{Q}}(P_{a,b,c}') + O(1) \\
&\leq N_S(D, P_{a,b,c}) + \varepsilon' h_{\mathscr{A}, \mathbb{Q}}(P_{a,b,c}') + O(1) \\
&\leq \sum_{p \mid abc} \log p + \frac{3}{n} h_{\mathbb{Q}}([a : b : c]) + \varepsilon'' h_{\mathbb{Q}}([a : b : c]) + O(1) \\
&\leq \sum_{p \mid abc} \log p + \frac{\varepsilon}{1 + \varepsilon} h_{\mathbb{Q}}([a : b : c]) + O(1),
\end{aligned}$$

and therefore $h_{\mathbb{Q}}([a : b : c]) \leq (1 + \varepsilon) \sum_{p|abc} \log p + O(1)$. Here we have used the fact that $N_S(D', P'_{a,b,c}) \leq N_S(D, P_{a,b,c})$, which follows from the fact that $\psi^* D - D'$ is effective.

This chain of inequalities holds outside of some proper Zariski-closed subset of Γ'_n, but it is possible to show that this set can be chosen so that it only involves finitely many triples (a, b, c).   □

The variety X_n admits a faithful action of $\mathbb{G}_m^{2n-2}$, by letting the first $n - 1$ coordinates act by $x_i \mapsto tx_i$ and $x_1 \mapsto t^{-i}x_1$ for $i = 2, \ldots, n$, and letting the other $n - 1$ coordinates act similarly on the y_i. This action respects fibers of the rational map (153), so the action extends to Γ_n, and the construction of Γ'_n can be done so that the group action extends there, too. Since $\dim X_n = 2n - 1$, the group acts transitively on open dense subsets of suitably generic fibers of ϕ. It is this group action that allows one to control the proper Zariski-closed subset of Γ'_n arising out of Conjecture 15.6b. The group action also provides some additional structure, and in fact Conjecture 15.2 can be proved in this context [93, Thm. 5.3].

32 An abc Implication in the Other Direction

The preceding sections give a number of ways in which some conjectures imply the abc conjecture. It is also true, however, that the abc conjecture implies parts of the preceding conjectures. While this is mostly a curiousity, since the implied special cases are known to be true whereas the abc conjecture is still a conjecture, this provides some insight into the geometry of the situation.

The implications of this section were first observed by Elkies [18], who showed that "Mordell is as easy as abc," i.e., the abc conjecture implies the Mordell conjecture. This was extended by Bombieri [6], who showed that abc implies Roth's theorem, and then by van Frankenhuijsen [85], who showed that the abc conjecture implies Conjecture 15.6b for curves. In each of these cases, the abc conjecture for a number field k would be needed to imply any given instance of Conjecture 15.6b. Here "the abc conjecture for k" means Conjecture 23.4b over k with $X = \mathbb{P}^1_k$ and $D = [0] + [1] + [\infty]$. It is further true that a "strong abc conjecture," namely Conjecture 25.3b with $X = \mathbb{P}^1_k$ and $D = [0] + [1] + [\infty]$, would imply Conjecture 25.1b for curves; in other words van Frankenhuijsen's implication holds also for algebraic points of bounded degree.

This circle of ideas stems from two observations. The first of these is due to Belyĭ [4]. He showed that a smooth complex projective curve X comes from a curve over $\overline{\mathbb{Q}}$ (i.e., there is a curve X_0 over $\overline{\mathbb{Q}}$ such that $X \cong X_0 \times_{\overline{\mathbb{Q}}} \mathbb{C}$) if and only if there is a finite morphism from X to $\mathbb{P}^1_{\mathbb{C}}$ ramified only over $\{0, 1, \infty\}$. For our purposes, this can be adapted as follows.

Theorem 32.1. (Belyĭ) *Let X be a smooth projective curve over a number field k, and let S be a finite set of closed points on X. Then there is a finite morphism $f : X \to \mathbb{P}^1_k$ which is ramified only over $\{0, 1, \infty\}$, and such that $S \subseteq f^{-1}(\{0, 1, \infty\})$.*

Proof. See Belyĭ [4] or Serre [74]. □

The other ingredient is a complement (and actually, a converse) to Proposition 25.2, using truncated counting functions.

Proposition 32.2. *Let k be a number field or function field, let $S \supseteq S_\infty$ be a finite set of places of k, let $\pi\colon X' \to X$ be a surjective generically finite morphism of complete nonsingular varieties over k, and let D be a normal crossings divisor on X. Let $D' = (\pi^* D)_{\mathrm{red}}$, and assume that it too has normal crossings. Assume also that the ramification divisor R of π satisfies*

$$R = \pi^* D - D' \tag{154}$$

(and therefore that π is unramified outside of $\mathrm{Supp}\, D'$).[4] Let $\mathscr{K}$ and $\mathscr{K}'$ denote the canonical line sheaves on X and X', respectively. Then, for all $x \in X'(\bar{k})$ not lying on $\mathrm{Supp}\, D'$,

$$
\begin{aligned}
N_S^{(1)}(D',x) &+ d_S(x) - h_{\mathscr{K}'(D'),k}(x) \\
&\geq N_S^{(1)}(D,\pi(x)) + d_S(\pi(x)) - h_{\mathscr{K}(D),k}(\pi(x)) + O(1),
\end{aligned}
\tag{155}
$$

with equality if $[\kappa(x) : k]$ is bounded. In particular, either part of Conjecture 25.3 for D' on X' is equivalent to that same part for D on X.

Proof. First, by (154), we have $\mathscr{K}'(D') \cong \pi^*(\mathscr{K}(D))$. Thus

$$h_{\mathscr{K}'(D'),k}(x) = h_{\mathscr{K}(D),k}(\pi(x)) + O(1),$$

so it suffices to show that

$$N_S^{(1)}(D',x) + d_S(x) \geq N_S^{(1)}(D,\pi(x)) + d_S(\pi(x)) + O(1), \tag{156}$$

with equality if $[\kappa(x) : k]$ is bounded.

Now let $Y_S = \mathrm{Spec}\, \mathscr{O}_{k,S}$ (or, if $S = \emptyset$, which can only happen in the function field case, let Y_S be the smooth projective curve over the field of constants of k for which $K(Y_S) = k$). Let $\mathscr{X}$ be a proper model for X over Y_S for which D extends to an effective Cartier divisor, which will still be denoted D. (This can be obtained by taking a proper model, extending D to it as a Weil divisor, and blowing up the sheaf of ideals of the corresponding reduced closed subscheme.) Let $\mathscr{X}'$ be a proper model for X' over Y_S. We may assume that π extends to a morphism $\pi\colon \mathscr{X}' \to \mathscr{X}$, and that D' extends to an effective Cartier divisor on $\mathscr{X}'$. We assume further that $\mathrm{Supp}\, D' = \pi^{-1}(\mathrm{Supp}\, D)$ (on $\mathscr{X}'$). Indeed, one can obtain $\mathscr{X}'$ and D' by blowing up the reduced sheaf of ideals corresponding to $\pi^{-1}(\mathrm{Supp}\, D)$.

---

[4] This condition is equivalent to (X',D') being *log étale* over (X,D) by [40, (3.12)], using the fact that (X',D') is log smooth over $\mathrm{Spec}\, k$ by (3.7)(1) of *op. cit.*

For $x \in X'(\bar{k})$, let $L = \kappa(\pi(x))$, let $L' = \kappa(x)$, and let $Y_S^{\flat}$ and Y_S' be the normalizations of Y_S in L and L', respectively. Then we have a commutative diagram

$$
\begin{array}{ccc}
Y_S' & \xrightarrow{\ i'\ } & \mathscr{X}' \\
\downarrow{\scriptstyle p} & & \downarrow{\scriptstyle \pi} \\
Y_S^{\flat} & \xrightarrow{\ i\ } & \mathscr{X}
\end{array}
\tag{157}
$$

in which the maps i' and i correspond to the algebraic points x and $\pi(x)$, respectively. We use the divisors D and D' to define $N_S^{(1)}(D, \pi(x))$ and $N_S^{(1)}(D', x)$, respectively. Then a place w of L contributes to $N_S^{(1)}(D, \pi(x))$ if and only if the corresponding closed point of $Y_S^{\flat}$ lies in $i^{-1}(\operatorname{Supp} D)$, and similarly for places w' of L'. By commutativity of (157) and the condition $\operatorname{Supp} D' = \pi^{-1}(\operatorname{Supp} D)$, it follows that

$$
N_S^{(1)}(D, \pi(x)) - N_S^{(1)}(D', x) = \frac{1}{[L' : k]} \sum_{w'} (e_{w'/w} - 1) \log \# \mathbb{F}_{w'},
$$

where the sum is over places w' of L' corresponding to points in $(i')^*(\operatorname{Supp} D')$, w is the place of L lying under w', $e_{w'/w}$ is the ramification index of w' over w, and $\mathbb{F}_{w'}$ is the residue field of w'. (In the function field case, replace $\log \# \mathbb{F}_{w'}$ with $[\mathbb{F}_{w'} : F]$.)

We also have

$$
d_S(x) - d_S(\pi(x)) = \frac{1}{[L' : k]} \sum_{Q'} \operatorname{ord}_{Q'} \mathscr{D}_{L'/L} \cdot \log \# \mathbb{F}_{w'},
$$

where the sum is over nonzero prime ideals Q' of $\mathscr{O}_{L',S}$, and w' is the corresponding place of L'. This sum can be restricted to primes corresponding to points in $(i')^*(\operatorname{Supp} D')$, since the other primes are unramified over L by Lemma 24.10 applied to $\pi : \mathscr{X}' \to \mathscr{X}$ and the relevant local rings. Therefore the inequality (156) follows from the elementary fact that

$$
\operatorname{ord}_{Q'} \mathscr{D}_{L'/L} \ge e_{Q'/Q} - 1,
\tag{158}
$$

where $Q = Q' \cap \mathscr{O}_{L,S}$. (This inequality may be strict if Q' is wildly ramified over Q.)

Now if $[L' : k]$ is bounded, then the differences in (158) add up to at most a bounded amount, so (156) holds up to $O(1)$ in that case.

The last assertion of the proposition follows trivially from (155). $\square$

The implications mentioned in the beginning of this section then follow immediately from Theorem 32.1 and Proposition 32.2, upon noting that (154) always holds for finite morphisms of nonsingular curves.

Acknowledgements Partially supported by NSF grant DMS-0500512

References

1. Ahlfors, L.V.: The theory of meromorphic curves. Acta Soc. Sci. Fennicae. Nova Ser. A. **3** (4), 1–31 (1941)
2. Artin, E: Algebraic numbers and algebraic functions. Gordon and Breach Science, New York (1967)
3. Ashline, G.L.: The defect relation of meromorphic maps on parabolic manifolds. Mem. Amer. Math. Soc. **139**(665), 78 (1999). ISSN 0065-9266
4. Belyĭ, G. V.: Galois extensions of a maximal cyclotomic field. Izv. Akad. Nauk SSSR Ser. Mat. **43**(2), 267–276, 479 (1979). ISSN 0373-2436
5. Bloch, A.: Sur les systèmes de fonctions uniformes satisfaisant à l'équation d'une variété algébrique dont l'irrégularité dépasse la dimension. J. de Math. **5**, 19–66 (1926)
6. Bombieri, E.: Roth's theorem and the abc-conjecture. Preprint ETH Zürich (1994)
7. Borel, E.: Sur les zéros des fonctions entières. Acta Math. **20**, 357–396 (1897)
8. Brownawell, W.D., Masser, D.W.: Vanishing sums in function fields. Math. Proc. Cambridge Philos. Soc. **100**(3), 427–434 (1986). ISSN 0305-0041
9. Campana, F.: Orbifolds, special varieties and classification theory. Ann. Inst. Fourier (Grenoble) **54**(3), 499–630 (2004). ISSN 0373-0956
10. Carlson, J., Griffiths, P.: A defect relation for equidimensional holomorphic mappings between algebraic varieties. Ann. Math. **95**(2), 557–584 (1972) ISSN 0003-486X
11. Cartan, H.: Sur les zéros des combinaisons linéaires de p fonctions holomorphes données. Mathematica (Cluj) **7**, 5–29 (1933)
12. Cherry, W., Ye, Z.: Nevanlinna's theory of value distribution. Springer Monographs in Mathematics. Springer, Berlin (2001). ISBN 3-540-66416-5
13. Corvaja, P., Zannier, U.: A subspace theorem approach to integral points on curves. C. R. Math. Acad. Sci. Paris **334**(4), 267–271 (2002). ISSN 1631-073X
14. Corvaja, P., Zannier, U.: On integral points on surfaces. Ann. Math. **160**(2), 705–726 (2004). ISSN 0003-486X
15. Deligne, P.: Équations différentielles à points singuliers réguliers. Lecture Notes in Mathematics, vol. 163. Springer, Berlin (1970)
16. Dufresnoy, J.: Théorie nouvelle des familles complexes normales. Applications à l'étude des fonctions algébroïdes. Ann. Sci. École Norm. Sup. **61**(3), 1–44 (1944) ISSN 0012-9593
17. Eisenbud, D.: Commutative algebra (with a view toward algebraic geometry), vol. 150 In: Graduate texts in mathematics. Springer, New York (1995). ISBN 0-387-94268-8; 0-387-94269-6
18. Elkies, N.D.: *ABC* implies Mordell. Int. Math. Res. Notices, **1991**(7), 99–109 (1991). ISSN 1073-7928
19. Evertse, J.-H.: On sums of S-units and linear recurrences. Compositio Math. **53**(2), 225–244 (1984). ISSN 0010-437X
20. Evertse, J.-H., Ferretti, R.G.: A generalization of the subspace theorem with polynomials of higher degree. In: Diophantine approximation. Dev. Math., vol. 16, pp. 175–198. Springer, NewYork (2008). ArXiv:math.NT/0408381
21. Faltings, G.: Endlichkeitssätze für abelsche Varietäten über Zahlkörpern. Invent. Math. **73**(3), 349–366 (1983). ISSN 0020-9910
22. Faltings, G.: Finiteness theorems for abelian varieties over number fields. In: Arithmetic geometry (Storrs, Conn., 1984), pp. 9–27. Springer, New York (1986). Translated from the German original [Invent. Math. **73**(3), 349–366 (1983); Invent. Math. **75**(2), 381 (1984); MR 85g:11026ab] by Edward Shipz
23. Faltings, G.: Diophantine approximation on abelian varieties. Ann. Math. (2) **133**(3), 549–576 (1991). ISSN 0003-486X
24. Faltings, G.: The general case of S. Lang's conjecture. In: Barsotti Symposium in algebraic geometry (Abano Terme, 1991), Perspect. Math., vol. 15, pp. 175–182. Academic, CA (1994)

25. Ferretti, R.G.: Mumford's degree of contact and Diophantine approximations. Compositio Math. **121**(3), 247–262 (2000). ISSN 0010-437X
26. Fujimoto, H.: Extensions of the big Picard's theorem. Tôhoku Math. J. **24**(2), 415–422 (1972). ISSN 0040-8735
27. Gasbarri, C.: The strong *abc* conjecture over function fields (after McQuillian and Yamanoi). Astérisque, (326): Exp. No. 989, viii, 219–256 (2010). Séminaire Bourbaki. Vol. 2007/2008 (2009)
28. Gillet, H., Soulé, C.: On the number of lattice points in convex symmetric bodies and their duals. Israel J. Math. **74**(2–3), 347–357 (1991). ISSN 0021-2172
29. Goldberg, A.A., Ostrovskii, I.V.: Value distribution of meromorphic functions. In: Translations of mathematical monographs, vol. 236. American Mathematical Society, RI (2008). ISBN 978-0-8218-4265-2; Translated from the 1970 Russian original by Mikhail Ostrovskii, With an appendix by Alexandre Eremenko and James K. Langley
30. Green, M.: Holomorphic maps into complex projective space omitting hyperplanes. Trans. Amer. Math. Soc. **169**, 89–103 (1972). ISSN 0002-9947
31. Green, M.: Some Picard theorems for holomorphic maps to algebraic varieties. Amer. J. Math. **97**, 43–75 (1975). ISSN 0002-9327
32. Green, M., Griffiths, P.: Two applications of algebraic geometry to entire holomorphic mappings. In: The Chern symposium 1979 (Proc. Internat. Sympos., Berkeley, CA, 1979), pp. 41–74. Springer, New York (1980)
33. Grothendieck, A. et al.: Revêtements étales et groupe fondamental. Springer, Berlin (1971). Séminaire de Géométrie Algébrique du Bois Marie 1960–1961 (SGA 1), Dirigé par Alexandre Grothendieck. Augmenté de deux exposés de M. Raynaud, Lecture Notes in Mathematics, vol. 224; arXiv:math.AG/0206203
34. Grothendieck, A., Dieudonné, J.-A.-E.: Éléments de géométrie algébrique. Publ. Math. IHES, 4, 8, 11, 17, 20, 24, 28, and 32, (1960–1967). ISSN 0073-8301.
35. Gunning, R.C.: Introduction to holomorphic functions of several variables. vol. II (Local theory). The Wadsworth and Brooks/Cole Mathematics Series. Wadsworth and Brooks/Cole Advanced Books and Software, CA (1990). ISBN 0-534-13309-6
36. Hartshorne, R.: Algebraic geometry. Springer, New York (1977). ISBN 0-387-90244-9; Graduate Texts in Mathematics, No. 52
37. Hayman, W.K.: Meromorphic functions. Oxford Mathematical Monographs. Clarendon Press, Oxford (1964)
38. Hodge, W.V.D., Pedoe, D.: Methods of algebraic geometry, vol. II. Book III: General theory of algebraic varieties in projective space. Book IV: Quadrics and Grassmann varieties. Cambridge University Press, Cambridge (1952)
39. Iitaka, S.: Algebraic geometry, An introduction to birational geometry of algebraic varieties, Graduate texts in mathematics, vol. 76. Springer, New York (1982). ISBN 0-387-90546-4; North-Holland Mathematical Library, 24
40. Kato, K.: Logarithmic structures of Fontaine-Illusie. In: Algebraic analysis, geometry, and number theory (Baltimore, MD, 1988), pp. 191–224. Johns Hopkins University Press, MD (1989)
41. Kawamata, Y.: On Bloch's conjecture. Invent. Math. **57**(1), 97–100 (1980). ISSN 0020-9910
42. Koblitz, N.: *p*-adic numbers, *p*-adic analysis, and zeta-functions, Graduate texts in mathematics, vol. 58, 2nd edn. Springer, New York (1984). ISBN 0-387-96017-1
43. Kunz, E.: Kähler differentials. Advanced Lectures in Mathematics. Friedr. Vieweg, Braunschweig (1986). ISBN 3-528-08973-3
44. Lang, S.: Integral points on curves. Inst. Hautes Études Sci. Publ. Math. **6**, 27–43 (1960). ISSN 0073-8301
45. Lang, S.: Algebraic number theory. Addison-Wesley, MA (1970)
46. Lang, S.: Fundamentals of diophantine geometry. Springer, New York (1983). ISBN 0-387-90837-4
47. Lang, S.: Hyperbolic and diophantine analysis. Bull. Amer. Math. Soc. (N.S.) **14**(2), 159–205 (1986). ISSN 0273-0979

48. Lang, S.: Number theory III: Diophantine geometry, Encyclopaedia of mathematical sciences, vol. 60. Springer, Berlin (1991). ISBN 3-540-53004-5
49. Lang, S., Cherry, W.: Topics in Nevanlinna theory, Lecture Notes in mathematics, vol. 1433. Springer, Berlin (1990). ISBN 3-540-52785-0; With an appendix by Zhuan Ye
50. Levin, A.: The dimensions of integral points and holomorphic curves on the complements of hyperplanes. Acta Arith. **134**(3), 259–270 (2008). ISSN 0065-1036
51. Levin, A.: Generalizations of Siegel's and Picard's theorems. Ann. Math. (2) **170**(2),609–655 (2009). ISSN 0003-486X
52. Levin, A., McKinnon, D., Winkelmann, J.: On the error terms and exceptional sets in conjectural second main theorems. Q. J. Math. **59**(4), 487–498 (2008). ISSN 0033-5606, doi:10.1093/qmath/ham052, http://dx.doi.org/10.1093/qmath/ham052
53. Lu, S.S.-Y.: On meromorphic maps into varieties of log-general type. In: Several complex variables and complex geometry, Part 2 (Santa Cruz, CA, 1989), Proc. Sympos. Pure Math., vol. 52, pp. 305–333. AMS, RI (1991)
54. Mahler, K.: On a theorem of Liouville in fields of positive characteristic. Canadian J. Math. **1**, 397–400 (1949). ISSN 0008-414X
55. McQuillan, M.: Diophantine approximations and foliations. Inst. Hautes Études Sci. Publ. Math. **87**, 121–174 (1998). ISSN 0073-8301
56. McQuillan, M.: Non-commutative Mori theory. IHES preprint IHES/M/00/15 (2000)
57. McQuillan, M.: Canonical models of foliations. Pure Appl. Math. Q. **4**(3, part 2), 877–1012 (2008). ISSN 1558-8599
58. McQuillan, M.: Old and new techniques in function field arithmetic. (2009, submitted)
59. Neukirch, J.: Algebraic number theory, Grundlehren der Mathematischen Wissenschaften, vol. 322 [Fundamental Principles of Mathematical Sciences]. Springer, Berlin (1999). ISBN 3-540-65399-6; Translated from the 1992 German original and with a note by Norbert Schappacher, With a foreword by G. Harder
60. Nevanlinna, R.: Analytic functions. Translated from the second German edition by Phillip Emig. Die Grundlehren der mathematischen Wissenschaften, Band 162. Springer, New York (1970)
61. Noguchi, J.: Lemma on logarithmic derivatives and holomorphic curves in algebraic varieties. Nagoya Math. J. **83**, 213–233 (1981) ISSN 0027-7630
62. Noguchi, J.: On Nevanlinna's second main theorem. In: Geometric complex analysis (Hayama, 1995), pp. 489–503. World Scientific, NJ (1996)
63. Noguchi, J.: On holomorphic curves in semi-abelian varieties. Math. Z. **228**(4), 713–721 (1998). ISSN 0025-5874
64. Osgood, C.F.: Effective bounds on the Diophantine approximation of algebraic functions over fields of arbitrary characteristic and applications to differential equations. Nederl. Akad. Wetensch. Proc. Ser. A **78** Indag. Math. **37**, 105–119 (1975)
65. Osgood, C.F.: A number theoretic-differential equations approach to generalizing Nevanlinna theory. Indian J. Math. **23**(1–3), 1–15 (1981). ISSN 0019-5324
66. Osgood, C.F.: Sometimes effective Thue-Siegel-Roth-Schmidt-Nevanlinna bounds, or better. J. Number Theor. **21**(3), 347–389 (1985). ISSN 0022-314X
67. Roth, K.F.: Rational approximations to algebraic numbers. Mathematika **2**,1–20; corrigendum, 168 (1955). ISSN 0025-5793
68. Ru, M.: On a general form of the second main theorem. Trans. Amer. Math. Soc. **349**(12), 5093–5105 (1997). ISSN 0002-9947
69. Ru, M.: Nevanlinna theory and its relation to Diophantine approximation. World Scientific, NJ (2001). ISBN 981-02-4402-9
70. Ru, M.: Holomorphic curves into algebraic varieties. Ann. Math. (2) **169** (1), 255–267 (2009). ISSN 0003-486X, doi:10.4007/annals.2009.169.255, http://dx.doi.org/10.4007/annals.2009. 169.255
71. Schmidt, W.M.: Diophantine approximations and Diophantine equations, Lecture Notes in Mathematics, vol. 1467. Springer, Berlin (1991). ISBN 3-540-54058-X

72. Serre, J.-P.: Géométrie algébrique et géométrie analytique. Ann. Inst. Fourier, Grenoble **6**, 1–42 (1955/1956). ISSN 0373-0956
73. Serre, J.-P.: Corps locaux. Hermann, Paris (1968). Deuxième édition, Publications de l'Université de Nancago, No. VIII
74. Serre, J.-P.: Lectures on the Mordell-Weil theorem. Aspects of Mathematics, E15. Friedr. Vieweg, Braunschweig (1989). ISBN 3-528-08968-7; Translated from the French and edited by Martin Brown from notes by Michel Waldschmidt
75. Shabat, B.V.: Distribution of values of holomorphic mappings, Translations of mathematical monographs, vol. 61. American Mathematical Society, RI (1985). ISBN 0-8218-4514-4; Translated from the Russian by J. R. King, Translation edited by Lev J. Leifman
76. Shiffman, B.: Introduction to the Carlson-Griffiths equidistribution theory. In: Value distribution theory (Joensuu, 1981), Lecture Notes in Math., vol. 981, pp. 44–89. Springer, Berlin (1983)
77. Silverman, J.H.: The theory of height functions. In: Arithmetic geometry (Storrs, Conn., 1984), pp. 151–166. Springer, New York (1986)
78. Siu, Y.-T.: Hyperbolicity problems in function theory. In: Five decades as a mathematician and educator, pp. 409–513. World Scientific, NJ (1995)
79. Siu, Y.-T., Yeung, S.-K.: A generalized Bloch's theorem and the hyperbolicity of the complement of an ample divisor in an abelian variety. Math. Ann. **306**(4), 743–758 (1996). ISSN 0025-5831
80. Siu, Y.-T., Yeung, S.-K.: Defects for ample divisors of abelian varieties, Schwarz lemma, and hyperbolic hypersurfaces of low degrees. Amer. J. Math. **119**(5), 1139–1172 (1997). ISSN 0002-9327
81. Stewart, C.L., Tijdeman, R.: On the Oesterlé-Masser conjecture. Monatsh. Math. **102**(3), 251–257 (1986). ISSN 0026-9255
82. Stoll, W.: Value distribution of holomorphic maps into compact complex manifolds. Lecture Notes in Mathematics, vol. 135. Springer, Berlin (1970)
83. Szpiro, L., Ullmo, E., Zhang, S.: Équirépartition des petits points. Invent. Math. **127**(2), 337–347 (1997). ISSN 0020-9910
84. van der Poorten, A.J., Schlickewei, H.P.: The growth conditions for recurrence sequences. Macquarie Math. Reports, 82-0041 (1982)
85. van Frankenhuijsen, M.: The *ABC* conjecture implies Vojta's height inequality for curves. J. Number Theor. **95**(2), 289–302 (2002). ISSN 0022-314X
86. van Frankenhuijsen, M.: *ABC* implies the radicalized Vojta height inequality for curves. J. Number Theor. **127**(2), 292–300 (2007). ISSN 0022-314X
87. Vojta, P.: Diophantine approximations and value distribution theory, Lecture Notes in Mathematics, vol. 1239. Springer, Berlin (1987). ISBN 3-540-17551-2
88. Vojta, P.: A refinement of Schmidt's subspace theorem. Amer. J. Math. **111**(3), 489–518 (1989). ISSN 0002-9327
89. Vojta, P.: Integral points on subvarieties of semiabelian varieties. I. Invent. Math. **126**(1), 133–181 (1996). ISSN 0020-9910
90. Vojta, P.: On Cartan's theorem and Cartan's conjecture. Amer. J. Math. **119**(1),1–17 (1997). ISSN 0002-9327
91. Vojta, P.: A more general *abc* conjecture. Int. Math. Res. Notices, **1998**(21), 1103–1116 (1998). ISSN 1073-7928
92. Vojta, P.: Integral points on subvarieties of semiabelian varieties. II. Amer. J. Math. **121**(2), 283–313 (1999). ISSN 0002-9327
93. Vojta, P.: On the *ABC* conjecture and Diophantine approximation by rational points. Amer. J. Math. **122**(4), 843–872 (2000). ISSN 0002-9327
94. Vojta, P.: Correction to: On the *ABC* conjecture and Diophantine approximation by rational points [Amer. J. Math. **122**(4), 843–872 (2000); MR1771576 (2001i:11094)]. Amer. J. Math. **123**(2), 383–384 (2001). ISSN 0002-9327
95. Vojta, P.: Jets via Hasse-Schmidt derivations. In: Diophantine geometry, CRM Series, vol. 4, pp. 335–361. Ed. Norm., Pisa (2007). arXiv:math.AG/0407113

 96. Vojta, P.: Nagata's embedding theorem. arXiv:0706.1907 (2007, to appear)
 97. Vojta, P.: On McQuillan's tautological inequality and the Weyl-Ahlfors theory of associated curves. arXiv:0706.3044 (2008, to appear)
 98. Weil, A.: L'arithmétique sur les courbes algébriques. Acta Math. **52**, 281–315 (1928)
 99. Weil, A.: Arithmetic on algebraic varieties. Ann. Math. **53**(2), 412–444 (1951). ISSN 0003-486X
100. Weyl, H., Weyl, J.: Meromorphic curves. Ann. Math. (2) **39**(3), 516–538 (1938). ISSN 0003-486X
101. Wong, P.-M.: On the second main theorem of Nevanlinna theory. Amer. J. Math. **111**(4), 549–583 (1989). ISSN 0002-9327
102. Yamanoi, K.: Algebro-geometric version of Nevanlinna's lemma on logarithmic derivative and applications. Nagoya Math. J. **173**, 23–63 (2004). ISSN 0027-7630
103. Yamanoi, K.: The second main theorem for small functions and related problems. Acta Math. **192**(2), 225–294 (2004). ISSN 0001-5962

Index

List of Participants

1. Amoroso Francesco, France
2. Bagalkote Jayasimha, India
3. Baran Burcu, Turkey
4. Bhowmik Prasenjit, India
5. Bost Jean-Benoît, France (**lecturer**)
6. Bouw Irene, Germany
7. Canci Jung Kyu, Italy
8. Carrizosa Maria, France
9. Cobbe Alessandro, Italy
10. Colliot-Thélène Jean-Louis, France (**lecturer**)
11. Corvaja Pietro, Italy (**editor**)
12. Creutz Brendan, Germany
13. Demeyer Jeroen, Belgium
14. Duma Bertrand, France
15. Dvornicich Roberto, Italy
16. Fehm Arno, Israel
17. Ferretti Andrea, Italy
18. Galateau Aurelien, France
19. Gasbarri Carlo, Italy (**editor**)
20. Hoermann Fritz, Germany
21. Huayi Chen, France
22. Ih Su-ion, U.S.A.
23. Illengo Marco, Italy
24. Isklander Aliev, United Kingdom
25. Jossen Peter, Hungary
26. Languasco Alessandro, Italy
27. Litcanu Gabriela, Germany
28. Loughran Daniel, U.K.
29. Marcovecchio Raffaele, France
30. Matev Tzanko, Germany
31. Mayer Hartwig, Germany
32. Mistretta Ernesto, France
33. Mller Jan Steffen, Germany

34. Pacienza Gianluca, France
35. Paladino Laura, Italy
36. Perelli Alberto, Italy
37. Perucca Antonella, Italy
38. Posingies Anna, Germany
39. Rhin Georges, France
40. Rossi Antonella, Italy
41. Sabatino Pietro, Italy
42. Salgado Cecilia, Spain
43. Surroca Andrea Cristina, Swisserland
44. Swinnerton-Dyer Peter, United Kingdom (**lecturer**)
45. Tossici Dajano, Italy
46. Van Valckenborgh Karl, Belgium
47. Viada Evelina, Switzerland
48. Viviani Filippo, Sweden
49. Vojta Paul, U.S.A. (**lecturer**)
50. Von Pippich Anna-Maria, Germany
51. Wevers Stefan, Netherland
52. Wouters Tim, Belgium
53. Zaccagnini Alessandro, Italy
54. Zannier Umberto, Italy

Lecture Notes in Mathematics

For information about earlier volumes
please contact your bookseller or Springer
LNM Online archive: springerlink.com

Recent Reprints and New Editions

LECTURE NOTES IN MATHEMATICS

Edited by J.-M. Morel, F. Takens, B. Teissier, P.K. Maini

Editorial Policy (for Multi-Author Publications: Summer Schools/Intensive Courses)

1. Lecture Notes aim to report new developments in all areas of mathematics and their applications - quickly, informally and at a high level. Mathematical texts analysing new developments in modelling and numerical simulation are welcome. Manuscripts should be reasonably self-contained and rounded off. Thus they may, and often will, present not only results of the author but also related work by other people. They should provide sufficient motivation, examples and applications. There should also be an introduction making the text comprehensible to a wider audience. This clearly distinguishes Lecture Notes from journal articles or technical reports which normally are very concise. Articles intended for a journal but too long to be accepted by most journals, usually do not have this "lecture notes" character.

2. In general SUMMER SCHOOLS and other similar INTENSIVE COURSES are held to present mathematical topics that are close to the frontiers of recent research to an audience at the beginning or intermediate graduate level, who may want to continue with this area of work, for a thesis or later. This makes demands on the didactic aspects of the presentation. Because the subjects of such schools are advanced, there often exists no textbook, and so ideally, the publication resulting from such a school could be a first approximation to such a textbook. Usually several authors are involved in the writing, so it is not always simple to obtain a unified approach to the presentation.

   For prospective publication in LNM, the resulting manuscript should not be just a collection of course notes, each of which has been developed by an individual author with little or no co-ordination with the others, and with little or no common concept. The subject matter should dictate the structure of the book, and the authorship of each part or chapter should take secondary importance. Of course the choice of authors is crucial to the quality of the material at the school and in the book, and the intention here is not to belittle their impact, but simply to say that the book should be planned to be written by these authors jointly, and not just assembled as a result of what these authors happen to submit.

   This represents considerable preparatory work (as it is imperative to ensure that the authors know these criteria before they invest work on a manuscript), and also considerable editing work afterwards, to get the book into final shape. Still it is the form that holds the most promise of a successful book that will be used by its intended audience, rather than yet another volume of proceedings for the library shelf.

3. Manuscripts should be submitted either online at www.editorialmanager.com/lnm/ to Springer's mathematics editorial, or to one of the series editors. Volume editors are expected to arrange for the refereeing, to the usual scientific standards, of the individual contributions. If the resulting reports can be forwarded to us (series editors or Springer) this is very helpful. If no reports are forwarded or if other questions remain unclear in respect of homogeneity etc, the series editors may wish to consult external referees for an overall evaluation of the volume. A final decision to publish can be made only on the basis of the complete manuscript; however a preliminary decision can be based on a pre-final or incomplete manuscript. The strict minimum amount of material that will be considered should include a detailed outline describing the planned contents of each chapter.

   Volume editors and authors should be aware that incomplete or insufficiently close to final manuscripts almost always result in longer evaluation times. They should also be aware that parallel submission of their manuscript to another publisher while under consideration for LNM will in general lead to immediate rejection.

4. Manuscripts should in general be submitted in English. Final manuscripts should contain at least 100 pages of mathematical text and should always include
   - a general table of contents;
   - an informative introduction, with adequate motivation and perhaps some historical remarks: it should be accessible to a reader not intimately familiar with the topic treated;
   - a global subject index: as a rule this is genuinely helpful for the reader.

   Lecture Notes volumes are, as a rule, printed digitally from the authors' files. We strongly recommend that all contributions in a volume be written in the same LaTeX version, preferably LaTeX2e. To ensure best results, authors are asked to use the LaTeX2e style files available from Springer's web-server at

   ftp://ftp.springer.de/pub/tex/latex/svmonot1/ (for monographs) and
   ftp://ftp.springer.de/pub/tex/latex/svmultt1/ (for summer schools/tutorials).

   Additional technical instructions are available on request from: lnm@springer.com.
5. Careful preparation of the manuscripts will help keep production time short besides ensuring satisfactory appearance of the finished book in print and online. After acceptance of the manuscript authors will be asked to prepare the final LaTeX source files and also the corresponding dvi-, pdf- or zipped ps-file. The LaTeX source files are essential for producing the full-text online version of the book. For the existing online volumes of LNM see: http://www.springerlink.com/openurl.asp?genre=journal&issn=0075-8434.

   The actual production of a Lecture Notes volume takes approximately 12 weeks.
6. Volume editors receive a total of 50 free copies of their volume to be shared with the authors, but no royalties. They and the authors are entitled to a discount of 33.3% on the price of Springer books purchased for their personal use, if ordering directly from Springer.
7. Commitment to publish is made by letter of intent rather than by signing a formal contract. Springer-Verlag secures the copyright for each volume. Authors are free to reuse material contained in their LNM volumes in later publications: a brief written (or e-mail) request for formal permission is sufficient.

Addresses:

Professor J.-M. Morel, CMLA,
École Normale Supérieure de Cachan,
61 Avenue du Président Wilson,
94235 Cachan Cedex, France
E-mail: Jean-Michel.Morel@cmla.ens-cachan.fr

Professor F. Takens, Mathematisch Instituut,
Rijksuniversiteit Groningen, Postbus 800,
9700 AV Groningen, The Netherlands
E-mail: F.Takens@rug.nl

Professor B. Teissier,
Institut Mathématique de Jussieu,
UMR 7586 du CNRS,
Équipe "Géométrie et Dynamique",
175 rue du Chevaleret,
75013 Paris, France
E-mail: teissier@math.jussieu.fr

For the "Mathematical Biosciences Subseries" of LNM:

Professor P.K. Maini, Center for Mathematical Biology,
Mathematical Institute, 24-29 St Giles,
Oxford OX1 3LP, UK
E-mail: maini@maths.ox.ac.uk

Springer, Mathematics Editorial I, Tiergartenstr. 17,
69121 Heidelberg, Germany,
Tel.: +49 (6221) 487-8259
Fax: +49 (6221) 4876-8259
E-mail: lnm@springer.com